THE NEW (SO-CALLED) MAGDEBURG EXPERIMENTS OF OTTO VON GUERICKE

ARCHIVES INTERNATIONALES D'HISTOIRE DES IDÉES

INTERNATIONAL ARCHIVES OF THE HISTORY OF IDEAS

137

OTTO VON GUERICKE

THE NEW (SO-CALLED) MAGDEBURG EXPERIMENTS OF OTTO VON GUERICKE

THE NEW (SO-CALLED) MAGDEBURG EXPERIMENTS OF OTTO VON GUERICKE

by

OTTO VON GUERICKE

Translation and Preface by

MARGARET GLOVER FOLEY AMES

KLUWER ACADEMIC PUBLISHERS
DORDRECHT / BOSTON / LONDON

Ottonis de Guericke
EXPERIMENTA
Nova (ut vocantur) Magdeburgica
De
VACUO SPATIO.

OTTO De GUERICKE
Serenifs: ac Potentifs: Elector: Brandeb:
Confiliarius et Civitat: Magdeb. Conful:

THE NEW (SO-CALLED) MAGDEBURG

EXPERIMENTS OF

OTTO VON GUERICKE

ON EMPTY SPACE

First published by the Reverend Father Kaspar Schott, S. J.
Professor of Mathematics at the University of Würzburg
and in the present more complete edition by the author himself who has included a number of additional experiments in the text.

At the same time certain other subjects have been added to this edition concerning the Weight of Air around the Earth, the Mundane Virtues, the Planetary System as well as the Fixed Stars and Immense Space which extends within as well as beyond them.

AMSTERDAM

PRINTED BY JOHANNES JANSSON OF WAESBERG 1672
THROUGH THE PRIVILEGE OF HIS IMPERIAL MAJESTY

OTTONIS DE GUERICKE
EXPERIMENTA
NOVA (*ut vocantur*) MAGDEBURGICA
DE
VACUO SPATIO

Primùm à R. P. *Gaspare Schotto*, è Societate Jesu, & Herbipolitanæ Academiæ Matheseos Professore:

Nunc verò ab ipso Auctore

Perfectiùs edita, variisque aliis Experimentis aucta.

Quibus accesserunt simul certa quædam

De Aëris Pondere circa Terram; de Virtutibus Mundanis, & Systemate Mundi Planetario; sicut & de Stellis Fixis, ac Spatio illo Immenso, quod tàm intra quam extra eas funditur.

AMSTELODAMI,

Apud JOANNEM JANSSONIUM à WAESBERGE, ANNO 1672.

Cum Privilegio S. Cæs. Majestatis.

Library of Congress Cataloging-in-Publication Data

Guericke, Otto von, 1602-1686.
[Experimenta nova (ut vocantur) Magdeburgica de vacuo spatio. English]
The new (so-called) Magdeburg experiments of Otto von Guericke / translation and preface by Margaret Glover Foley Ames.
p. cm. -- (Archives internationales d'histoire des idées ; 137)
Translation of Experimenta nova (ut vocantur) Magdeburgica de vacuo spatio.
Includes index.
ISBN 0-7923-2399-8 (alk. paper)
1. Physical sciences--Early works to 1800. 2. Astronomy--Early works to 1800. 3. Pneumatics--Early works to 1800. 4. Madgeburg experiments. I. Ames, Margaret Glover Foley. II. Title. III. Series.
Q155.G9313 1993
500--dc20 93-5121

ISBN 0-7923-2399-8

Published by Kluwer Academic Publishers,
P.O. Box 17, 3300 AA Dordrecht, The Netherlands.

Kluwer Academic Publishers incorporates
the publishing programmes of
D. Reidel, Martinus Nijhoff, Dr W. Junk and MTP Press.

Sold and distributed in the U.S.A. and Canada
by Kluwer Academic Publishers,
101 Philip Drive, Norwell, MA 02061, U.S.A.

In all other countries, sold and distributed
by Kluwer Academic Publishers Group,
P.O. Box 322, 3300 AH Dordrecht, The Netherlands.

Printed on acid-free paper

TABLE OF CONTENTS

BOOK II

EMPTY SPACE

BOOK IV

MUNDANE VIRTUES AND OTHER ALLIED SUBJECTS

BOOK V

THE SPHERE OF LAND AND SEA AND ITS COMPANION, THE MOON

BOOK VI

OUR PLANETARY SYSTEM

PREFACE

The contemplation of nature, to which the blessed Basilius bears witness, is a threshold of joy, an eternal source of pleasure for the spirit, a portal of tranquility, a bridge from the celestial to the earthly and the peak of man's happiness. Attaining this, the soul seems to be aroused, as it were, from a deep lethargy and enters a region of light, forgetful of self, and assumes the role, not so much of a man touched by divinity, as that of an earth-bound deity.

That well-known verse, indeed, is true:

"If mortal souls were to understand the innermost workings of the cosmos,
Royal power and might would be nothing more to them than meaningless froth."

To attain such knowledge of nature, the art of oratory, an elegance of speech, or the ability to debate incisively are of little account. "For in seeking this a thousand Demosthenes, a thousand Aristotles can be laid prostrate by a single man of mediocre talent who has seized upon a better way to find the truth. Such a hope, therefore, must be removed: for indeed, men, more learned and superior to us in book-learning, will be found who, to the shame of nature itself, can make that which is, in fact, false, true." Galileo, *Dialogus Cosmicus*, page 35.

Therefore theories which are demonstrated by experiment and visual perception must be preferred to those derived from reasoning, however probable and plausible, for many things seem true in speculation and discussion which in actual fact defy reality.

"It follows from this that all science is empty, deceptive, and pointless unless it is supported by experiment. What inconsistencies otherwise successful and perceptive scholars bring forth without its help! It is experimentation that dissolves all doubts, reconciles difficulties, is a unique teacher of truth, furnishes a torch in darkness and instructs us how to determine the true causes of things by disentangling knotty problems." Kircher, *Ars Magnetica*, page 570.

Thus scientists who rely upon their conjectures and hypotheses alone, repudiating experimentation, can make no conclusive statements about the nature of the world, for when man does not rely upon experiments, he often wanders so far from the truth as the sun which we see is distant from the earth.

Gilbert Clerke very recently acknowledged this fact in the Preface of his book, *De Plenitudine Mundi*, where he writes:

"Natural science at one time was dependent upon debatable and dubious assertions, (expressed in sesquipedalian words), designed to deceive true scientists and other scholars rather than to enlighten them.

* * *

"So it has happened that a man who, thanks to his argumentative powers would acquire for himself a meaningless reputation for knowledge and would be looked upon as an outstanding scholar, a man who knew very little about himself but seemed to feel he had much more insight than anyone else, in the paroxysms of his madness, would chatter about whatever nonsense came into his head.

"Finally, however, there shone forth, with God's help, certain men of more exalted talents who led the way in a new kind of research, using both reason and experimentation in combination.

"From these men there was conceived the hope that scholars would not embrace some sort of shadow or delusion of truth, but truth itself and that, putting aside controversy, natural science would come to be allied with mathematical science. This is not the task of one age nor a single individual to bring to fruition, but neither need we doubt (when one scholar after another supports this same kind of reaearch) that finally with the new mysteries of nature being removed daily and the dark night of errors dispelled, truth, in all its purity, (insofar as the innate weakness of humankind allows) will appear and the marriage torches lighted.

"Indeed, in the present state of affairs, it pains me that up to this time, I have seen a certain need for dissention and arguing back and forth while the leaders of this new science dispute about the composition and plenitude of the world and trip over the threshold of knowledge, etc."

Because, in truth, scholars have sharply disputed with one another for a long time the question of the vacuum (does it exist or not? or what is it?) and each one has defended his theory like a determined soldier defending a stronghold against an attacking enemy, for this reason I felt a burning desire to investigate the truth of this question which has been a source of dispute, not being able to lay my feelings to rest nor extinguish them until I should find time to carry out an experiment involving the vacuum.

I performed this experiment in a number of ways and my efforts were not unrewarded: indeed, I designed several pieces of apparatus to prove the existence of that vacuum whose existence had always been denied.

Later on, when I had been sent on behalf of the state to the Imperial Diet being held in Regensburg in 1654 several enthusiastic followers of these kinds of investigations had heard about my aforementioned experiments: they succeeded in convincing me to demonstrate some of them and I endeavored to do this insofar as I could, considering the limits of my capabilities.

At the conclusion of the meetings of the Diet and when it was breaking up, my experiments were brought to the notice of his Imperial Majesty as well as the Electors and several Princes who were on the point of leaving but wished to see my experiments demonstrated before their departure. I could not refuse their request under the circumstances.

These experiments were of particular interest to His Eminence the Noble Elector Johann Philipp, Archbishop of Mainz and Bishop of Würzburg, who of all the spectators present, persuaded me to make a similar piece of apparatus for himself. Because of the limitations of time, however, this apparatus could not be reproduced by my workmen and so, at the request of his Eminence, I handed over the experimental instruments I had brought with me to Regensburg for a financial consideration.

As soon as the Reverend Fathers of the Society of Jesus and the professors of the public University of Würzburg, under the chairmanship of the same Eminent Elector, had verified these experiments of mine, they wrote out a draft of their findings and communicated them to the scholars in Rome and elsewhere and at the same time sought their opinions. Among this group of particular importance was the Reverend Father Kaspar Schott, Professor of Mathematics at the same university, who wrote me regarding the experiments and sought more information about them. Ultimately he added, as an Appendix to his book, *De Arte Mechanica Hydraulico-Pneumatica*, published in 1657, these "Experimenta Nova" of mine and called them "Magdeburgica:" these were set up in type so that they might be available for many to read and study.

After these experiments had been exposed to public scrutiny, many others that I had hitherto carried out were added. These, in turn were published by the aforementioned Father Schott in Book I of *Technica Curiosa* which dealt with the *Magdeburgica Mirabilia*, published in 1664, together with the *Antiqua Experimenta Magdeburgica* and the *Nova Experimenta Magdeburgica*. In addition to Father Schott, other scholars were found who wrote on this same subject and, indeed, almost every one who saw the apparatus and the results stemming from the experiments was seized with astonished admiration. And just as the Reverend Father Schott attests in the Foreward to his *Technica Curiosa*, page 3, (and other places as well), "I do not hesitate to speak openly and declare without qualification that I have never seen nor heard of any experiments of this kind more worthy of admiration nor read nor even imagined such, and I do not believe that the sun ever shone upon anything like them from Time Immemorial. And I might add, this is the judgment of powerful princes and learned scholars as well, with whom I have been in communication as regards these experiments, etc." So these feelings will be made abundantly clear by everyone else everywhere who has brought forth treatises on this subject.

Although, in fact, it was never my intention to have my experiments printed and published, nevertheless the very diverse judgments of the "De Vacuo" of which some approve and others attack the contents, in order that no one could be surprised by so many different and often remarkably confused concepts and at the same time satisfy those who urgently asked for a clarification of these experiments, I undertook the publication of the treatise, *De Spatio Vacuo* in its entirety for its advantage in providing a deeper understanding of the field of natural science. I finished the work on the 14th of March, 1663; I did not seek to correct nor refute the numerous and frequently inconsistent opinions of philosophers (except briefly in Chapters 35 and 36 of Book III where at least the usual and typical objections of the Reverend Fathers and professors, who are named explicitly in the Appendix of Schott's *Ars Hydraulico-Pneumatica*, are discussed). Indeed, such a treatise would be too long-winded and tedious for the reader to encompass. Rather its aim is to recall from ingrained and poorly conceived imaginings all who do not labor under preconceived opinions but regard the experiments dispassionately and weigh them on the fair scale of truth, having acquired from them rich experience and knowledge. Indeed, when there is evidence from facts, words are not necessary. One need not dispute and take up arms against an opponent who denies experiments that are completely reliable; let such a one keep for himself his own opinion and dig in the dark like a mole. An exact science does not march into battle but celebrates its victory and rests in the deep tranquility of truth. Other branches of science, however, are subject to controversy because they are devoid of the rational certitude in which an exact science is rich. So it comes about that the human spirit, after it has wandered for a long time in the circle of arts and science finally comes to find rest in the certain knowledge of an exact science.

As for the form in which this treatise is presented, it is not intended to be eloquent nor elegant in verbiage: however, if there is any fault in my use of words, I should like to be excused, because we are concerned with facts, not words which are only in the service of facts rather than the other way around. Indeed, all things could not be adequately described verbally, but often for the sake of brevity, had to be omitted or reduced to the common idiom as the old proverb says, "Speak with many words but think with a few."

And although this book had been completed seven years ago, as I mentioned, I was hindered in its publication in part by illness and in part by a series of business commitments. This work would have remained in the desk for a still longer time, however, had not several distinguished men, together with Lukian, thinking it a wretched state of affairs that knowledge was being drained from the books alone without being substantiated by any published experiments, put an end to my vacillation, encouraging me not to hesitate and refuse any longer in bringing out this work for general use. For this reason I did not want to disregard their wishes for its publication.

Just as all things are not pleasing to everyone, so one can easily surmise that opponents will not be lacking, as the proverb says, "Let no one be persuaded otherwise, for he is going to have both enemies and followers who embarks upon great undertakings."

And indeed, (to quote from Seneca), there is no mortal man who does not suffer from some sort of ignorance; "this is an age-old ill of mankind. For as often as a man errs, he errs like a man. To reproach a man for his erring, however, is to reproach humankind itself."

For this reason, one should not expect to be so fortunate as to be able to bring forth any work that is flawless. Indeed, for our part, we sometimes commit errors in pointing out the faults of others and in the same vein must readily accept the criticism thoughtful and honest men make of results found in this work which are not substantiated by experiment. In this situation, when we learn or have to admit another conclusion that is better and perhaps correct, we must be ready to recognize the truth.

Before all else, however, it is our wish that the contents of this work be encompassed by the cloistered walls of natural science and not slip out from these confines into other fields that are perhaps mixed with faith or belief: rather, this treatise must be limited by the principles of natural science alone made manifest by experimentation. But should some thoughtless or inadvertent remark slip out contrary to our intention, however, we should like to retract this statement herewith. We grant to everyone the freedom of dissent but for our part we are prepared to follow truth rather than agreement for the sake of harmony. As for the rest, we believe that there will be no lack of keen and perspicacious minds, being stimulated by my experiments to arrive at perhaps better and deeper insights, who are going to undertake this work at some future time. And so farewell, gracious reader, and regard our efforts in good part.

Written in Magdeburg, the 14th of March, 1670

DEDICATED TO THE "EXPERIMENTA NOVA MAGDEBURGICA" OF THE MOST DISTINGUISHED AND EXCELLENT GENTLEMAN, OTTO VON GUERICKE

To delve into the manifold mysteries of nature
is the task of an inquiring and fertile mind.
To follow the tortuous paths of nature's wondrous ways
is work more difficult and not designed for everyone.
You, Distinguished Sir, Magdeburg knows as its Burghermaster
as well as an outstanding researcher in the field of science.
Whether one speaks with you informally or studies your work alone,
he soon will confirm your genius openly and without a feeling of doubt.
May I make a small joke? While you prove quite clearly that a vacuum exists
in your Book, there is not a vacuum to be seen!

In this insignificant poem, I have set down no insignificant proof of my eternal gratitude to my most excellent patron with warm wishes for a long life.

Johannes von Gersdorf
Nobleman of Lausitz

SHORT BIOGRAPHY OF O. VON GUERICKE

Guericke (Gericke), Otto von (born in Magdeburg, Germany, 20 November 1602; died in Hamburg, Germany, 11 May 1686.

Otto Guericke, the son of Hans Gericke and Anne von Zweidorf, was a descendant of an aristocratic family of Magdeburg who had occupied positions of authority and prominence there for three hundred years.

Destined to enter political life as a young man, he was registered as a student in the Faculty of Arts at Leipzig from 1617 to 1620. When the opening campaign of the Thirty Years War threatened to engulf Leipzig, however, he was sent to Helmstedt, in Brunswick, to attend the university there. Later he studied law at Jena in 1621 and 1622, continuing his studies in this field in Leyden where he also attended lectures on astronomy, mathematics, and physics.

Upon completing his university studies, Guericke returned to Magdeburg and as a young man of twenty-four (1626) was elected to the City Council and elevated to the Bench of Magistrates. Four years later he assumed the additional duties of City Contractor in which capacity he supervised the building, maintenance, and defense of the city's fortifications. Little is known of Guericke's personal life except that in 1626 he married Margarethe Alemann, the mother of his only surviving child, a son also named Otto. Several years after her death in 1645 he married Dorothea Leutke, but was again left a widower. Shortly after his second wife's death, he and his son went to live with Guericke's mother on their manorial estate. In 1681, after he had retired from public office, he went to live in Hamburg with his son who had been elected a magistrate of Brandenburg. He spent the rest of his life there until his death in 1686.

Because Magdeburg was regarded as a city of considerable strategic importance, it was inevitable that it play a significant role in the Thirty years War which was spreading over Germany. Inasmuch as Guericke was vitally involved in the affairs of Magdeburg at this time, particularly as supervisor of the fortifications, he was an important participant in the two successive sieges of Magdeburg which resulted in its destruction in 1631.

Guericke left Magdeburg at this time and joined the army of Gustavus Adolphus of Sweden in his victorious sweep through Germany. He served with the Swedish forces as Quartermaster General and Chief Siege Engineer when they laid siege to the city of Erfurt.

Four years later he found service with the Saxon Army. In these dual capacities Guericke was able to continue to serve his native city by acting as envoy to the changing occupation powers.

He returned to Magdeburg after these foreign commitments and devoted much of his time and resources to the rebuilding of the city. Feeling respect and gratitude for his devotion, his fellow citizens sent him to Dresden and Leipzig for three years where he helped formulate thc treaty which brought peace to Germany.

From 1642 until 1666 diplomacy consumed much of his time as did his duties as Burghermaster, a position he held for twenty years until 1676. He was likewise elevated to the nobility by being appointed Counsellor to the Elector of Brandenburg. It was at this time that Guericke assumed the noble "von."

Throughout his varied philanthropic career, however, beginning in 1632, Guericke had devoted his brief and limited leisure time to scientific experimentation. Thus, congruent with his duties as diplomat, he attended international congresses and princely courts where he could take part in the exchange of scientific ideas. He was able to present some of his own experiments on several occasions at Regensburg in 1653–54 and again in 1663 at the court of the Great Elector of Berlin. Although the politically important statesmen and scientists who comprised his audiences willingly examined and evaluated his work and found it worthy of notice, nonetheless they questioned its validity. He was regarded, for the most part, as an enthusiastic amateur rather than as a scholar to be reckoned with, because he was neither affiliated with a university nor was he a member of any learned society.

BOOK I

THE SYSTEM OF THE WORLD, ACCORDING TO THE MORE COMMON PHILOSOPHICAL THEORIES

CHAPTER 1

WHAT THE WORLD IS, OR WHAT IS USUALLY UNDERSTOOD BY THE TERM, WORLD

Before we begin a discussion of empty Space (in which the system of the world is held and which generally is regarded as some sort of celestial matter), it will be necessary to explain rather fully in this First Book certain concepts which must be presented in sequence so that they can be more readily understood; astronomers in particular, but philosophers as well both ancient and modern, set forth and pass judgment on these theories about the earth and the disposition, order, course, distance and size of bodies associated with it.

Aristotle's definition of the world.

Aristotle, who is regarded as the most eminent of the philosophers, defines and describes the world to Alexander the Great in his book, *De Mundo*: "It is a joining together of sky and earth and of the elements contained therein." Others give the following definition: "It is the order and arrangement of all things, the central position of which the earth holds, motionless and stable and endowed with the richness of life. Its outermost boundaries are self-contained and the most elevated region is called the sky (the domicile of God) which is filled with celestial bodies and stars, etc. . . ."

Another definition.

The world is commonly defined as "an elegant and adorned scaffold of sky and earth, together with the sun, moon, planets, and fixed stars and whatever exists beyond them, whether on a higher level or a lower (among which is the Ninth Heaven, likewise the First Mobile and the Fiery Heaven)."

A more common description of the world.

But how the system of this mighty temple of God's creatures is actually assembled, set up, and organized in itself, that is, how are these great bodies of the world arranged relative to one another in their order, proportion, distances, situations, locations, connection, and difference? And how do they maintain their courses and complete their motions? Are they joined together by a bond of substance, like the pages of a book? Does a state of continuity exist between them or are they openly disassociated, etc.? Philosophers in antiquity vigorously disputed this question and even now they continue to do so.

The origin of astronomical studies.

Whatever we know with certainty at the present time, however, this surely was not known to man at the very beginning, but only with the passing of time, and after much investigation and observation did it come to be more apparent and convincing to later generations. But since complete knowledge is possessed by no man, mathematicians and astronomers, observing the stars, have been in existence for centuries and their successors have added to the store of knowledge associated with these sciences from century to century. Among this number the Chaldaeans and Egyptians were particularly active. It is believed, moreover, that Abraham, the Patriarch, was particularly learned in these disciplines. Furthermore, according to the testimony of Philo, Moses had absorbed a knowledge of astronomy from the Chaldaeans and Egyptians. St. Stephen readily confirms this statement when he says in the Acts of the Apostles that "Moses was learned in all the wisdom of the Egyptians," of which the most important element was astronomy, as Diodorus, Diogenes Laertius, Philo Judaeus, Jamblichus, and Eusebius testify.

It is not known, however, how far they had progressed in this knowledge, for nothing more than general opinions has been handed down to posterity about them in the sacred records nor in the writings of the pagans. Thus no definite nor distinct mention of planets or solar and lunar eclipses is found in these writings.

The oldest of the extant observations are those of the Babylonians, for Ptolemy quotes Hipparchus as saying that several eclipses had been observed in Babylonia a little more that seven hundred years B.C. Whatever is recorded probably is a fabrication. Then came the Greek philosophers who yielded to no one in their knowledge of astronomy and geography. They engaged in such vigorous controversies and debates with one another in describing and making conjectures about the world, its system, nature, and origin, however, that each fought in defense of his own opinion with the pen as forcefully as enemies attack one another with the sword.

Conjectures about the world.

Thales of Miletus, born in 640 B.C., was the first among the Greeks to forecast a solar eclipse. He held that there was only one world which was derived from God. Metrodorus of Chios, an astronomer, however, contended that the world had always been in existence. Democritus, Epicurus, and Leucippus postulated the existence of many worlds. The Pythagoreans held that the world is one thing, the universe another and that the former, like an animal that breathes, sucking the vacuum outside of it, draws it into itself. They also believed that the sun is in the middle of our world and that the earth revolves around it with the other planets. Aristotle contended that the world has its beginning and end in eternity. He considered it to be unique, however, and equated it with the universe. Plato, however, held that the world itself had had a beginning but would not have an end.

These, then, were the diverse conceptions held of our earth. That it seemed to consist of an essentially round globe surrounded by water and air and that inhabitants dwelt upon it seems to have been unknown to the ancients who lived 2300 and more years ago.

Diverse opinions about our earth.

Now since the outer regions of Asia, Africa, and Europe (and much less America, which we call the New World) had now been known in those early times, and since man saw nothing beyond himself except boundless seas, many thought for this reason that the earth was either (1) surrounded by water which in turn served as its foundation or (2) had foundations reaching to an infinite depth or (3) that because of its very extensive bulk, it pressed upon the lower air which thus prevented it from falling. Thales of Miletus and a number of others subscribed to the first theory and the words of the 24th Psalm also seem to voice this popular concept, "For He hath founded it upon the seas, and established it upon the floods." Xenophon of Colophon and his followers favored the second theory to which it is generally believed that the words of the 104th Psalm refer, "Who (God so) laid the foundations of the earth, that it should not be removed forever." Democritus, Anaxagoras, and others of the same belief defended the third hypothesis.

The philosophers of antiquity did not know that the earth was a sphere and existed freely in the atmosphere.

Now since these theories were generally accepted, no one at that time believed that the earth is a sphere whose surface is inhabited throughout and on all sides; instead they held to divergent opinions. Thus Anaximander stated that the earth was like a column, Leucippus, a cylinder or war drum; Cleanthes, a cone or top; Heraclitus, a boat-shaped vessel; Democritus, a concave disc; Anaximemes and Empedocles, etc., a level table. Thereafter Parmenides and Epicurus came nearer actual fact, as did Heraclitus, Ponticus, and Ecphantus. The former believed that the earth was round like a ball while the latter not only attributed the shape of a sphere to the earth but also some movement (not such that it could move forward and change place, however, as a wheel does). When Plato was still a youth, moreover, he declared that the earth was situated in the center of the world and turned about its own axis toward the east in the space of twenty-four hours and therefore that the sun and other stars within the same period of time seemed to turn towards the west. Pythagoras and his pupils, furthermore, attributed to the earth not only a movement like that of a wheel turning around its own axis within a space of twenty-four hours, but also such a motion whereby it could move forward around the sun during the year, changing its position. For this reason Philolaus is said to have expressed the belief that "the earth is one of the stars." On the other hand, Aristotle and Ptolemy after him, although they attributed a spherical shape to the earth (particularly because of the apparent circular shadow of the earth which is manifest and projected upon the moon during a lunar eclipse), nevertheless, did not acknowledge that it moved. This opinion is commonly held even to this day. Now, as for the antipodes, the Church Fathers, including Chrysostome, Jerome, Augustine, Theodoretus, Theophilactus, Lactantius, likewise Bede and several others, were unwilling to grant their existence, but contended that this belief was not only contrary to the Holy Scriptures, but also human reason. For just as a house cannot stand without a foundation and a roof, so of necessity, they held, the earth must stand on an immovable foundation and have the sky as a roof. See Johann Ludwig Gottfried's *Historia Antipodum et Americae*.

Those who postulate the existence of the antipodes are labelled as heretics.

For this reason they label those who assert that there are antipodes as stupid men who have been deluded by groundless error: in fact, they have condemned them as heretics. We have as an example, Virgilius, Bishop of Salzburg, who in 745 A.D. or thereabouts, was charged with heresy by Bonifacius, Archbishop of Mainz, for no other reason than that he held there were antipodes. Now when this information was transmitted to the King of the Bohemians and thence an appeal was directed to Pope Zaharias, Virgilius was condemned because of his "heretical antipodes," for the very reason that these "absurd, ill-founded deductions" were not considered to be in accord with the Holy Scriptures. See Aventinus, Book III *Historia Bavariorum* and Bartholomaus Keckermann's *Commentarius Nauticus*.

From these and other sources it is clear how little scientific knowledge our ancient forebears living 2,000 years before us had about the true system of the world, for it is obvious that they had observed nothing more than the simple courses of the sun, moon, and fixed stars, which were regularly completed within twenty-four hours and had merely noted the years and months in a general way. As for the planets, Saturn, Jupiter, Mars, Venus, and Mercury, as well as the whole system of the world, they had little or no information.

CHAPTER 2

THE MOTION OF STARS, THE WANDERERS OR PLANETS, AS WELL AS THE STATIONARY OR FIXED STARS

Planets were first written about 2500 years ago.

It can he ascertained from early writings that twenty-five hundred years ago several philosophers came to observe movements in certain individual stars other than their aforementioned twenty-four hour cycle. By reason of this unusual movement, they called these stars wanderers or planets, and for this reason they began to formulate certain hypotheses according to which they calculated the motions that were made, to analyse these motions, and to divide them arithmetically into certain periods expressed in years, months, days, hours, and minutes. They were of the opinion, moreover, that these planets rise and set, now in these regions of the horizon, now in those, and that they swing from the direction of the south up higher into the north and decline again lower into the south, both movements being, however, within certain bounds. Similarly at times the planets move more rapidly, at others more slowly, and appear now larger and again smaller. In appearance, moreover, they differ from one another as well as from fixed stars and at times are close together and at others are separated. Occasionally they hide one or another of the fixed stars and then in turn are themselves covered. Furthermore they are not equally visible to all inhabitants of the earth.

Observations about the movements of the planets.

The paths of Mercury and Venus.

The philosophers in particular observed that the planets which they called Mercury and Venus are followers, as it were, of the sun, for they never wander far from it. Venus, indeed, is never more than a sign and a half away and Mercury is not even a full sign. They also precede the sun at times and move forward (that is, according to the series or succession of the signs, as from Aries into Pisces), and occasionally they remain stationary (that is, for a period of time they do not seem to move either forward or backward). Similarly they move into the south and again into the north, now to a greater and now to a lesser degree than the sun.

What are the forward and backward motions and the state of rest?

The paths of the more distant planets.

Furthermore, as for the planets which they named Mars, Jupiter, and Saturn, they remarked first of all that these were not bound to the sun to as great a degree as the two aforementioned planets, but so departed from it as occasionally to be opposite it, namely six whole constellations away. They likewise observed that these planets annually progressed toward the more easterly stars, at times proceeding forward, at times backwards, and again remaining stationary, and also that these planets moved backwards very rapidly and were visually of the greatest size when they were opposite the sun. They further remarked that Mars moved backward over a larger area than Jupiter, and Jupiter than Saturn, but also that Saturn covered this distance in a longer time than Jupiter, and Jupiter than Mars. Finally they discovered that these planets move into the south and north, now to a greater, now to a lesser degree than the sun, etc.

Observations about the sun.

As regards the sun, they discovered that when it sets, visible stars will set after it, which, after several days, are no longer to be seen; and that stars, invisible before the rising of the sun, after several days rise with it and eventually appear before it. They recorded, moreover, that there were occasional total eclipses and frequent partial eclipses of the sun which occurred only during the new moon and never when it was full. Similarly they remarked that the greatest declination of the sun decreases as much in the north as in the south and that it is not as great at one time as another. They also made similar observations. As regards the moon, they were aware that from the time of the new moon, it becomes more and more removed each day from the sun with the result that it moves in an increasingly easterly direction until it completes its circuit. They also noticed that it appears periodically in various phases as it grows horned or sickle-shaped and then becomes a half-moon or semi-circle, and then a hump. After it has attained its full orb, it wanes again into a hump, a semi-circle, and a horned shape. Occasionally it is totally eclipsed and frequently partially eclipsed, but only during the full moon. The ancients similarly recorded that it moves on occasion into the north as well as the south, at times to a greater, at times to a lesser, degree than the sun, etc.

Observations about the moon.

The motions of the fixed stars.

As for the fixed stars, they discovered that apart from their daily motions (wherein they saw the stars turn each day from rising to setting), they did not always maintain the same distance from the points of the equinox, but in their natural motion travelled more slowly and, (as it has seemed to some), irregularly, that is, rapidly and again more slowly. For Spica, in the constellation Virgo, to cite an example, which Timochares observed shortly after the death of Alexander the Great to precede the point of the autumnal equinox by eight degrees, was seen by Hipparchus, two hundred years later, to precede by only six degrees and by Ptolemy, two hundred and sixty years later by not more than three degrees. At the present time, namely fifteen hundred years later, it has been observed that this star follows the same point by nearly nineteen degrees. In the same way the star, α in the constellation Aries, which in Timochares' time was only two degrees distant from the vernal equinox, is now discovered to be more than twenty-eight

Precession of the equinoxes.

(ERRATUM read 28 for 29) degrees distant. This movement is commonly termed the precession of the equinoxes.

And because this motion is completed above the poles of the ecliptic very slowly in the manner described, Ptolemy calculated that one circuit is completed within 36,000 years (if one reckons that only one degree is covered every one hundred years). According to the Alphonsine Tables, however, the circuit is made in 49,000 years, while contemporary scholars set the time as more nearly 25,000 years, with one degree's being completed within seventy years.

Various scholars have proposed different explanations as to the cause of these motions and others of this kind, and to account for their appearances (let us not speak here of the wavering motion "trepidatio" of approach or recession, or of the oscillating motion "libratio", arising from a certain irregularity in the precession of the equinoxes, as long as the point at which celestial longitudes are usually calculated, namely the vernal equinox, is uncertain, making its appearance earlier at one time than another). The older scientists were commonly of the opinion that even though the celestial movements appeared uneven and irregular to us, they were, nevertheless, regular and uniform in themselves since they believed that immortal celestial bodies cannot appropriately admit of irregularity. Therefore they sought to account for these appearances by postulating circular and even motions, based either on the assumed inactivity of the earth or its motion. Indeed a great part of mankind at that time wished to place an immobile earth in the center of the world.

The motion of "trepidatio".

CHAPTER 3

THE TWO SYSTEMS OF THE ANCIENTS: THAT BASED UPON THE PRESUMED INACTIVITY OF THE EARTH AND THAT BASED UPON ITS MOVEMENT

The first exponents of the hypothesis of an immobile earth were Anaxagoras, Democritus, and others, who declared that stars moved freely in Space and that they were not bound to solid spheres. Nor did they believe that there was a First or Second Mobile by which they were carried to the west. Rather the stars proceeded in a simple movement toward the west, and only appeared to be moved toward the east when in fact they were being carried slowly toward the west.

But they also should have explained why they move not only slowly, but also rapidly at times, why they frequently appear either larger or smaller, why they now move backwards and again are stationary, etc. For this reason, Pythagoras of Samos, a renowned Greek philosopher (who at the time of the first siege of Jerusalem, which sets his birthdate about 600 B.C., left his homeland because of the enemy presence in that part of Italy known as Magna Graeca, and settled at Crotona) devised another system whereby the appearances of the planets and the theories developed about them correspond admirably with their movements. According to his system, the sun is situated in the center of the world and the earth is mobile, a theory which his students and their successors took up later. Among his disciples was Philolaus (who is believed to have lived in the Italian town, Tarentum, and on whose account Plato visited Italy). He held that the earth revolves and is one of the stars. Likewise Aristarchus of Samos was one of his pupils (who embraced the Pythagorean doctrine in his writings and handed it down to posterity), as well as Ecphantus, Nicetas, Seleucus, Cleanthes, and others.

Pythagoras lived 600 years B.C.

Inasmuch as the belief in the immovability of the earth is an old one and has commonly been held to be true because it seems to be so, many philosophers have rejected the the doctrine of Pythagoras. Among this number was Aristotle as well as Claudius Ptolemy, a renowned astronomer and others of his contemporaries, who began his astronomical observations at Alexandria in Egypt around 130 A.D. He wrote the *Almagestum*, a comprehensive treatise dealing with the structure of the world. His death occurred in 147 A.D. and Tycho Brahe asserts that he had been of the royal stock of the Ptolemys. Thereafter John of Holywood of Yorkshire, the Alphonsine Tables, Peurbach, Christoph Clavius, and others, in order to account for the different appearances of the stars, but still to maintain the immovability of the earth, decided that the stars were attached or connected, as it were, to solid spheres which carried them about in their diverse and varied motions. For example: just as when a ship is moving south, a sailor can walk north from the bow to the stern, so can a planet move from west to east and vice versa when it is being carried by its solid sphere from east to west during a twenty-four hour period.

Ptolemy and his contemporaries.

This belief has been held by almost everyone up to this time and is concisely summed up in the following Chapter.

CHAPTER 4

A BRIEF RESUME OF THE SYSTEM OF THE WORLD BASED UPON PTOLEMY'S THEORY OF THE IMMOVABILITY OF THE EARTH

"The earth holds the central position in the world and rests immobile with its water and surrounding layer of air. Fire encompasses it and separates it from the rest of the world.

The moon passes nearest the earth in its orbit and its curved smooth surface is not injured by its fire. Mercury, which revolves about the moon, is in turn circled by Venus. These three wanderers are called "the lower planets". The sun, king, as it were, and ruler of all, occupies a middle place on its throne. It is attended on one side by the earth, moon, Mercury, and Venus and on the other side by its attendants Mars, Jupiter, and Saturn, at a distance, which lie in "the upper reaches of the world". These three are called "the upper or higher" planets hecause early astronomers had placed their orbits and courses above that of the sun. The region of the stars or firmament extends above these seven heavens, glittering with its many lights. At the time of Aristotle, the entire expanse of the world was included in these eight regions. With the passage of time, however, other movements were observed which led to the designation of new heavenly regions. The concept of the equinoxes introduced the Ninth Heaven when the movement of "libratio" was attributed to the starry firmament. Thereafter these nine heavenly regions were surrounded by a tenth called the First Mobile and this, in turn, by the Fiery or Empyrean Heaven.

Furthermore they held that all the spheres of the planets were solid and compact but that the individual motions of each were not interrupted by one another inasmuch as their polished surfaces allowed only of a glancing contact. Thus the moon completed its orbit without interference by Mercury and in the same way Venus did not interrupt the orbit of Mercury. And so all spheres were permitted to move without injury in an orderly and reasonable manner. In this way collisions were absent and stability reigned in the heavens, etc." Thus wrote Philolaus in his treatise *De Systemate Mundi*.

Pierio Valeriano of Belluno, moreover, described the system of the world to Cardinal Alexander Farnasius thus: "First of all you must bear in mind that this earth which we inhabit, together with the water surrounding it and adhering to it, within, above, and below, should be regarded as an essentially round ball which is

PLATE I

encompassed on all sides by air. This air, in turn is warmed by a layer of fire surrounding it. Now in these four simple bodies, namely earth, water, air, and fire, all elements are present which are later transformed into various kinds of matter. The ingenuity of the astronomers has disclosed that above these spherical bodies the seven orbits of the planets or wanderers lie, much indeed, like the skins of an onion. These are the moon, Mercury, Venus, the sun, Mars, Jupiter, and Saturn. It is believed that the orbit of the fixed stars lies above the sphere of Saturn and that the whole expanse of the heavens is transfixed by a number of fiery nails. Human vision seems to bear out this assumption. Above the stars, moreover, there is thought to be another heaven which the wise men of old called the First Mobile, inasmuch as it was moved by the power of God Almighty (for there is no other explanation). It rushes in a constant and rapid rotation from east to west and then completes its cycle in the reverse. It draws along with it the sphere of the fixed stars and those of the wandering stars. Some are drawn along more slowly and others more quickly according to their respective nearness to or distance from this very rapidly moving sphere. More contemporary astronomers have postulated a Tenth Heaven. Some have located the perpetual and tranquil abode of divinity there, as being a place where divine spirits and the souls of the devout stand before God and enjoy eternal peace. From the foregoing you can see that there is little agreement among the philosophers and astronomers about the heavenly spheres as well as the position and order of the wandering stars. We are merely setting forth here the various hypotheses that have been given by numerous scholars: as for us, we will defer our discussion of these questions until another time."

See Plate I.

Many astronomers, as well as the entire school of the Peripatetics, leave no stone unturned to save or defend and maintain this system, as is amply evident in their numerous books and writings *De Theoriis Planetarum ac Firmenti*. Here one can see how they regarded the varied and different spheres of the planets (exceeding fifty in number), both in their totality and individually; likewise how many and varied were the planetary circles, positions, epicycles, deferents, lines, centrics and eccentrics, and motions, etc. they devised, with the result that the planets, with their individual spheres, solid and compact, were made not only to revolve irregularly about their own axes, but with regularity about some other point. And thus by contrary motions, by spirals, as it were, these planets in an extraordinary way curved up and down and turned from rising to setting in the same time as from setting to rising for all to see. Consequently, they now moved swiftly, now slowly, at times were even stationary, and again after an interval, proceeded backwards. As time went on, and the more accurate the observations of astronomers became, the more spheres, circles, lines, centrics and eccentrics they were forced to devise by reason of new and hitherto unknown phenomena. For this reason, the hypotheses in *De Theoriis Planetarum* came to be so complicated that it was not only difficult to understand, but was clearly contrary to human reason. For surely it follows according to this system of the Ptolemaics that one and the same planet should be said at one time to be lower, but at another time higher. Take Mars as an example. This planet, although it has been established

The difficulties arising from this system.

elsewhere in the Ptolemaic system as heing above the sun, when it moves downwards to such a degree that its sphere breaks through that of the latter and takes a position below the sun (an occurrence which is a manifest impossibility because of the aforementioned solidity of the spheres), at this time approaches nearer the earth than the solar body itself. A short time later, however, it rises again and is found to be a great distance above it.

Let us not speak now of those diverse and varied motions of fixed stars that have been pondered over and recorded previously in Chapter 2. From the above-mentioned, as well as numerous other similar inconsistences inherent in this system, uncertainty, gross irregularity, and a complete disregard for reason are clearly apparent.

There is another system, however, which should be investigated, wherein the appearances of the courses of stars and planets are more in agreement with reason and nature.

CHAPTER 5

A REVISED PYTHAGOREAN SYSTEM OF THE WORLD WHEREIN, ACCORDING TO COPERNICUS, THE SUN IS HELD TO BE IN ITS CENTER

Since, then, the Ptolemaic system presented so many incurable defects and unreasonable hypotheses, Nicolaus Copernicus of Torun, Doctor of Medecine and Canon of Frauenburg in Prussia, was moved about one hundred years ago to resurrect the system of Pythagoras of Samos, Philolaus, Aristarchus, and their followers, who were mentioned previously in Chapter 1. They had made the sun the center of the world and claimed that the earth, together with the other planets, rotated about it. (Even before the time of Copernicus, in about 1440, Cardinal Nicolaus of Kues also held firmly to this conviction.) Copernicus assigned to the earth, therefore, the place previously occupied by the sun between the planets, Mars and Venus, thereby introducing a remedy for those difficulties and inconsistencies that arose from the Ptolemaic theory. His concept is illustrated in Plate II. In this diagram, the region of the fixed stars bounds the world and is completely without movement, and insofar as we can perceive, orbicular or spherical. The sun, moreover, is in the center of this heavenly vault and likewise appears stationary. With these two regions comprising immobile boundaries, as it were, the moving planets are arranged in the space between and carry out their various courses both around the sun and below the region of

See Plate II.

The sun is the center of the world.

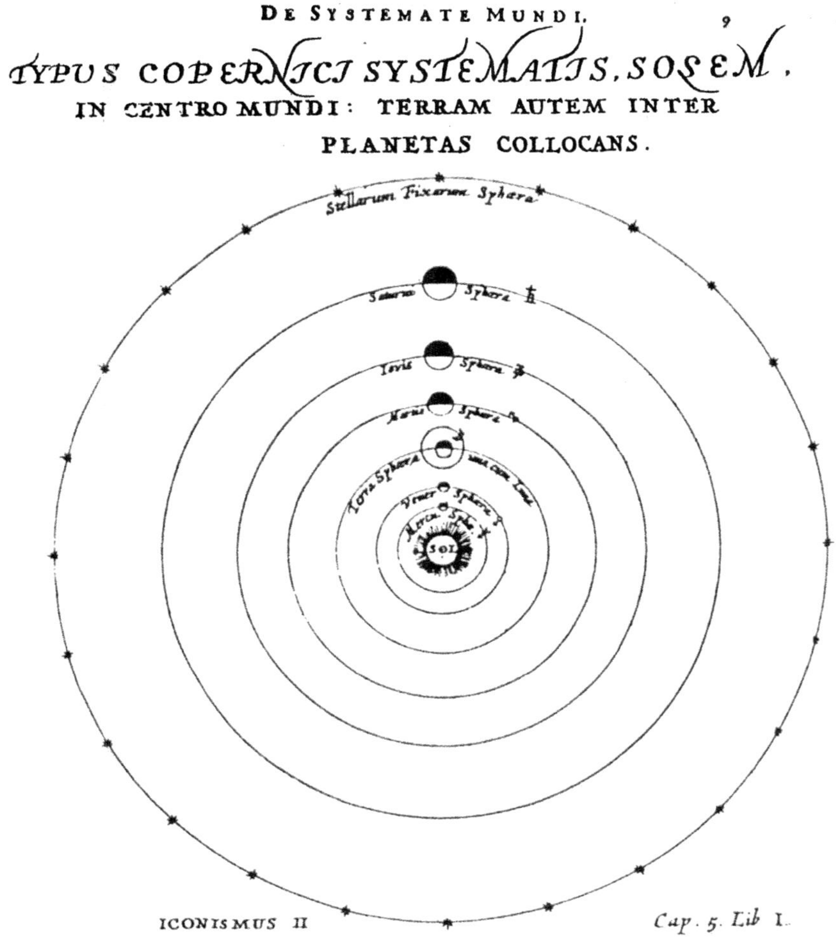

PLATE II

the fixed stars. First of all the planets, Mercury is situated nearest the sun with the result that it describes the shortest orbit of all about the sun and completes it most quickly, essentially within three months. Venus holds the second position. It has a larger orbit than Mercury and moves with greater speed than the more distant planets, thus completing its orbit within seven and a half months.

The order of the planets.

In the third place is the earth itself which embraces Venus in its

The planet earth carries the moon along with it.

orbit and thus completes its course more slowly, namely within twelve months or one year. It passes between Venus and Mars with so much distance on both sides that in the space between, it carries along the moon which follows in attendance. The moon, although it is carried about the sun every year in company with the earth, circles the earth by itself every month. Mars is fourth and in travelling about the earth, completes its circuit in essentially two years. Jupiter, in the fifth place similarly circles Mars and takes about twelve years to complete its orbit. The sixth is Saturn whose circuit embraces the foregoing and requires almost thirty years to accomplish its circuit.

The sphere of the fixed stars.

Finally, the space extending from Saturn to the fixed stars is vast and essentially boundless. To be sure the distance of the fixed stars from the earth is estimated to be so great that not only does the sphere of the earth, when compared to the region of fixed stars, have dimensions of a point, and indeed, all astronomers will concede this without contradiction, but even the great circle in which the earth follows its course around the sun and whose radius is the distance between the earth and sun, even this circle is of the dimensions of a point. All of these questions are thoroughly dealt with in the six books of Copernicus *De Revolutionibus Orbium Celestium* which was published in Nürnberg.

CHAPTER 6

THE DAILY AND ANNUAL MOVEMENT OF THE EARTH ACCORDING TO COPERNICUS

Having postulated that the earth, together with the moon, has its place between Venus and Mars, Copernicus attributed to it a twofold motion, daily and annual, to which a third is added, namely "inclination", although this cannot properly be termed motion.

The daily motion of the earth.

Daily motion is the revolution of the earth about its own axis and in travelling towards the east, it completes its circuit within twenty-four hours. In this way that portion of the earth which is turned toward the sun enjoys daylight, while that which is turned away, endures night and the regions of the heavens which appear in succession, as it were, come into sight, or rise, and then, after a time, pass away and become hidden, or set.

The annual motion of the earth which

Annual motion is the progression of the earth through the zodiac; that is, its movement between Venus and Mars and progress towards the east under the signs of the zodiac. During this time, the earth passes around the sun and completes its circuit within a year, that is, when the surface of the earth is turned about a center, namely its own axis. This center, in turn, is gradually advanced in accordance with the succession of the zodiacal signs. For example, it is like a globe turned above a plane whose center is moved forward down the length of the plane while its surface is turned about its center or axis. Because of this motion, one finds that when the earth lies between the sun and one zodiacal sign, the sun then covers the opposite sign and is said to be "in it". For example, when the earth is in Libra, the sun appears in Aries, and when it passes from Libra into Scorpio, the sun appears to pass from Aries into Taurus, and so on successively. The result is that it is the earth itself which passes through the zodiac and describes an ecliptic, although it is the sun which seems to pass through it.

should rather be termed "latio" or translation.

Finally the motion of "inclination" is not in fact motion, but a modification, as it were, of annual motion. Indeed, if in the great circle (which is the aforementioned circle the earth describes about the sun in its annual motion in accordance with the succession of the signs) you imagine both the ecliptic and the equator whose plane is thought to extend to the fixed stars, while at the same time the axis passing through the sun is thought to extend into the same region until it ends at the borders of the heavens, and then if you picture that the earth moves through the ecliptic and first comes into Aries with its equator coinciding with the plane of the equator of the world, at this time it is clear that the axis of the earth is also parallel to the axis of the world. Thus it is evident that the equator of the earth is always parallel to the equator of the world, although the presence of winter and summer and the variation in the length of days and nights might seem to indicate otherwise. If these hypotheses are accepted, then many problems which before and after the time of Copernicus had taxed the resourcefulness of the astronomers will be resolved harmoniously. This will be discussed more fully in Book V.

The motion of "inclinatio".

CHAPTER 7

OBJECTIONS OF THE ASTRONOMERS AND NATURAL PHILOSOPHERS TO THE COPERNICAN SYSTEM

Since, however, almost everyone has been of the conviction that the earth is immobile since it is a heavy body, the dregs, as it were, of the universe and for this reason situated in the middle or the lowest region of the heavens, and since they are of the opinion that the Sacred Scriptures also confirm this belief, the additional criticisms and contradictions of the astronomers themselves as well as those of the natural philosophers and theologians continue to persist. Some of the early critics, like Aristotle, Ptolemy, and others added force to these criticisms that have been brought forward. Another group added new objections of their own and when they saw themselves refuted by fundemental mathematical and philosophical arguments, they called upon the authority of the Scriptures. The arguments which Ptolemy relied upon to prove that the earth is stable, are essentially as follows. The first is based upon the injuries to be feared as a result of the daily motion of the earth. "Animals and whatever else the earth sustains, will fall into the air and leave the sphere of earth; for the earth must move 5400 miles in one day, or 1/6 of a mile per second, and for this reason, (so he believes), it will disperse, dissipate or scatter everything into the heavens with utmost speed etc." (Book IV, Chapter 5 shows, however, that this is prevented by the conserving virtue of the earth.)

The first objection based on the injuries to be feared as a result of the daily motion.

The second objection is "If the earth moves towards the east in its daily motion, everything that is not on its surface will seem to move westwards. Clouds, rain, hail, snow, birds, arrows, stones, and other objects projected upwards and sent downwards show, however, that this is not the case since their movement is not only easterly but in every other conceivable direction in the heavens. The earth, therefore, is not moved from west to east but is at rest." (The refutation of this objection is to be found in Book V, Chapter 12.)

The second objection from objects projected upwards.

Ptolemy bases his third objection upon the weight of the earth "For this reason he concluded that it is completely incapable of a twenty-four hour movement." (According to Book V, Chapter 15, nevertheless, a heavy body is more suited to a rotating motion than any other.)

Third objection.

Others present these arguments to prove that the earth is stable: "No stone, arrow, or any other body that is projected upwards or cast down from on high would touch the spot of earth lying directly beneath it. Rather, while this object is in the air, the point on the ground towards which it would tend would withdraw from it." Likewise, "A cannon ball exploded in a westerly direction would travel farther than if it were projected into the east." Likewise, "A necessary result of the motion of the earth would be that nature at no time would use a straight line. It would prefer instead a spiral, arc, or circle." (As a matter of fact, there is a great difference between motion and translation. Concerning this, see Book V, Chapter 16. Here one finds that objects are not moved forward but are borne directly to the center of the earth, although they are, at the same time, carried around by the air, according to Book V, Chapter 12).

Because these and other similar objections have been carefully answered by our experiments as well as by Galileo in his Dialogues, *De Systemate Mundi*, pages 90, 94, 136, and 139, Philip Lansberg, in his *Commentationes in Motum Terrae Diurnum et Annum*, pages 7 and following, and Daniel Lipstorp in *Copernicus Redivivus* pages 110, 128, 134, 140, 150, and following, we do not think it necessary to repeat the answers here in greater detail. Instead we refer the reader to the references given.

CHAPTER 8

OBJECTIONS TO THE COPERNICAN THEORY FROM THE SACRED SCRIPTURES AND THEIR REFUTATIONS

I

Objection to the mobility of the earth.

The words of Psalm 104 cited above in Chapter 1 are commonly quoted as an argument against the mobility of the earth. "Jehovah (Who) laid the foundations of the earth, that it should not be removed forever." Or to quote from other books. "He founded the earth upon her bases and it will not be moved for ages to come." Likewise in Ecclesiastes, Chapter I, "But the earth abideth forever." (In Hebrew עֹמֶדֶת is the equivalent of "permanet.") The followers of Copernicus, however, reply that these and similar phrases are metaphorical and if we insist on interpreting them so simply, countless other absurdities can be drawn from the Sacred Scriptures. Indeed, the Scriptures do not purport to explain the system of the universe or to decide whether it is the earth or sun which occupies the center position or which of them is immobile. They rather make known to man in the honor of God the perfection of his work and its permanence and incorruptibility. Those who are eager, therefore, to prove that the words of the Scripture reject the natural orbit of the earth or its daily motion should first show where the Scripture denies the movement of the earth and condemns those who support this theory. But this is nowhere! Well then, the critics answer, the Bible attests to the immobility of the earth from its durability, its quality of endurance, and its perfection. Isaiah, Chapter 66, to be sure, speaks in this sense when he says, "For as the new heavens and the new earth, which I make to stand before me, saith the Lord, so shall your seed and your name remain." One should not conclude from these words that either the sky or the earth stands still and does not move itself, any more than assume that men from the seed of Abraham stand still and do not move. One should understand the immovability of the earth to refer only to its durability and its quality of endurance.

II

Objection to the stability of the sun.

An objection to the immovability of the sun is cited in Psalm 19 (ERRATUM read 19 instead of 18) "In them hath He set a tabernacle for the sun, which is as a bridegroom coming out of his chamber, and rejoiceth as a strong man to run a race. His going forth is from the end of the heavens, and his circuit unto the ends of it: and there is nothing hid from the heat thereof."

The Scriptures speak of what we commonly perceive.

Likewise in Ecclesiastes, Chapter I, there are the words, "The sun also ariseth, and the sun goeth down, etc.". Now since the Sacred Scriptures were not given to us by God so that we might learn to philosophize from them or become mathematicians and natural philosophers, but rather to impart instructions for eternal salvation, for this reason scholars misunderstand the scope of the Scriptures when they judge geometrical and astronomical questions according to its precepts. For the Scriptures speak of these things only as they commonly appear to men and pay little heed to their innate and actual qualities. One may ask, for instance, whether or not the smallest stars shining in the heavens are of greater brilliance than the moon. Now, to be sure, the Scriptures say that "the moon sheds great light" because it seems to be greater than all the stars as a result of its nearness to the earth. As a matter of fact, however, any fixed star, no matter how small, is larger than the moon.

In the same way men, when they approach a city in the course of a journey and for the first time see the tops of its spires and then proceed down the road from which the view is somewhat obstructed so that they can no longer see the spires, say, "The spires have fallen from sight, have disappeared, or have hidden themselves. When, however, they came through a clearing, they say, "The towers appear, rise, or show themselves", which is not actually the case, but only what the towers appear to do.

Similarly when we emerge from some narrow valley, we say that a wide plain opens up. In the same way Christ said to Peter, "Lead into the depths, etc.," but it is not as though the small pond of which he was speaking were deeper in one place than another. Thus, the followers of Ptolemy speak of the solstice, although they deny that the sun stands still. The followers of Copernicus, moreover, say that the sun rises and sets, although they are well aware that the rising of the sun is the result of the daily motion of the earth itself towards the sun and that the setting of the sun results from the earth's turning away. Similarly we speak of the eclipse of the sun when in fact it is an eclipse of the earth, etc.

Now, indeed, because others have treated of this subject, among whom is Paulo Antonio Foscari, the Carmelite, who wrote a letter, published in Naples in 1615, to the head of the Carmelite order, with the approval of the College of Papal Theologians, we beg the reader to turn to this author as well as the *Introductio in Martem* by Johann Kepler, the Imperial Mathematician. Here he will find a resolution to the contradictions of these and other authorities of the Scriptures who commonly oppose the motion of the earth and the stability of the sun.

III

Objection from the sun's standing still in Gibeon.

Critics employ the miracle of Joshua, Chapter 10 to refute the theory of an unmoving sun in the center of the world: "And Joshua spoke in the sight of Israel, 'Sun, stand thou still upon Gibeon; and thou, moon, in the valley of Ajalon.' And the sun stood still, and the moon stayed, until the people had avenged themselves upon their enemies. Is not this written in the book of Jasher, the Upright? So the sun stood still in the midst of heaven, and hasted not to go down about a whole day. And there was no day like that before it or after it, that the Lord hearkened unto the voice of a man." But if a person wishes to interpret these words simply and to extend this miracle beyond the region of Gibeon or Ajalon to the heavenly world and the sphere of earth, then many difficulties arise. (1) For the text says that the sun had stood in Gibeon and likewise in the middle of the heavens. Unless we are averse to common sense, it follows that Gibeon and the middle of the heavens lie in one and the same place or are one and the same thing. (2) The text says that the sun had stood in Gibeon and the moon in the vale of Ajalon. But since these places are not a German mile distant from one another, how can one perceive the difference either from Ajalon or Gibeon in the position of bodies so far away as are the sun and moon? For just as the Israelites saw the sun in Gibeon, so must they have they have seen it in Ajalon, as everyone knows who is aware that one side of this triangle measures one German mile, while the other contains 1,040,600 miles (Ptolemy's computation of the distance of the sun from the earth). It is quite apparent, therefore, that the sun and moon did not stand still in the great expanse of heaven and earth but only in the eyes of the sons of Israel, and indeed in the two aforementioned places. Thus had Joshua spoken in their sight, "Sun, stand thou still upon Gibeon, and thou, moon, in the valley of Ajalon, etc.". Thus the miracle should be confined within their own boundaries, or else what need would there have been to speak of the sun and moon's positions in Gibeon and Ajalon if they had actually stood still in the whole world? (3) Because mention was made not only of the position of the sun, but also of the moon, and since with the sun shining, no light could be shed on the earth by the moon, one must assume that at the time Joshua bid the sun stand still, it had already set and the moon had been about to rise. This miracle, therefore, should so be interpreted that when the sun had in fact set and the moon was beginning to rise, a great glow from the sun remained in the center of the heavens. For otherwise, if the sun had remained in the middle of the heavens, the people would not have seen the moon, nor would they have had need of its light. (4) Now if the sun had stood still in the company of the moon, then all the other planets had either remained with them, as well as the vast multitude of fixed stars or they had departed from their usual course. If the latter were true and they did this, then the order that had been set up from the beginning of the world was rendered completely confused and disordered as well as the relation of the sun and moon to the fixed stars and the planets. Had they stood still with the sun and moon, however, then in order that the sons of Israel might overcome their enemies in this tiny restricted region of Gibeon and the vale of Ajalon, thousands of huge astral bodies at a great distance from the earth, in comparison with which the earth could not be as large as the smallest grain of sand in the sea, stopped in their course. But indeed this is unthinkable and beyond the imagination of mortal man, nor has it been recorded in the Scriptures that so extensive a miracle had taken place, one that far surpasses the wonder of the sun's standing still. This is quite apparent from the vast number of the stars, their great distance away, and their size. (5) If the whole expanse of heaven stood still, it would follow that all the children of the earth as well as the animals conceived and brought to birth within this space of about twenty-four hours would have had one and the same horoscope for their conception and nativity or the same zodiacal sign, which in fact is a principal governing force in an individual's nature and future life. It would also follow that the whole expanse of months and years had been interrupted and destroyed in one day, an event mentioned nowhere by any writer or the Scriptures themselves. (6) Someone may wonder when the sun stood still in the heavens for two days, whether or not all people, animals, trees, and plants, especially those found in southerly regions and in this land of Gibeon, would not have suffered damage from burning or scorching because of the heat from an immobile sun, and contrarily whether the rest of the animals on the other or opposite side of the earth had not suffered two nights of hunger, cold, and other great trials. (7) The length of the day also makes it clear that this miracle did not extend beyond Gibeon and Ajalon: "For this day was as two." Ecclesiastes, Chapter 46. Now the longest day in Palestine has 14 (ERRATUM read 14 for 13) hours and the shortest, ten hours. Consequently, this day could not be longer than 28 hours. We know, nevertheless, that in polar regions summer days are usually much longer than twenty-eight hours, in fact, they last whole months! Thus it is patently clear that this double day should be considered as two consecutive days in the region of Gibeon, and in consequence, that this miracle should not be extended to all lands, but should include only the land of Gibeon, where there never was before or after so long a day. But should one wish to interpret that day as being as two in any region whatsoever of the earth at the time of the miracle, it follows then that in the polar regions, where days during the summer last weeks and even months, these two had been doubled in length. This clearly is impossible, however, for the engagement of the Israelites was completed in that day, which was as two (that is, a day of twenty-four or twenty-six hours). (8) This miracle is taken from the book of Jasher which in turn had probably received it from another source. Since this Scripture, therefore, serves only to transmit the record of this miracle, one can ask whether the book of Jasher had been a canonical or an apocryphal writing, and if the former, whether the canonical text had heen lost. (9) It is not likely that at that time this miracle was intended to define a theory for the true system of the world, and thereby to affirm or deny that the sun or earth stood still. Rather it seems much more likely that the miracle was meant to reveal to mankind what God had done for Joshua. It is of no concern whether this had been done through the sun's actually standing still, or its appearance of doing so, or through the immobility of the turning earth, or through the brightness of the sun. Likewise it is immaterial whether God appeared in the form of fire, a fiery column, a cloud, three men, a dove, the face of Stephen, or that of an angel, etc.

IV

Objection from the sundial of Ahaz.

We should likewise investigate the miracle that God performed in curing the King, Hezekiah (Kings II, Chapter 20). "By a sign Isaiah was to know that the king was going to be cured. And Isaiah asked the king whether the shadow should go forward ten degrees or go back ten degrees. And Hezekiah answered, 'It is a light thing for the shadow to go down ten degrees: nay, but let the shadow return backward ten degrees.' And Isaiah the prophet cried unto the Lord and He brought the shadow ten degrees backward by which it had gone down in the dial of Ahaz." We also find this recorded in Isaiah, Chapter 38, verse 8 in these words, "Behold I will bring again the shadow of the degrees, which is gone down in the sundial of Ahaz, ten degrees backward. So the sun returned ten degrees, by which degrees it was gone down." Now this miracle of the sun's having moved backward in the heavens can be accepted by those who commonly make superficial judgments and are scientifically ignorant of the true system of the world. When this text, however, is examined from a twofold point of view, that is, from an assessment of the true situation as well as from actual appearances, and when one makes a careful distinction between that which concerns the orbit of the earth in its entirety and that which concerns the people of Judaea in particular, it is quite clear that this miracle of the sundial had indeed occurred only on the sundial of Ahaz and not on all the sundials of the earth. For as God had given a sign to Hezekiah when he was ill and prayed to Jehovah for help and not to the world as a whole (Chronicles II, Chapter 32, verse 24), so this divine sign was evident on the dial of Ahaz alone. Had the sun actually moved backwards, the sign would have been made for all men, but this clearly, was not the case. This fact is further confirmed in the aforementioned reference in Chronicles: "Howbeit in the business of the ambassadors of the princes of Babylon who sent unto him to enquire of the wonder that was done in the land –" (That is, in the land of Judaea, and not in the heavens, for there had been no occurrence of this kind in Babylon. Otherwise, it would not have been necessary for the Babylonians to ask the Judaeans about the sign had this occurred in Babylon or had the sun turned back in the sky, for then the whole world would have seen it). Let us assume, however, that the sun had turned back, and we shall see what necessarily follows from such an assumption. Either the sun turns back alone and by itself or it does this together with the First Mobile, that is, with the other planets and all fixed stars, as the followers of Ptolemy assert. Now if the sun turns back alone it can only do this in its own ecliptic.

Furthermore if ten lines make ten hours, the sun turned back 150 degrees, which constitute five whole signs of the zodiac. Thus, summer was either made from winter or winter from summer, both of which are unlikely occurrences. But if, in fact, the sun turned back with the First Mobile through a twenty-four hour cycle, all the heavenly spheres, those of the fixed stars as well as the planets, were similarly turned back so that the shadow might recede on the dial of Ahaz. This clearly would have upset all nature, its order and sequence of things, the rise and fall of stars, the ephemerides, and astronomical tables; none, however, noticed that this occurred. Hezekiah, moreover, sought nothing more than the turning back of the shadow on the dial of Ahaz and thus God seems not to have granted more than was desired nor to have multiplied miracles of such magnitude to no purpose. Indeed, that day would have been almost three times longer than other days, for after the sun had completed ten lines or hours, and had again turned back ten lines, it should again have run forward through ten lines, thus completing a day of thirty hours which is the equivalent of almost three days in Palestine. But this would be contrary to Joshua's words cited above, "that there was no day like that before it or after it" and again "that day was as two", Ecclesiastes, Chapter 46. At the time of Hezekiah, therefore, this day was an ordinary day and consequently it was not the sun, but a shadow that receded on the dial of Ahaz. For clearly the day was not lengthened as the time involved was too short, a fact which the text itself confirms when it says, "when Isaiah cried unto Jehova, and he brought the shadow ten degrees backward, by which it had gone down etc.". In other words, God brought the shadow backward at that moment or hour when this miracle was expected and while the assembled people watched, and not within ten hours, which constitute a whole day in Judaea in wintertime.

This miracle occurred, then neither in the sun itself nor on all the sundials throughout the earth, but only on the dial of Ahaz. For surely the sun does not run back in a moment and return again along the same course which it had completed in a full day. Indeed, had this been the case, the sun would have made winter from summer and should have completed a circuit of 5,450,760 German miles in a scarcely perceivable moment of time, provided one grants that the sun is mobile and calculates with Ptolemy that it covers 272,538 miles an hour in its ecliptic. Such a miracle would clearly have been contrary to all human understanding, perception, and reason in view of the required alteration of time and the sun's retrograde movement, not to mention that of the dependent planets and fixed stars. There is little doubt that had this miracle occurred, it would have been investigated a great deal more thoroughly than it was and the writers in Babylon and elsewhere would not have passed over it in silence.

Miracles, therefore, should be accepted within limits, for inasmuch as God made man a rational creature, so do we function rationally and decline to defend such miracles which subvert all reason and disparage the power and order of God. For what could weaken His power more than to conclude that He was unable to make one shadow recede on the dial of Ahaz without disturbing so many thousand heavenly bodies and the whole fabric of time?

The argument of the author.

Now these are the arguments which I found on this subject presented by writers who support the Copernican system. Certainly it is not my intention to defend these arguments or to make their controversy mine, much less to write against the Sacred Scriptures (if this should arise). No one will deny, I am sure, that the Scriptures must be interpreted in many places not with bald literalism, but with rationality and good sense. Otherwise it would follow from the words of the Psalms quoted above that the earth "should not be removed forever" and likewise that "the earth abideth forever," remarks which would give false credence to Aristotle's theory concerning the imperishability and eternity of the earth. It would likewise follow from Job, Chapter 26 where the "column of the heavens" is discussed and Psalms 2 and 22, where the "uttermost parts" of the earth are mentioned, that the heavens are supported by posts and beams, thereby indicating that an end is foreseen for the spherical body.

Hereupon with the words of Johann Kepler in his aforementioned work, I exhort my reader "by awareness of the divine goodness conferred upon mankind, to praise and celebrate with me the wisdom and bountifulness of the Creator, which I disclose to him in a thorough explanation of the design of the world and an investigation into first causes, once error has been dispelled, and thereby to recognize not only the great protection He affords all earthly creatures through the abiding power and stability of the earth (for which God be praised), but also through its movement, so well-concealed and worthy of wonder. If a person be too dull-witted to comprehend his knowledge or too enfeebled by his wounded piety to believe Copernicus, I encourage him to dismiss the school of astronomers and, if he wishes, even those among the condemned philosophers he favors, and pursue his own investigations and cease his earthy wanderings. With his eyes raised to the heavens above, for with these alone he sees, and with his whole heart, let him pour out his praise of God, the Creator. Meaning to offer up no less reverence to the Lord for his discoveries than the astronomer to whom God granted clearer vision with the eye of his mind, he can and so desires to honor his God. So much for the authority of the sacred writings. As for the opinions of the saints regarding natural phenomena, I herewith give a brief reply. In theology, to be sure, authority should carry weight, but in philosophy, the decrees of reason should be heard. Therefore although it was St. Lactantius who said that the earth was not round, St. Augustine who denied the existence of the antipodes although he admitted that the earth was round, and contemporary authorities holding holy office, who, though conceding that the earth is small, nonetheless refuse to admit its movement, still sacred reason carries more weight with me than these. I can demonstrate from philosophy with due respect to the Doctors of the Church that the earth is round, is inhabited in the antipodes, is of a smallness worthy of greatest scorn, and is carried in its orbit about the heavens."

Authors who reconcile the words of the Scriptures.

So much for Kepler, the highly distinguished mathematician of the Emperor Rudolph II. If one should require further discussion of this matter, he should look into Philip Lansberg's *Commentationes in Motum Terrae Diurnum et Annuum*, pages 11 ff., Peter and Jacob Lansberg's *Apologia*, Peter Cruger's *Prognosis Astrologica* for the year 1639, Chapter 4, Thomas Wite"s *De Mundo Dialogus*, Book 2, page 132, Campanella's *Pro Galilaeo*, David Frölich's *Prognosis* for 1640, Daniel Lipstorp's *Copernicus Redivivus*, the last chapter, Michael Havemann's *Astrea*, Chapter 6, pages 25 ff., (herein he explains opposing fundemental testimony from the Scriptures based on Hebraic and Chaldean sources and rabbinical commentaries), and many others. But since indeed, Tycho Brahe saw with discerning eyes, among other things, the extraordinary error of the Ptolemaic system in designating the location of Mercury and Venus, and recognized that Mercury is nearer the sun than Venus, that Venus is nearer the sun than the earth, that they never come into opposition, and for this reason that the Ptolemaic system of necessity requires correction, nevertheless, he did not wish to subscribe to the Pythagorean or Copernican system. Thus he proposed another system, perhaps to procure for himself the support of those who denied the motion of the earth. This theory we will now discuss.

CHAPTER 9

THE WORLD SYSTEM ACCORDING TO TYCHO BRAHE, THE NOBLE DANE. HE ASSIGNS THE EARTH TO THE CENTER OF THE WORLD AND THE SUN TO THE CENTER OF THE PLANETS WHICH REVOLVE AROUND IT (WITH THE EXCEPTION OF THE MOON)

Tycho Brahe (a man heretofore not sufficiently praised, for by reason of his ingenious perception, his enthusiasm, and his expenditures verging on the kingly we have more precise observations of celestial motions and canons more compatible with the heavens than ever before), in order to retain the earth in the center of the world, as popular opinion commonly held, devised a system of his own, altering the Ptolemaic and Copernican systems. By following the middle course, he allowed the sun to have motion, as did the Ptolemaic theory, but assigned it to the center of the remaining planets, those other than the earth and moon, as did the Copernican theory. He proclaimed further that the heavens were fluid. See Plate III.

Description of Tycho's system.

In Tycho's system, the earth is in the center of the heavens, that is, in the center of the moon, sun, and fixed stars. Among this number the sphere of the moon is first, then that of the sun, and finally that of the Eighth Sphere. The orbits of all these spheres are concentric with the earth. The sun is reckoned to be the center of the five planets, Mercury, Venus, Mars, Jupiter, and Saturn, which circle it as their own particular leader or king and always look to it in the midst of their revolutions. Thus within a space of twenty-four hours, the whole planetary system together with the sun, turns about the earth every day. In addition to this circuit, however, each planet completes its own individual orbit every year, either more or less rapidly according to its size and distance from the sun. Mars circles these at a very large interval (bringing it nearer the earth than the sun itself when it rises at sunset and sets at sunrise). Then Jupiter and Saturn circle this planet in turn. From these movements (wherein the sun draws along these five planets in its course around the earth each day in the manner described, like an independent planetary system), other phenomena are derived. Now indeed, although this system of Tycho's has been ingeniously devised, nevertheless, it will become evident from what follows that this theory actually is not valid. To be sure, Tycho sees the particular flaw of the Ptolemaic system which locates Venus and Mercury near the earth when, as a matter of fact, they are nearer to and always attendant upon the sun, never deviating far from it, as we said above in Chapter 2. Nor do they come between the earth and sun in opposition, an occurrence which would necessarily arise if they passed around the earth. For these reasons, then, no one can deny that they complete their orbits about the sun and not, indeed, about the earth. Furthermore, he also sees that the Ptolemaics are gravely in error, when to preserve and defend their system, they assert that the spheres of the planets are dense and compact, something, however, that Tycho does not admit. He denies the presence of solidarity in the heavens and substitutes instead a fluid matter through which the sun, like a charioteer, drives the rest of the planets about the earth every day. But however this may be, Tycho's system is contrary in many ways to human reason as well as to mathematical proof.

Objections to Tycho's system.

First of all, in this system the planets wander everywhere and their paths are more intricate than the webs of spiders. For, contrary to the laws of nature, Mars' course is projected through the circular orbit, or ecliptic of the sun as well as through the paths of the rest of the planets, whereby their revolutions, through dislocation, as it were, become mutually involved and completely confused.

In the second place, it is contrary to reason and intelligence that so extensive a solar system including the five planets, should complete a circuit of the earth in the relatively short space of twenty-four hours, indeed, even pass through matter as well, though it be of the most transparent and fluid kind. For this system is so extensive that the whole sphere of the earth in comparison should not be regarded as even the size of a point, let alone that of a pea.

The third question now arises as to how this planetary system which likewise embraces the earth, can be moved without the intervening light matter or substance, which Tycho calls aetherial air, being pulled along and the earth's being separated from elements of this substance that are contiguous to it. For while this planetary system, in which the earth holds not the center position, but only the middle place in respect to the sun, is carried around, of necessity the earthly sphere which Tycho established to be outside of the center of this system, must be carried at the same time in it and with it.

Fourthly, no reason can be given for the sun's having such power or the capacity for pulling about with it an extensive system of this kind in which the bodies are so widely separated and huge, within the space of twenty-four hours and indeed accomplishing this in such a way that all the individual bodies, nonetheless, keep to their own courses each year.

Fifthly, it is not reasonable that Tycho describe the orbit of ☉ as being much smaller that those of ♄, 4, ♂, for in this way, he makes

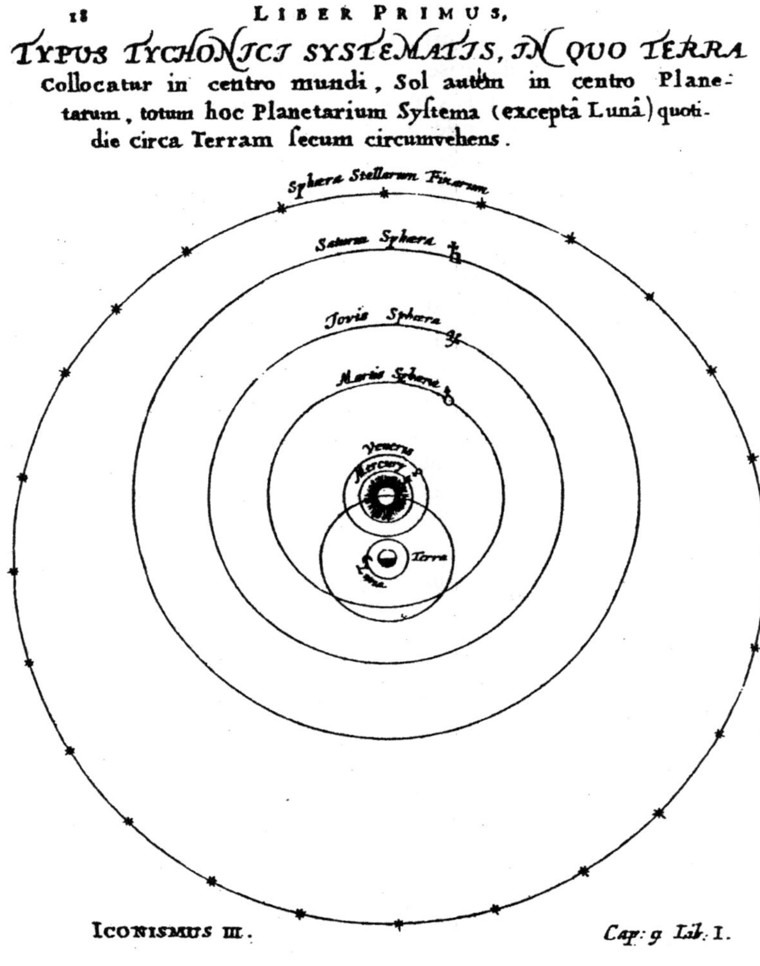

PLATE III

them confused. Ptolemy was more careful in this respect for in assigning to each of the planets its own deferent, he so enclosed their orbits within spheres that they could not go beyond them.

Authors who have disputed Tycho's system.

But inasmuch as others have treated of this matter, for example Kepler in his *Epitome Astronomiae Copernicanae*, Book 4, Philolaus in his *De Systemate Mundi*, Chapters 2, 3, 4 and Book 4, Chapter 38, Daniel Lipstorp, in *Copernicus Redivivus* Chapter 7, Michael

Havemann in his *Epistola Sideralis Scientiae*, etc. it is not our intent, therefore, to discuss the matter further here. We shall only add the particular circumstances which moved Tycho to devise a new system such as this. In the Ptolemaic system, he could not countenance the absurd distribution of the heavenly bodies, the unnecessary postulation of so many and so great epicycles to account for the phenomena of retrogradation and absence of motion, the incongruous assumption that in this system the uniform circular motion of a body should be described not about its own center, but about that of another eccentric, and an acceptance of the solidity of the spheres, etc. In the Copernican theory, he objected most to the fact that even though nothing was in conflict with mathematical principles, to ascribe motion to the earthly sphere, a heavy body in itself, and slow and unwieldy of movement, was contrary to the laws of physics, as well as against the Scriptures which confirm, so he believes, the stability of the earth. And finally, he thought that the space (empty straight up to the stars) between spheres ħ and 8 was essentially immeasurable.

The reasons why Tycho devised his system.

CHAPTER 10

ANOTHER SYSTEM WHEREIN THE EARTH IS SITUATED IN THE CENTER OF THE WORLD AND MOVES ABOUT ITS OWN AXIS IN THE SPACE OF TWENTY-FOUR HOURS

There is, in addition, a fourth theory of the world system about which the first writers were Heraclides Ponticus, likewise Ecphantus and the youthful Plato. These were followed by Tost, Longberg, and most recently by Michael Havemann, Senior Officeholder of the State Ministry in his *Astraea sive Sideralis Scientia*, and others. These place the earth in the center of the world, and although they do not allow of its moving forward on its central axis, they assign, nonetheless, daily motion to it, that is, setting and rising within a period of twenty-four hours. It is in accordance with the following optical principal that the sun (ERRATUM read "ut sol caeteraque" for "caeteraque") and other stars are seen within the same space of time to set: "Those bodies visible to the eye and which move with equal velocity as the eye, seem to be at rest. Those which move more slowly, however, seem to move backward, and those which move more rapidly, seem to be carried forward." Vitelo, Book 3, propositions 135, 138, Euclid, *Optica*, theorem 59). To each of the planets, however, they attributed individual motion in free space and to the firmament, the slow motion of the equinoctal precession, that is, one revolution in 25,000 years, a subject discussed above in Chapter 2. They call his motion secondary motion and declare that it is free of all impetuosity, contrariety, error, and fluctuation so that no planet in the heavens circulates more quickly on one day and more slowly on another or that a man who is now standing perfectly still, would be made to move forward moments later, and shortly thereafter to move backward in crab fashion, etc., as the Ptolemaics have taught. This theory poses several questions, however, that must be addressed. (1) Why it is that the planets now appear larger to us, now smaller and consequently do not always maintain the same distance from the earth, for at times they approach very near to it and at others recede at a distance from it. They should likewise explain why they seem to move now more slowly and again more quickly. They say that these appearances are the result of the eccentric. For example, according to their theory, when a planet revolves in a circle which is not described about the center of the world or earth, but about another point, of necessity it is at one time nearer, and at another, more removed from us. Furthermore even though it travelled uniformly through the signs of the zodiac in a uniform time, nevertheless, because of its eccentric motion, it would not appear uniformly the same to us. (2) When one asks why one or another of the planets seems to us not only to move more quickly at times and again more slowly, but at one time to stand still, and again move forward and then backwards crabwise, at this point they are compelled to devise epicycles in the eccentric orbits of the planets whose centers they represent to be on the circumference of the eccentrics. In this way, they seek to account for these anomalies of the planets, for they say that the epicycle moves into opposite regions from the eccentric when these circles are opposite to one another, as it were, and the planet in an epicycle appears to us either to stand still or to move backward.

But indeed, it is completely contrary to reason to assume that planets have individual eccentrics in the world. Now what would this center be? For what good and to what end? And by what virtue, I ask, are these stars moved about such an imaginary eccentric in the fluid heavens? Or how is it possible for a planet to complete not only its own circuit about an imagined eccentric, but even to run through its own individual course in an epicycle and indeed to accomplish this in free space? There is little doubt that the Ptolemaic hypothesis is more tenable than the foregoing (for tò each planet Ptolemy assigned its own individual solid sphere, and in so doing, these anomalies in the heavens can be accounted for more reasonably). Inasmuch as other writers have discussed these matters with sufficient completeness, we shall turn from this question for the sake of brevity and proceed to other problems.

CHAPTER 11

THE CORRECTED AND IMPROVED COPERNICAN THEORY OF THE WORLD

After students of optics had invented the telescope in this century, the truths of the old Pythagorean and Copernican theories became much clearer than before. For once again they served as a firm foundation for these investigations and opened a wide field for discovery and speculation. At this time, Galileo Galilei, a professor at the University of Pisa and the astute mathematician of the Grand Duke of Tuscany, discovered in the year 1610, with his famous telescope, the four moons around Jupiter, which were called the satellites of Jupiter or the Medici stars after the city of the Medici where they were first discovered. Simon Marius, the first of the Germans to observe them, called them the Brandenburg stars (because the margraves of Brandenburg were his lords and he received their patronage).

The satellites or the companions of Jupiter.

Subsequent investigators thereafter brought to light certain new and hitherto unknown facts about the rest of the planets which will be dealt with specifically at the proper time and place. Suffice it to say, however, that with the help of telescopes they discovered that all planets are rough spheres, not unlike our own earth, without light of their own, but reflecting the rays of the sun to our eyes. Inasmuch as the sun is the focal point of our system, our discussion should of necessity begin here.

CHAPTER 12

THE SUN

The sun is visible in the middle or center of our world and controls the group of planets surrounding it. Like a king seated on his royal throne, he not only illuminates all the planets with his light-giving rays, serving as the lamp of the world, but even regulates their motions and allows them to share in his properties of warmth and life. For indeed, without his vital force in the reproduction of animals, all generative efforts on our earth would be unavailing. Here, then, in this Creation marvelous beyond all others, can the unique omnipotence of God properly be seen.

The sun is continually turning.

The sun turns in place about its own axis and when it is observed with a powerful telescope equipped with colored glass to prevent the viewer's being blinded, it appears to rotate continually, seething with its extremely hot vapors and luminous flames. As a matter of fact, this continual revolving motion can clearly be seen by the naked eye in fair weather when the sun is either on the point of rising or of setting.

The planets rotate in the same direction as the sun.

Since other planets always seem to rotate in the same direction as the sun and its spots, there can be little doubt that its impelling motor and rotating virtue helps move all the planets periodically within a year. Furthermore they participate in this solar motion to a greater or lesser degree and thus increase or diminish their speed according to their relative distance from the sun.

See also Book VI, Chapter 11.

It follows, consequently, that the slower these planets rotate, the more distant they are from the sun; contrariwise the more quickly they move, the nearer they are to it. Since, moreover, the transmitting and motive virtue of solar light rays is imparted to the planets surrounding them with great intensity, it is accordingly the case that this power is also less intense the farther and more removed a planet is from its source, namely the sun.

CHAPTER 13

THE SUN SPOTS

See Plate IV, Chapter 19 *of this Book.*

So much for our brief discussion of the sun: now we must consider the spheres around it. Among this number the spheres called the sun spots, cc, are nearest to it. Discovered (ERRATUM read "detectae" for "detecta") during our lifetime after the invention of the telescope, they have been accounted for in different ways by various scholars. (Their movements and courses tend from the easterly region of the sun and proceed to the west, like the uninterrupted orderly course of Mercury and Venus.)

Some believe that the sun like the earth is colored by vapors and is sometimes covered by clouds which are produced very easily and subsequently disperse. Some scholars like Father Christoph Scheiner and others think that these spots, or islands wandering over the sun, are attached to its body and turn with it around its own axis, a revolution which is completed within twenty-seven days.

The sun spots are planets.

Others hold that these spots are stars, a reasonable conclusion since we should not postulate the existence of spots or unsightly bodies on the surface of the sun, the eye of the whole world. And thus, even though they seem to be so near to the solar disc that they cannot be separated from it in any way and therefore appear as spots to us, they are stars, nonetheless, being carried about the sun in their own orbits not unlike Mercury and Venus. For indeed these planets do not appear to us to be moving about the sun: rather they seem to be nothing more than spots, when they move between the sun and our line of vision.

These bodies, then, revolving about the sun, very effectively reflect upon the sun the light they have received from it: this occurs in part because they are very near it. By reason of this reflection, they produce small bright patches, or "faculae" on the sun's disc, or so it seems. These bright patches are never evident without the spots since, in fact, they have their origin in them and move along with them. These spots, however, are located and move, not on the surface of, but outside and around the sun and are actually stars or very small planets which, because of their proximity to the sun, as Mercury occasionally is, can only be seen by us when they are in opposition, namely running between the line of the sun and our vision. Then they appear to our telescopes as sun spots.

The nearer these spots are to us, the larger they appear. The more distant they are from the sun, the more time they require to complete their periodic revolutions about it. On the other hand, the smaller spots turn in a smaller circle and for this reason revolve around the sun more quickly. Because of this diversity in their courses, therefore, they at one time mutually approach, at another time recede from, and occasionally blanket one another.

The reason that the spots at times appear oval.

Why then, do these spots at times appear oval and again of other different shapes? One reason is that when they revolve on the outermost rim of the sun's disc, they necessarily appear more contracted and smaller because the portion lighted by the sun is invisible to us while only the small darkened part which is left is seen. The nearer they are to the middle of the sun's disc and the more in our direct line of vision, the greater and rounder they always appear to us. A second reason for this variation in their shape is that when they almost coincide with one another in the line of our vision and partially cover one another, then they appear oval. Furthermore, because there are many of these bodies quite often, together with their moons, which congregate in one mass and then separate again, it appears to us that new spots are at times brought to birth, which later on are destroyed or disappear. This occurs when spots of different size and distance from the sun, or to put it another way, spots which follow the same course at different rates of speed, mutually blanket one another, and when those that are smaller and consequently nearer the sun are partially eclipsed by one or another larger spot and subsequently are again uncovered. What makes this theory even more credible is the fact that this occultation is particularly evident around the middle of the sun's disc.

In the great distance between the sun and Mercury there are many very small planets with their satellites.

Furthermore if one recognizes the existence of sun spots, it is apparent that there are as many solar eclipses, so to speak, as there are stars of this kind which intercept our line of vision and the sun. They frequently appear before the sun in various and uncertain numbers, for sometimes (although this is unusual), there are as many as thirty at one time, while at others, twenty, or occasionally only one or two, and sometimes none. Solar eclipses, therefore, occur in greater numbers and much more frequently than is commonly believed. As for the effects of these spots upon our earth, no one has come to any conclusion as yet. Because of them, however, astrologers recognize an element of uncertainty in their predictions.

CHAPTER 14

MERCURY

Mercury revolves in the orbit nearest the sun, except for the path of the sun spots. The Peripatetics held that this planet was provided with its own light and natural fire. In our century, however, the telescope reveals that at one time it is totally lighted, then humped, then of half-form and finally horned or sickle shaped like our moon. (See (ERRATUM, read "Iconism. sequent" for "Iconism.") Plate III, figure 1.) Now, however, it is indisputable that Mercury is lighted by the sun in the same way as the moon and earth. Furthermore, it is spherical and covered with rugged mountains.

In its course, Mercury does not deviate from the sun more than thirty degrees and at no time does the earth pass between it and the sun. Thus, one must necessarily conclude that it is nearer the sun than any other planet and for this reason describes the shortest orbit of all the planets about the sun, requiring less than three months (or more precisely 87 days, 23 hours, and 45 minutes) to accomplish this. Because of its proximity, Mercury is frequently blanketed by the sun's rays and cannot be seen by us except when it is at its greatest distance from the sun. When it does appear, however, we of necessity see it in the form of a hemisphere. This was experimentally confirmed with the new telescope invented by Francesco Fontana who published his new *Observationes Coelestium* in Naples in 1646.

It is evident from the *Annales Francorum in Vita Caroli Magni* in the library of Pierre Pithou published in Paris in 1588 that this planet appeared to be a dark spot on the sun on March 22, 807 A.D. Scaliger writes of a similar observation in his *Exercitio 72 Contra Cardanum*, as did Kepler in the year 1607, May 28. Pierre Gassendi refers to Mercury as a spot on the sun on Nov. 7, 1631 in his *Institutiones Astronomica*, Book II, Chapter 14. The famous and celebrated Danzig mathematician and town-councillor, Johann Hewelke, together with Herr Büthner made a similar observation on May 3/13, 1661 that Mercury appeared as a spot on the sun.

Mercury at times appears as a sun spot.

CHAPTER 15

VENUS

The planet Venus follows Mercury and never deviates more than 45 degrees from the sun, nor is it ever in opposition to the sun. By reason of a greater orbit than Mercury's, it completes its periodic motion somewhat more slowly within seven and a half months, or more exactly, 224 days, 17 hours, and 26 minutes. Its speed, however, is greater than that of the planets surrounding it.

Viewed with a telescope, it appears dark having no light of its own, hilly, and of irregular surface like the moon. This is particularly true when it appears horned. We see it as round and very bright when it is above (or beyond) the sun, but when it is below (or on the near side of) the sun, only a portion of it is seen illuminated, making it appear half-full or sickle-shaped. One can can conclude, then, with fair conviction that (1) the nearer this planet is to us, the smaller it seems because less of its surface catches the light of the sun. (2) The orbits of Venus and Mercury do not pass around the earth, as the Ptolemaics held, but rather around the sun. (3) When Venus and Mercury are in conjunction with the sun and lie in our direct line of vision, they appear to us as spots on the sun.

Venus is an opaque body like the moon.

Furthermore, the aforementioned Francesco Fontana wrote that two small globes or spheres revolve around Venus which he calls "courtiers or attendants". His first observation found Venus to be horned and similar to the moon in size when observed by the naked eye without a telescope. In his second observation he saw a small

Francesco Fontana ascribed two attendants to Venus.

sphere, dark purple in color, in the middle of its disc. His third observation found two separate stars of the same dark purple color adjacent to it, as it were, in the horns of the planet. His fourth observation revealed not two, but only one small sphere on the upper convex portion of the planet. With his fifth observation he found one small sphere near the concave section of Venus. (See Plate III ERRATUM, read "Iconism. sequent." for "Iconism"), figure 2.)

CHAPTER 16

THE EARTH AND THE MOON

The earth's position lies in the midst of the planets.

The tropical year.

The earth's position and that of the moon lie outside of Venus in the orbit which Ptolemy and Tycho assigned to the sun. Taking the moon with it as a companion, the earth completes its periodic course around the sun in a year (the tropical year which consists of 365 days, 5 hours, 48 minutes, and 45 seconds, and the sidereal year of 365 days, 6 hours, 9 minutes, and 26 seconds, with a time difference of 20 minutes, 41 seconds). Furthermore, the earth completes its full orbit every day in twenty-four hours, turning about its own axis from west to east and repeating this revolution 365 times in a year. During the same interval the moon turns around the earth twelve times. But how and by reason of what innate capacity does the moon accomplish this? And what causes this rotary movement? Why does the earth turn about its own axis and at the same time turn about the sun annually? Why does not the moon turn about its own axis like the earth rather than complete its orbit as an epicycle about the earth in 29 days? We shall discuss these and other questions in Book V.

CHAPTER 17

MARS

Mars circles the earth in a larger orbit, indeed, than the earth's (ERRATUM read "at" for "ut") but in a smaller one than Jupiter's. Its course around the sun takes almost two years (or one Egyptian year which is about 321 days, while its sidereal year is completed on the following day or night.)

When Mars is opposite the sun (especially when the sun is in the sign of Leo and Mars is in Aquarius and Pices), this planet is much

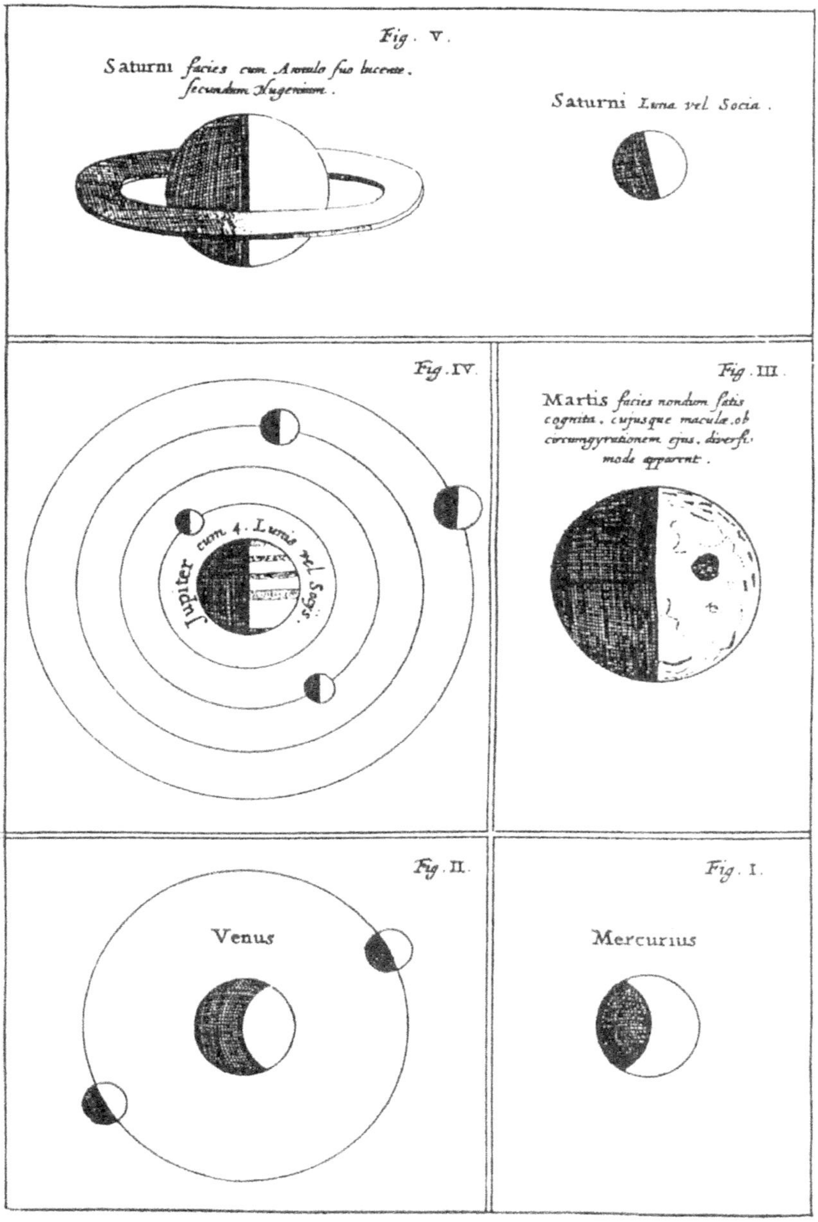
Fig. V.
Saturni facies cum Annulo ſuo lucente. ſecundum Hugenium.
Saturni Luna vel Socia.
Fig. IV.
Jupiter cum 4. Lunis vel Sociis.
Fig. III.
Martis facies nondum ſatis cognita, cujuſque maculæ, ob circumgyrationem ejus, diverſi mode apparent.
Fig. II.
Venus
Fig. I.
Mercurius

PLATE IV

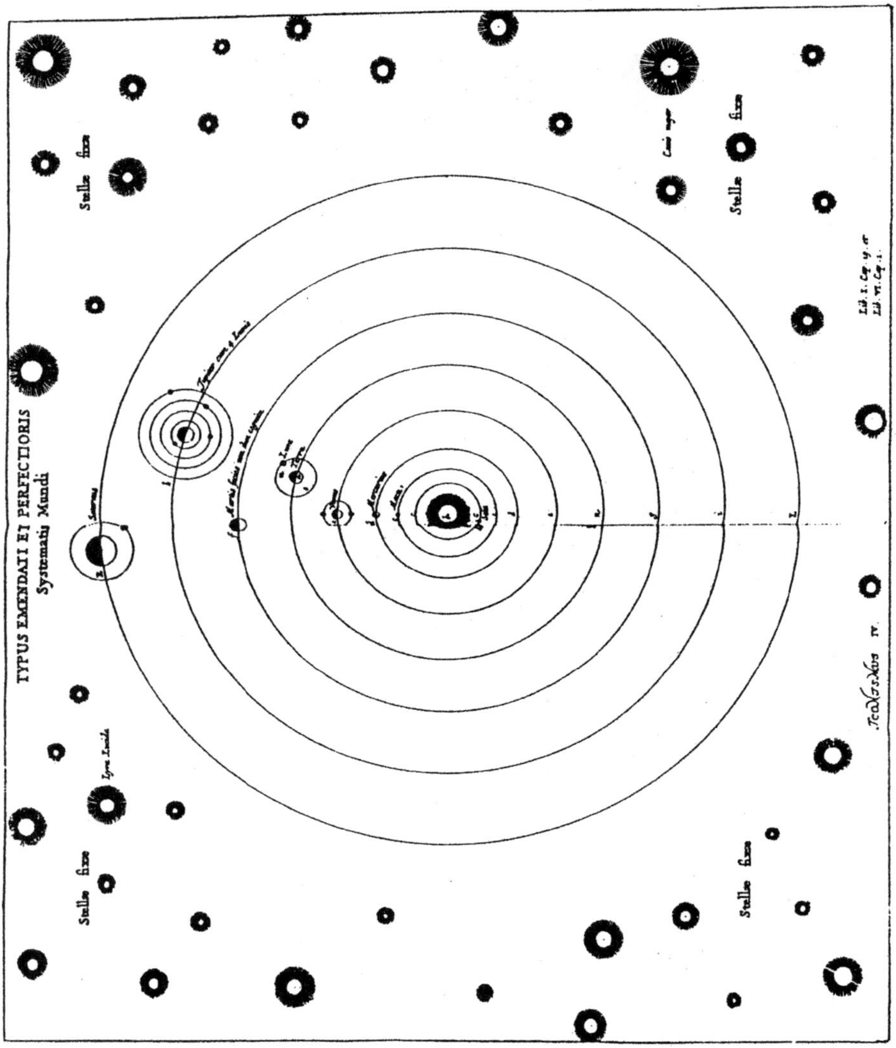
TYPUS EMENDATI ET PERFECTIORIS
Systematis Mundi
Stellæ fixæ
Stellæ fixæ
Stellæ fixæ
Stellæ fixæ

nearer the earth than when it is in conjunction with the sun. At this time it appears to be much larger than Venus, an observation which has led the uninitiated to believe that a new star is rising.

Is this planet circled by attendants or not? Our observations are limited, but there are those who assert that Mars is accompanied by satellites. When Francesco Fontana first observed Mars with his new telescope, he discovered it to be a perfect sphere but with a black cone in the middle, similar to a small, extremely dark sphere. Could this be its satellite? Or is there a very large hollow in the planet's surface? He left this judgment to others. In his second observation, Fontana saw with great clarity the dark sphere or cone become progressively less distinct as the planet became proportionately less round. He believed that this might be caused by Mar's revolving around its own axis. See also Plate II, (ERRATUM read "Iconism. sequent." for "Iconism."), figure 3. Father Anton Maria Schyrläus de Rheita in his *Oculus Enoch et Eliae*, Book IV, Chapter 5 observed that Mars' form was never perfectly round, but rather rugged and covered with many attendant spots. He could furnish no further information as to the existence of satellites about this planet.

Why it is that this planet and Venus do not seem to us as disparate in size as their diverse distances might seem to warrant, is due in part to our eyes in whose moisture light is broken down. For this reason brightly shining objects in the distance do not appear to us with simple clarity. See Book IV, Chapter 14 "De Differentia Aspectus" on this subject. Rheita, Kircher, Zupi, the Neapolitan mathematician and others are of the opinion Mars circles or revolves about its own axis.

CHAPTER 18

JUPITER

Jupiter lays claim to the place next to Mars and completes its passage through the zodiac in eleven Egyptian years and almost 315 days or eleven sidereal years and essentially 318 days.

Jupiter is accompanied by four moons.

Astronomers of old always believed that this planet was solitary. In our century, however, the telescope has revealed that it is accompanied by four satellites or moons. Its sphere, moreover, is divided into three or four zones by dark or blackish bands which vary in appearance. Our vision informs us that the moons are borne about Jupiter in their own epicycles and can be distinguished from one another by the time required for their revolutions. Schyrläus found that (1) the moon nearest to Jupiter completes its orbit in about 42 hours, (2) the next one in $3\frac{1}{2}$ ordinary days, (3) the third in 7 days, and (4)the one on the outside in 17 days. Furthermore, all the moons above Jupiter seem to revolve from west to east and those below in the opposite direction. Thus they appear larger to us when they approach Jupiter from the east, and smaller when they come from the west. They seem to move more slowly in quadrature and more rapidly in conjunction with the planet. What's more, there is little doubt that they at times are hidden from view behind Jupiter, again reappear, and from time to time eclipse one another so that when they are situated between Jupiter and the sun, they produce a kind of eclipse or shadow on the planet itself. Galileo, Mayr, Schikhard, Gassendi, and Schyrläus have made computations as to the distances of these moons from Jupiter and their masses as well. It is believed that the first or nearest moon to Jupiter is of the size of our moon and is three diameters of Jupiter distant from the planet. It appears two times larger if one stands on the surface of Jupiter than our moon appears to us on the earth. The above-mentioned writers state that the second moon is of essentially the same size as Mercury and is four diameters distant from Jupiter. The third is only slightly smaller than Venus and is six diameters of Jupiter distant from the planet. The fourth and outermost moon is not much smaller than our earth and is ten diameters of Jupiter distant. All these moons appear relatively larger when one is on Jupiter than the sun and moon appear to us.

Although many may regard these calculations as questionable, nevertheless, these distances and masses of the moons can be calculated from the presumed mass of Jupiter and its distance from the earth.

As for the five other satellites of Jupiter, in addition to the four already mentioned, which Father Anton Maria de Rheita and Francesco Fontana say they had seen, one should look into the *Almagestum* of Riccioli, page 489. For there he says: "Jupiter is irregular around its edges, apparently because of the mountains and rolling ground of its terrain". One should also see Kircher's *Itinerarium Ecstaticum*, page 267.

CHAPTER 19

SATURN AND PLANETS IN GENERAL

The outermost planet which is farthest from the sun is Saturn. This moves in the largest orbit, as one would expect, and completes its circuit in essentially thirty years (or 29 Egyptian years and about 174 days) and returns to its point of departure within that time. It completes the same sidereal year, however, in 29 years and essentially 184 days.

The appearance of Saturn.

In our century, the use of the telescope has disclosed several noteworthy phenomena associated with these heavenly bodies which seem to have escaped the notice of the schools of ancient philosophers. From many of their observations modern scholars now conclude that Saturn has one or two satellites and furthermore ascribe to it an encircling luminous ring. Although they do not, in fact, agree on all points (see Rheita, folium 287, Fontana, page 126, Borel, page 53, Lipstorp, page 25, and Riccioli, page 487), nonetheless their texts can be used in conjunction with the treatise of Johann Hewelke which discusses the actual appearance of Saturn and its recurrent phases. Christiaan Huygens in 1659 also wrote a treatise which particularly dealt with the Saturnian system.

Planets in general.

It can generally be remarked insofar as planets and their satellites are concerned, that is, if good telescopes are used for observation, that they are (1) dark bodies in themselves which are only partially illuminated, significantly on the side which is turned toward the sun. Their light is thus derived from the sun and its borrowed rays strike our eyes. (2) They are essentially round, but are marked with excrescences and mounds like our moon and earth which do not conform to the geometric norm of perfect roundness. Nonetheless, they appear round to our senses. (3) These planets, like the earth, are composed of hard, solid, and dense matter which reflects the light of the sun. They embody the four so-called elements, an observation which even Schyrläus, Copernicus' adversary, supports in his *Oculus Enoch et Eliae*, Book IV, Chapter 1, page 175. (4) Each planet is enveloped in its own atmosphere and clouds. And (5) they complete daily revolutions about their own axes, an observation confirmed by the aforementioned Fontana, page 108 and Schyrläus, Book IV, Chapter 2, page 203. A diagram of an emended and more accurate world system can be seen in Plate IV.

Plate IV.

CHAPTER 20

A GENERAL DISCUSSION OF THE DISTANCES OF THE STARS

Here one should remark that it is generally accepted that although

The great size and distance of earthlike bodies is unbelievable.

the distances of the sun and moon as well as the planets cannot really be compared with the distance of the fixed stars, nevertheless, one can visualize their immensity to be such that in comparison, the distance of our earth from the outermost planet, Saturn, is of the proportions of a mere pea. Book V, Chapter 2 will discuss this question further: elsewhere it has been well confirmed by many who have been trained in the mathematical sciences.

Thus even though the distances and proportions of these bodies or their spheres relative to one another can be calculated numerically, nevertheless, they cannot be reproduced in their true proportions by mechanical instruments or delineated or described on any chart or flat surface. Thus it is impossible for the human intellect to capture these distances and magnitudes.

For what mortal man can actually comprehend or conceive of the actual size of our earth whose diameter contains 1,718 (or 1,720 in round numbers) German miles, 860 in radius and 5,400 in circumference, with fifteen miles being the equivalent of one degree?

Several noteworthy considerations relative to the great distance of the stars.

So that an amateur delving in these sciences may perceive all the more clearly that the extent of this planetary system (we are not yet discussing the far vaster distance of the fixed stars) is so great that our own sphere of earth in comparison is of absolutely no consequence, he should contemplate the following fundamental and noteworthy considerations. First, it was known to the designers of the sundial that its base corresponded with the center of the earth, and its stilus represented the axis of the world projected through the center of the earth. Thus every sundial was constructed for use not on the surface of the earth but as though we were in its center. Thereby the calculated distance from the center of the earth to the surface of the precise place we happen to be is judged to be of little account and is virtually imperceptible. In the second place we know that astronomers have calculated the movements and positions of the planets in the heavens so precisely that they can predict their oppositions and conjunctions (for example, the eclipses of the sun and moon). Still we perceive no difference in the accuracy of our astronomical calculations whether they are made relative to the center or the surface of the earth, even though we use extremely delicate astronomical instruments. we have the same results as if the observations had been carried out in the center of the earth. Thirdly, wherever we are on land or sea we see the whole hemisphere of the heavens and perceive that when one of the fixed stars which is diametrically opposite to another is in the process of rising, the other is seen to be setting. In the fourth place,one can see from Riccioli's computation in Chapter 25 that the radius of Saturn's sphere, in the middle distance either from the sun or earth, contains 73,000 radii

The earth's radius is of negligible proportions compared to the distance of the planets.

of the earth. If the earth's radius is calculated as 1, then (according to Book 5, theorem 19 or Book 12, the last theorem of Euclid) the cube of its sphere will be to the cube of Saturn's sphere as 1 is to 389,017,000,000,000 (ERRATUM read "389,017,000,000 000" for "38,901,703,000,000,000").

This single digit, however, is infinitesimal when compared to so large a number and accordingly the earth is of the same minute proportions when compared with the volume of Saturn's sphere. In the same way that the earth cannot be compared with the sphere of Saturn, neither can the latter be compared with the sphere of one of the fixed stars, even though this star should be one of the nearest to us. These stars will be discussed more fully in Book VII.

The first writers to record the distance of the stars.

The first to turn his attention to this problem of investigating the distance and sizes of the stars was Aristarchus, the Samian, who lived four hundred years before Ptolemy and recorded for posterity the theories of Pythagoras. Hipparchus of Rhodes followed him: then Ptolemy with his *Magnum Opus* of five books, Albattani and Alfargani, the Arabs, and more recently Müller (or Königsberger), Copernicus, Tycho Brahe, Kepler, Wendelin, Lansberg, Riccioli and a number of others.

We do not know with accuracy the distances and magnitudes of stars.

Aristarchus in his small treatise on the distances and sizes of the sun and moon very wisely did not attempt to determine their limits, but preferred to set forth certain outermost bounds for them, for indeed, it must be admitted that we mortals do not know with any precision their distances and sizes. Although the distance and size of the moon is established with some certainty from its parallax and apparent diameter, still the parallaxes of the sun and the other planets cannot be determined with such certainty and consequently their distances are more a matter of conjecture. The parallaxes of the fixed stars, moreover, are even much less certain than these and for this reason we cannot conjecture their sizes and distances, except within the limits of certain calculations. Thus we cannot hope to know with precision the extent of the heavenly bodies although there are certain astronomers who dare to define their distances, thus amazing their readers with their conclusions, when, for example, they state that the New Comet is 373,807 2111/4097 German miles from the earth.

From so precise a computation which involves such trifling amounts, ill-informed people are apt to believe that it is impossible for such an astronomer to be in error who computes with one digit in his calculations. To be sure, such writers are in a position to deceive ignorant men at times by hundreds and often thousands and more miles, either too many or too few. See Galileo in this regard. We make these statements here not to bring into question all such measurements, but to make people aware of the true situation who wish to measure the apogee, perigee, or eccentrics of the sun and planets and to criticize others for some measurement or other which they have found to be a slight amount smaller or larger. They themselves should be aware that they too may make incorrect measurements of the heavens.

The basis of the calculation of the sun and moon.

The basis upon which depends the calculation of the distances of the moon and sun as well as those of the rest of the planets and even the fixed stars, however, is the parallax.

What the parallax is.

The parallax is defined as the distance or difference between the true position of a certain star as seen from the center of the earth, and its apparent, or observed, position. This is also termed the "variation" or "difference" in appearance.

The horizontal parallax.

The parallax has a twofold aspect: one is the longitudinal parallax, or the parallax of height which we use in lunar measurement. The other is the horizontal parallax which is necessary for measuring the distance of the sun from the earth.

The parallax of height is the difference between the true height of a particular star (that is, as viewed from the center of the earth) and the height observed by the eye on the earth's surface. From these two heights an angle is made in the center of a star. It is formed by two straight lines, the one leading from the center of the earth and the other from the eye on the surface of the earth. This is termed the angle of the parallax of height.

The horizontal parallax is the angle made in the center of a star by two straight lines, the one proceding horizontally from the center of the earth when a particular star rises, the other from the eye on the surface of the earth as directed to the center of the same star. These two lines in the center of the star form an angle which is called the horizontal parallax.

The nature of vision through triangles.

Furthermore, one must be aware of the fact that whatever we see, this appears, as it were, in a triangle. Thus the more distant we are from a particular star, the smaller this visual angle becomes until it finally is imperceptible to the eye. Then the whole triangle merges into a line, with the result that the eye sees nothing more. This will be demonstrated in Book IV, Chapter 13.

CHAPTER 21

THE DISTANCE OF THE MOON FROM THE CENTER OF THE EARTH

The distance of the moon from the center of the earth is calculated from its height at mid-day, both its apparent and true height, and the radius of the earth.

The apparent mid-day height is observed from a particular place, Rome, for example, by some instrument like a large brass quadrant which has been divided with precision into 90 degrees and into minutes of degrees. This is termed the apparent height or the observed position of the moon.

The true mid-day height (which is measured by a presumed observation from the center of the earth) is calculated from the true position of the moon in the sky with respect to longitude and latitude. For since the position of the moon, that is, its latitude as well as longitude in the ecliptic, has been tabulated from lunar motion, one can calculate from the latitude of its motion and the angle of greatest latitude, how great its true height should be for a given position in the heavens at a given time. These two observations give the parallax of height between the observed (at Rome, that is) and the true height of the moon (from the center of the earth). This is the basis for reckoning the distance of the moon from the earth.

It is difficult to arrive at the parallax correctly.

As a matter of fact it is difficult to arrive at this parallax of height (1) for the angle of parallax is obviously very sharp in view of the moon's great distance from the earth and the proportionately small diameter of the earth. Thus the angle contains only a few minutes of one degree and even the most exacting scholar can easily make some error or inadvertent mistake in calculation which might result in a great discrepancy in the final result. (2) Astronomers are frequently deceived by their instruments. (3) A change in atmosphere may make the moon seem now higher, now lower, then greater and again smaller than it in fact is. We shall demonstrate this observation in Book V, Chapter 10. (4) This calculation is particularly dependent upon the reckoning derived from tables as to the true position of the moon and to its conformity with the position that has been observed.

Various parallaxes found by astronomers.

Because astronomers for these and other reasons calculate parallaxes varying from 40′ to 66′ and thereby find that the moon is not always at the same distance nor of the same size, but at various times is now farther away and now nearer; thus, in addition to the various motions that they attribute to the moon, they now postulate the motion in height whereby the moon is at one time higher and again lower, They ascribe this movement either to an eccentric, an epicycle, the eccentric of an epicycle, or a simple elliptic eccentric. In view of the different parallaxes of the moon's height that are evident at different times, they conclude that the moon's height varies in its distance from the center of the earth and that this is apparent from its diameter appearing now greater and again smaller.

What the apogee and perigee are.

They call the most distant point the apogee and the nearest the perigee.

Thus Ptolemy and Albattani and Alfargani after him, declared that the moon is in apogee, that is, the greatest distance from the earth, when it is full or new, and is in perigee, that is, the smallest distance when it is half.

Others held that the greatest and the smallest distance of the moon from the earth occurred during new moon and full moon, and that it was midway during half moon. Kepler was of this number although he later made both equal.

Others decided that the greatest and smallest distance of the moon occurred during halfmoon and that it was half of this during new and full moon. Of this number were Copernicus, Magini, Tycho, Longberg, Lansberg, Bouillau, Riccioli, and others. The variations in the astronomers' computations can be seen from the following Table.

The Distance of the Moon from the Center of the Earth Measured in Radii of the Earth

	In Full Moon	In New Moon	In Half Moon	
	Apogee	Perigee	Apogee	Perigee
Ptolemy	64–10	53–50	43–53	33–33
Copernicus	65–30	55–8	68–21	52–17
Tycho	58–8	54–52	60–36	52–24
Longberg	57–38	54–23	60–4	51–17
Lansberg	64–10	54–0	66–58	51–12
Bouillau	61–40	56–30	64–15	53–55
Riccioli	64–15	53–45	66–42	51–20
Wendelin	63–56	56–4		
Kircher	62–50	57–24		
The Author See Book V Chapter 22.	64–0			

CHAPTER 22

THE SIZE OF THE MOON

Once the distance of the moon from the center of the earth had been found, astronomers sought to calculate its size in relation to the earth from the radii of both these bodies.

As we have already indicated, the radius of the earth is 860 German miles. The apparent radius or diameter of the moon (according to observations made with the quadrant) is found to vary in size with different astronomers because of the variation in their calculations for the apogee and perigee. This is apparent in the following Tables.

The apparent Diameter of the Moon is Found by:

	In Full Moon	In New Moon	In Half Moon	
	Apogee	Perigee	Apogee	Perigee
Ptolemy	31′–20″	35′–20″	42′–8″	55′–0″
Albattani	29′–30″	35′–20″		
Alphonsine Tables and Peurbach	29′–0″	36′–8″		
Copernicus, Rheinhold, and Magini (earlier)	30′–0″	35′–38″	28′–45″	36′–44″
Magini in his "Supplementae Ephemerides"	30′–30″	34′–40″		
Clavius and Hérigone	30′–0″	36′–8″		
	In conjunction			
	25′–36″	28′–48″		
	In opposition			
Tycho, Tost, Muliers	32′–0″	36′–0″	32′–32″	36′–0″
Longberg	32′–0″	34′–0″		
Kepler	30′–0″	34′–0″		
Lansberg	30′–0″	32′–44″ 35′–38″		
Gassendi	26′–36″	31′–8″		
Bouillau	31′–10″	34′–0″	29′–56″	36′–4″
Wendelin	28′–40″	32′–42″		
Kircher	29′–22″	34′–40″		
Riccioli	28′–0″	33′–30″		
The Author See			27′–0″	35′–50″
Book V, Chapter 22	30′–0″	30′–0″	30′–0″	30′–0″

If one assumes that the radius of the moon is 15 minutes and that its distance from the center of the earth is 64 radii of the earth, this

To find the size of the moon.

distance is computed to be 55 ,040 German miles. Thereby it follows that just as (in the Tables of Sines, Tangents, and Secants) 100,000 parts of a radius are to the distance of the moon, namely 55,040 miles, so are the 435 parts of a tangent of an angle of 15′ to the radius of the moon, namely 240 German miles. Inasmuch as 'The volume of spheres are the cubes of their diameters", according to the last theorem, Book 12 of Euclid, by cubing the diameter of the earth and moon, or according to the 19th theorem, Book 5 of Euclid, their radii, one will arrive at their volume. Thus, for example, the cube of the moon's radius of 240 miles is 13,824,000 and the cube of the earth's radius of 860 miles is 636,056,000. When the cube of the earth's radius is divided by the cube of the moon's radius, the result is $46\frac{152}{13824}$, or the number of times greater the earth is than the moon.

A Table of the Comparative Sizes of the Moon and Earth According to the Various Astronomers

The Earth is larger than the moon according to	Ptolemy	40	times
	Copernicus	43	
	Tycho	42	
	Longberg	51	
	Kepler	52	
	Lansberg	45	
	Maestlin	44	
	Bouillau	45	
	Wendlin	53	
	Schyrläus	43	
	Riccioli	55	
	The Author See Book V, Chapter 22.		

CHAPTER 23

THE DISTANCE OF THE SUN FROM THE EARTH

The distance of the sun is calculated from the parallax of height. It is extremely difficult to compute this parallax since one must reckon in advance the true altitude of the sun from its motion so as to distinguish the difference between its true and apparent height. The true motion of the sun and its position in the heavens, however, cannot be found with such precision (because the sun's great distance allows no proportion to be made with respect to the earth's radius) and inaccurate calculations in minutes and seconds are bound to result. Astronomers, therefore, found other means and used the moon for this at first.

Thus Aristarchus of Samos, a fellow-countryman of Pythagoras, handed down a method which employed the distance of the moon from the center of the earth and its deviation from exact quadrature when it shines as a half moon.

Hipparchus and his popularizer, Ptolemy, devised a third method which their later imitators followed. These were Albattani, Müller, Maurolico, Magini, Lansberg, and Bouillau who held that three numbers were required to measure the distance of the sun from the earth, first the radius of the sun as well as the observed radius of the moon, secondly, the distance of the moon from the center of the earth by means of its parallax, and thirdly, the radius of the earth apparent on the moon at the time of a lunar eclipse. From these numbers one can calculate the horizontal parallax of the sun.

The Distance of the Sun from the Earth Calculated in Radii of the Earth by

	In Perigee	In Apogee
Hipparchus	1357	1429
Possidonius	1357	13147
Ptolemy with Maurolico, Clavius Barozzi, Galileo, Maestlin, and others who agree. According to his calculation, in apogee the sun is 1,040,600 German miles distant, its daily orbit is 6,540,914 German miles and it covers 272,538 German miles every hour (if one accepts the sun's movement), 4542 (ERRATUM read 4542 for 4524) German miles every minute and 76 German miles every second (which is a 3600th part of an hour and, as Mersenne asserts, corresponds in length most nearly to the slow throbbing of a pulse or heartbeat).	1126	1210
Albattani	1068	1146

The distance of the sun from the earth according to Ptolemy, calculated in German miles, as well as its orbit in days, minutes and seconds.

Lansberg, however, in his "Uranometria" accused Albattani of error, for the latter like Ptolemy, calculated the radius of the sun at apogee to be 15′–41″ and a full moon's maximum distance from the earth as 64 1/6 radii of the earth. He further calculates that the apparent radius of the moon is no greater than 14′–45″ and that the apparent radius of the shadow of the earth is no greater than 38′–20″. From these calculations he believes it follows that the horizontal parallax of the sun is 3′ and its distance 1146 radii of the earth, when

The miscalculation of Albattani.

		In Perigee	In Apogee
	it should follow that the solar parallax is 26″ and the sun's distance $7,936\frac{1}{2}$ radii of the earth. Copernicus in Book IV of his *Revolutiones* discovered this error before Lansberg.		
	Alfargani	1210	1220
Copernicus' error (ERRATUM, This and following marginal note should be interchanged, as has been done).	Copernicus and Magini	1105	1179
	Lansberg accused Copernicus of a similar error because the latter calculates the distance of a full and new moon in apogee from the center of the earth as $65\frac{1}{2}$ radii of the earth, its apparent radius at 15′, the radius of the shadow of the earth as 40′–18″, and finally the radius of the sun in apogee as 15′–50″. He believes that it follows from these calculations that the horizontal parallax of the sun is 2′–55″, and its distance from the center of the earth, calculated in radii of the earth, as 1179 (ERRATUM read 1179 instead of 1172), although it should follow from the foregoing computations that the solar parallax is 3′–39″ and the distance of the sun in apogee as 15′–50″. He believes that it follows from these calculations that the horizontal parallax of the sun is 2′–55″, and its distance from the center of the earth, calculated in radii of the earth, as 1179 (ERRATUM read 1179 instead of 1172), although it should follow from the foregoing computations that the solar parallax is 3′–39″ and the distance of the sun in apogee as $941\frac{53}{60}$ radii of the earth.		
	Tycho Brahe	1117	1182
Tycho's error.	Tycho also comes in for criticism by Lansberg when he states that the distance of a new and full moon at apogee is $58\frac{8}{60}$ radii of the earth, and its apparent radius at 16′, the radius of the sun at apogee as 15′, and the radius of the earth's shadow as 43′. It follows, however, from these calculations that the angle of the half cone of the earth's shadow comes to 16′–8″, a number greater than the radius of the sun, when in fact it should be smaller because the edge of the shadow ought to be at a greater distance from the center of the sun than the center of the earth.		
	Longberg	1242	1334
Longberg's error.	Lansberg finds the same error in this computation as he did in Tycho's. Longberg postulates the distance of a full and new moon as $57\frac{1}{2}$ radii of the earth, the radius of the sun as 15′, of the moon as 16′, and the shadow as 43′, which is similar to Tycho's computations.		

Kepler's error.

	In Perigee	In Apogee
From these calculations it follows that the angle of the half cone of the shadow is 16′–38″, a number greater than the radius of the sun, when in fact it should be smaller.		
Kepler in his *Ephemerides*	1736	1800
"Epitome Astronomiae"	--	3469
The Parallaxis in the "Rudolphinae Tabulae"	3327	3438
Lansberg did did not find Kepler's calculations valid either, because the latter makes the radius of the shadow when the moon passes in apogee as 44′–22″ and contends that this figure has more validity than that which Tycho derived, namely 43′, from lunar eclipses, He is in error, however, Landsberg says. The latter asserts that the radius of the shadow when the moon passes in apogee is greater than Tycho's computation. Furthermore, Riccioli accuses Kepler of inconsistency in his measurement of the sun since in *De Nova Stella*, Chapter 16, he asserts that the sun is distant 1432 radii of the earth while in the *Ephemerides*, page 2 he sets this distance as 1800. He is not consistent, therefore, in his computation of the horizontal parallax of the sun because he states in the same place that this parallax is 2/3 smaller than that of Tycho, and if this is the case, the parallax is 1′–2″; but if, indeed, the distance of the sun is 1800 radii, the parallax will be 1′–55″. In his preface to the "Ephemerides" in the year 1619, he says that he has confirmed from the smallness of the parallaxes and the harmonic intervals of Mars that the polar parallax is no greater than 1′–0″. If this is assumed, then the distance of the sun should be 3438 radii of the earth.		
Galileo	1208	1208
Lansberg	1446	1550
Bouillau	1433	1485
Wendelin	14407	14905
Schyrläus	1927	2073
Riccioli	7000	7600
Langren	3420	3420
Our calculation, according to Book VI, Chapter 8	2644	2644

CHAPTER 24

THE SIZE OF THE SUN

The sun is greater than the earth, according to	Ptolemy (with whom Maurolico, Clavius, Barozzi, etc. are in agreement)	166
	Albattani	186
	Copernicus	334/162
	Tycho	140
	Longberg	196
	Kepler	3375
	Lansberg	434
	Bouillau	343
	Wendelin	262/144
	Schryläus	1,000
	Riccioli	38,000
	Our computation, Book VI, Chapter 9	1,521

The reason for disagreement as to the distance and the size of the sun.

These disagreements both as to the distance of the sun as well as its size derive from the fact that (1) these men, otherwise great scholars, have been very much preoccupied (ERRATUM, read "torserint" for "forcerint") with the apogees and perigees of the two luminaries, namely the moon and sun, when, as a matter of fact, these bodies, especially the sun, are always at approximately the same distance from the earth, as will be seen in Book V, Chapter 11. (2) Then the fact that they have found the diameters of the sun and moon, as well as the shadow of the earth, to vary so much in size at different times (for it has been discovered that the orbs of the stars, even though they remain at the same distance from the earth and the same height above the horizon, now appear smaller, now greater. For the darker the night is, the larger they seem, and when rain has fallen and the air has cleared, they seem smaller. The atmosphere itself, moreover, is not always of the same height, and thus, when it is higher, they seem greater and nearer but when it is lower, they seem smaller and more distant. This will be demonstrated in Book V, Chapter 10. (3) Finally it is of greatest importance how the measuring of these heavenly bodies is carried out and what instruments are used, for if the measuring is not uniform, a variety of answers always results.

They Found the Apparent Diameter of the Sun

Authors	In Perigee	In Apogee
Aristarchus, Archimedes,	30′–0″	30′–0″
Ptolemy, Albattani, Alfargani, Alphonsine Tables and Thabit	33′–20″	31′–20″
Peurbach, Müller, Clavius, and Santbech	34′–0″	31′–0″
Copernicus	33′–54″	31′–40″
Moestlin	31′–45″	29′–36″
Tycho, Longberg, Magini, Muliers and Biancani	32′–0″	30′–0″
Kepler and Herigone	31′–4″	30′–0″
Galileo	30′–0″	30′–0″
Lansberg	35′–58″	33′–34″
Wendelin	30′–33″	29′–30″
Bouillau	33′–30″	32′–18″
Gassendi and Fournier	31′–6″	30′–12″
Schyrläus	34′–41″	34′–41″
Langren	30′–24″	29′–51″
Riccioli	33′–8″	30′–50″
Our computation, Book VI, Chapter 9	30′–0″	30′–0″

The Apparent Radius of the Earth's Shadow at the Point of the Moon's Transit at the Time of a Lunar Eclipse

Authors	Sun–Apogee		Sun–Perigee	
	Moon-Apogee	Moon Perigee	Moon-Apogee	Moon-Perigee
Ptolemy	40′–44″	51′–3″	39′–58″	50′–15″
Albattani	38′–20″	51′–11″	39′–0″	50′–15″
Alphonsine Tabs.	37′–42″	46′–57″	36′–46″	46′–1″
Copernicus	40′–18″	47′–52″	39′–21″	46′–55″
Tycho and Longberg	43′–0″	47′–0″	42′–4″	46′–4″
Kepler	44′–22″	49′–41″	43′–52″	49′–11″
Lansberg	39′–0″	46′–19″	38′–2″	45′–21″
Wendelin	39′–1″	46′–33″	38′–30″	46′–2″
Bouillau	41′–51″	47′–5″	41′–16″	46′–30″
Riccioli	38′–32″	49′–2″	37′–30″	47′–45″
Our computation, Book VI, Chapter 9	40′–0″	40′–0″	40′–0″	40 –0″

Here one should also take note that frequently contradictions occur in the postulated radius of the earth's shadow by reason of the hypotheses put forward by these authors as to the size of the radii of the sun and moon as well as the parallaxes they have given. In some cases these differences are large, in others small.

Horizontal Parallaxes of the Sun

Authors	Sun–Apogee	Sun–Perigee
Possidonius	1′–40″	1′–50″
Ptolemy	2′–51″	3′–4″
Albattani	3′–0″	3′–14″
Alfargani	2′–50″	3′–3″
Fernel	2′–36″	2′–26″
Copernicus	2′–55″	3′–6″
Neander	2′–54″	3′–5″
Moestlin	2′–51″	3′–2″
Tycho	2′–54″	3′–7″
Longberg	2′–35″	2′–45″
Kepler in the "Rudolphinae Tabulae" but calculated from a distance of 1800	0′–58″	1′–2″
Lansberg	2′–13″	2′–23″
Bouillau	2′–19″	2′–24″
Kircher	1′–41″	1′–51″
Bettini	3′–0″	3′–0″
Rheita	1′–39″	1′–49″
Wendelin	0′–14″	0′–14″
Langren	0′–59″	0′–59″
Riccioli	0′–27″	0′–29″
Our computation, Book VI, Chapter 19 (ERRATUM, read VI instead of V)	1′–18″	1′–18″

Unbelievable error results from the smallest difference in astronomers' calculations.

Finally, in computing the sun's distance, one should bear in mind that it is of great importance to arrive at its true parallax. This parallax is so small, however, in view of the enormous distance of the sun, that it is doubtful whether it is perceptible. Thus if the parallax is uncertain, the distance of the sun is also uncertain. As a matter of fact, unbelievable error results from the smallest difference in calculation in its size, for

If its horizontal parallax should be	3′–0″	the distance	1146	
	2′–0″	of the sun	1719	
	1′–0″	will be	3438	radii of
	0′–30″		$6875\frac{1}{2}$	the
	0′–15″		15470	earth

CHAPTER 25

THE DISTANCES OF THE PLANETS FROM THE EARTH AND THEIR MAGNITUDES

First of all, we must bear in mind that the distances and magnitudes of the planets had not been observed by the early astronomers, Hipparchus and Ptolemy.

Secondly, the followers of Copernicus base their calculations on the radius of the great orbit of the earth (we shall deal with this more fully in a subsequent chapter) and employ it as the basis for the beginning and completion of their computations. For indeed, this orbit can give a more accurate parallax than the radius of the earth and thus they discovered that the nearer the planets are, the greater their parallax is, and the more distant they are, the smaller it is. Finally, we must also remember that the relative distances of the planets as well as their positions are regarded differently by the followers of Copernicus and those of Ptolemy because with the former, the sun occupies the central position whereas with the latter, it is the earth which occupies this place, as can be seen from the Plate of the Copernican system.

The distance of Mercury calculated in radii of the earth.

	Perigee	Middle	Apogee	Magnitude		
Albattani	65	115	165		21952	
Alfargani	64	115	167		19000	
Longberg	820	1311	1812			
Copernicus	629	1142	1655		12	
Tycho	630	1150	1660	Mercury	19	times
Kepler	1736	3049	4342	is		
Lansberg	821	$1498\frac{1}{2}$	2176	smaller	12	
Wendelin		5713		than		
Schyrläus		764		the	8	
Riccioli	5246	8057	10868	earth	256	
Our computation Book VI, Chapters 13 and 14	1348	2644	3940		4	

The distance of Venus from the earth as calculated by

Albattani	167	618	1121			
Alfargani	167	641	1115		28	
Longberg	332	1263	2194			
Copernicus	297	1142	1987	Venus	$3\frac{1}{2}$	
Tycho	300	1150	2000	is	$3\frac{1}{6}$	
Kepler	859	3207	5555	smaller		times
Lansberg	399	$1498\frac{1}{2}$	2598	than	$3\frac{1}{2}$	
Wendelin		10552		the		
Schyrläus		1438		earth	2	

		Perigee	Middle	Apogee	Magnitude
	Riccioli	2241	7580	12919	$\frac{2}{3}$
	Our computation Book VI, Chapters 13 and 14	595	3020	4693	$\frac{2}{3}$
The distance of Mars from the earth as calculated by	Albattani	1216	4584	8853	Mars is
	Alfargani	1216	5032	8848	greater than the earth $1\frac{1}{12}$ times
	Longberg				
	Copernicus		2275		smaller than the earth 8 times
	Tycho	410	5471	3080	smaller than the earth 13 times
	Kepler		5150		
	Lansberg	556	2275	3994	smaller than the earth 8 times
	Wendelin		22257		
	Schyrläus		3047		greater than the earth 3 times
	Riccioli	2373	11000	21005	smaller than the earth 14/100 times
	Our computation Book VI, Chapters 13 and 14	1095	3739	6383	greater than the earth 2 times
The distance of Jupiter from the earth as calculated by	Albattani	8237	10704	13171	95
	Alfargani	8853	10423	14378	81
	Longberg	5048	6605	8163	
	Copernicus	4538	8091	10798	Jupiter is $25\frac{2}{5}$ times greater than the earth
	Tycho	2564	3990	5416	14
	Kepler	13395	17569	21740	
	Lansberg	6222	8091	9960	$25\frac{2}{5}$
	Wendelin		75039		
	Schyrläus		10799		42
	Riccioli	26441	36500	47552	685
	Our computation, Book VI, Chapters 13 and 14	13100	9150	11803	12
The distance of Saturn from the earth as calculated by	Albattani	13100	15800	18500	91
	Alfargani	14378	16800	$20612\frac{1}{2}$	79
	Longberg	10303	12354	14405	
	Copernicus	8678	10580	12600	Saturn is $46\frac{2}{3}$
	Tycho	8760	10550	12340	22 times greater than the earth
	Kepler	27027	32156	37285	
	Lansberg	12534	14880	$17227\frac{1}{2}$	$46\frac{2}{3}$
	Wendelin		139677		

	Perigee	Middle	Apogee	Magnitude	
Schyrläus		20000		Greater	165
Riccioli	57743	73000	90155	than the	891
Our computation, Book VI, Chapters 13 and 14	11716	14360	17004	earth	$29\frac{1}{2}$

Although it is quite evident that there is no method for measuring these bodies that warrants being called exact in view of the great discrepancies and variations among astronomers in calculating distances and magnitudes of the planets wandering in the sky, still there is no more certain or effective method than through the proportionate relationship of the planetary orbits. This is demonstrated in Book VI, Chapter 14.

CHAPTER 26

THE HEIGHT OR DISTANCE OF THE FIXED STARS FROM OUR EARTH, ACCORDING TO THE FOLLOWERS OF PTOLEMY AND ARISTOTLE

Essentially all astronomers have held that the fixed stars, wheresoever they may be in the sky, are encompassed in one sphere and have a common center with that of the world, namely the earth.

They agree unanimously that their distance should be set far beyond the planets. For even if the distance and magnitude of the planets were so great that the human mind could not conceive of them, nevertheless the distance of the fixed stars clearly surpasses all human comprehension, perception and power of imagination, for experimental evidence through the ages has disclosed to us that fixed stars make no parallax with respect to the radius of the earth. Furthermore (if Copernicus' hypothesis is accepted), these stars make only an imperceptible parallax with respect to the diameter of the great orbit of the earth. Therefore Hipparchus, Ptolemy, and all those who preceeded them, left this measurement completely unexplored for they thought, almost to a man, that the earth was at rest in the center of the world with the result that the path of investigation based upon the diameter of the great orbit of the earth was closed to them.

The distance of the stars clearly surpasses all human comprehension and faith and power of imagination.

Astronomers of old left these distances unexplored.

Albattani and Alfargani, astronomers of the first rank among the Arabs, as well as Longberg, Tycho Brahe and others, declared that the distances of the fixed stars can be measured if one assumes that

The distance of the stars according to Alfargani.

the radius of the earth makes a parallax. Thus, according to Alfargani's calculation (with which the Ptolemaics, Clavius, Magini, Barozzi, Maurolico, etc. generally agree), the outermost bounds of the starry heavens lie distant 40,000 or 45,000 radii of the earth, the equivalent of 38,848,275 German miles if the radius of the earth is presumed to be 859 German miles.

The course of the fixed stars according to the Ptolemaics in a day, hour, minute and second.

If this calculation is accepted, then the diameter of the starry heavens will be 77,696,550 German miles and according to the proportion of the diameter to the circumference, namely 7 to 22, the circumference of the fixed stars at the equator will be 244, 189,157 German miles, If this is divided into twenty-four hours, in a single hour any star at the equator will pass through 10,174,548 German miles, in a minute, 169,565 German miles, and in a second, a wink of an eye, 2,826 German miles, provided that one assumes that it is the sphere of the fixed stars and not the earth which travels about in this twenty-four period.

The parallax of the fixed stars.

From this distance computed by Alfargani, it follows that the fixed stars have a parallax of 4″ 3″, a calculation supported by Scipione Chiarmonti in his *De Universo*, Book IV, Chapter 10, page 139.

The distance of the stars according to the Ptolemaics, a more common assessment.

Now there are other astronomers who do not believe that the fixed stars are so far away and therefore attribute a larger parallax to them (one of 10", as it were). Thereby they demonstrate their height to be only 22,000 or 20,000 radii of the earth distant from us, the equivalent of 17,637,740 German miles or 35,275,480 German miles for the diameter of the starry sphere. Furthermore, according to the proportion of the diameter to the circumference (that is, 7 to 22), the circumference of the heavens at the equator contains 110,865,794 German miles. If this is divided by the daily revolution of twenty-four hours, it follows that any star on the equator passes through 4,619,408 German miles an hour, 76,990 German miles a minute and 1,280 German miles a second, which is the equivalent of the pulsing of a human artery.

Riccioli's calculation of the distance of the fixed stars.

Giovanni Riccioli in his *Almagestum Novum*, Book VI, Chapter 7, number 17 and Book VII, Chapter 11, section 6 ascribed a parallax to the fixed stars of two seconds (according to which the stars are distant 100,000 radii of the earth, or of one second, whereby he assigns the distance of the fixed stars as being 210,000 radii of the earth, the equivalent of 180,600,000 German miles. Accordingly, the total diameter of the starry sphere is 420,000 radii of the earth or 361,200,000 German miles and its circumference at the equator 1,320,000 radii of the earth or 1,135,200,000 German miles. Consequently in an hour's time, a star at the equator passes through 55,000 radii of the earth or 47,300 German miles, and in a minute $916\frac{2}{3}$ radii of the earth or 13,138 German miles. According to a statement in Chapter 7, number 18, however, he states that a star passes through 549,500 radii of the earth in an hour, 9,158 in a minute, and 152 in a second or 130,720 German miles. The error in his calculation here is apparent, however, in his addition of an extra zero.

According to Riccioli, a fixed star passes through 130,720 German miles in any given second.

CHAPTER 27

THE HEIGHT OR DISTANCE OF THE FIXED STARS FROM THE EARTH ACCORDING TO THE FOLLOWERS OF TYCHO

There are others, among whom is Tycho Brahe in his *Progymnasmata*, Book I, page 481 who contend that the fixed stars are at a smaller distance from the earth, and are only 14,000 radii of the earth away. Longberg and others subscribe to this hypothesis. According to this calculation wherein the sphere of stars is 12,040,000 German miles distant from the earth, the circumference of this great sphere is 88,000 radii of the earth or 75,680,000 German miles. Thus it is necessary that the fixed stars near the equator progress 3,153,333 German miles every hour, 52,555 German miles every minute (which is 1/60 of an hour), and 875 German miles every second (which is 1/3600 of an hour), or the equivalent of a moment.

Now indeed, since these measurements of Tycho's have no mathematical basis whatever, their source seems to rest upon an assumption more uncertain than certain. For when Tycho states that Saturn at the time of its greatest distance from the earth, is 12,340 radii of the earth away, perhaps he goes on to assume that the fixed stars are 14,000 radii distant. The distance of the fixed stars, however, is not as small as this since no one can show any parallax of the more distant planets, much less the fixed stars, from the radius of the earth. Now what if, for example, the fixed stars showed a parallax of only one second, their distance would be 268,965 radii of the earth or 231,309,900 German miles, to use the calculation of Chiaramonti, referred to in the preceding chapter. (Riccioli in this chapter, however, arrived at another answer for this calculation.)

Tycho's measurements have no mathematical basis.

Since, however, the stars admit of no parallax, it follows that the altitude given by Tycho and others of his group is completely inexact. When Father Anton Maria Schyrläus, who was a follower of his, realized this, he declared in his book, entitled *Oculus Enoch et Eliae*on page 194 that the distance of the fixed stars could be obtained

The calculation of Father Anton Maria.

more appropriately from his own theoretical proportion which follows.

(1) He sets the sun's diameter as ten times greater than that of the earth, that is, as 1 is to 10 or as 10 is to 100. He tries as the same time to prove this result using the Table of Sines. Thus he assigns to the diameter of the sun's disc at its so-called middle distance (ERRATUM delete period after"distantia") in the spring a value of 34′41″ for which, according to the Table of Sines, 100 corresponds to the base number 10,000 which is the square root of the distance of the sun, namely 10,000 (ERRATUM read 10,000 for 1,000): but to the earth he assigned 10 units, 10 being the cube root of 10,000 of this distance in diameters of the earth. Since, however, 10 units in the table of Sines corresponds with 3′24″ or essentially 3′30″, he concludes that the diameter of the earth from the degrees of the ecliptic contains 3′30″ and its radius 1′45″, as many minutes and seconds as he believes the three highest planets have in their orbits.

(2) In his opinion, the square of the diameter of the sun is the earth's distance from the sun; that is, the earth is 100 diameters of the sun distant or 1,000 diameters of its own distance.

(3) The cube of the sun's diameter is the distance of Saturn from the sun. Thus Saturn is 1,000 solar diameters distant from the earth.

(4) He believes that the heavens are the square of Saturn's distance away, that is, the starry skies are 1,000,000 solar diameters distant, or 10,000,000 diameters or 20,000,000 radii of the earth.

According to Maria, the heavens move 200,000 German miles in a second.

Inasmuch as the radius of the earth is 860 German miles, the distance of the heavens must be 17,200,000,000 German miles. Thus, if it is the starry sky and not the earth in its daily movement that completes this course in twenty-four hours, any star that happens to be on the equator will pass through 716,666,666$\frac{2}{3}$ German miles in one hour, 11,944,444 German miles in a minute, and 199,074 German miles in a second in its course through the heavens, which is the equivalent of 36 circumferences of the earth or essentially 4674 German miles. The renowned mathematician, Father Christoph Clavius in his *In Sphaeram Johannis de Sacrobosco* (i.e. John Holywood), a defender of the Ptolemaic system, is forced to concede that a speed of motion as great as this disturbs all common sense and intelligence and although he realizes that the distance of the stars is extraordinarily great, nonetheless he remains firm in his opinion that the fixed stars revolve each day. For he writes, "This space is of such a great dimension that in one hour, a star on the equator covers as much distance as a man could run in 3,000 years, that is, if he covers ten German miles a day."

Father Christoph Clavius ascribes to the stars almost the greatest speed of motion.

clusions associated with the Ptolemaic and Tychonic theories, the followers of Copernicus revised their system and attributed motion to the earth.

CHAPTER 28

THE DISTANCE OF THE STARS ACCORDING TO THE FOLLOWERS OF PYTHAGORAS OR COPERNICUS WHO LOCATE THE SUN IN THE CENTER OF THE WORLD

The Pythagoreans, in placing the sun in the center of the world and denying that the fixed stars move, base their measurements of the fixed stars not on the radius of the earth (because they look upon the radius of the earth as virtually nothing when compared with the unbelievable distance of the heavens) but rather on the radius of the great orbit of the earth, from which they seek a parallax. Regarding this, we must take note that:

Method of measuring according to the Copernicans.

(1) The radius of the great orbit of the earth is the distance of the earth from the sun which, according to Ptolemy, is assumed to be 1210 radii of the earth or 1,040,600 German miles. See preceding Chapter 23.

What the radius of the great orbit of the earth is.

(2) The great orbit of the earth (which, according to the Ptolemaic or Tychonic theory is the orbit or path of the sun) is the circuit which the earth makes about the sun every year. According to Ptolemy, this is found to be 7,606 radii of the earth or 6,540,914 German miles.

What the great orbit of the earth is.

(3) The sphere of the great orbit of the earth (or, in the Ptolemaic system, the sphere of the sun or sphere of the sun's path) is said to be that spherical body (or, in our opinion, that empty Space) of which the mind has conceived and which the earth makes around the sun in its annual circuit.

What the sphere of the great orbit of the earth is.

According to Euclid, Book V, theorem 19 and Book XII, the last theorem, spheres are intrinsically the cubes of their radii and therefore if we cube the radius, 1, of the earth, the result is l; and if we cube the radius of the great orbit of the earth which is 1210 radii of the earth, the result is 1,771,561,000. Likewise if we cube 220 (for the radius of the great orbit contains the radius of the sun this many times), the answer will be 10,648,000 solar spheres which the sphere of the great orbit can hold. From this it follows that the orbital sphere of the earth is 1,771,561,000 times greater than the earth and 10,648,000 times greater than the sun; that is, the capacity of the

The capacity of the sphere of the great orbit of the earth according to Ptolemy's computation of the distance of the sun from the earth.

sphere of the great orbit, according to Ptolemy's theory, is so great that it can encompass the earth 1,771,561,000 times and the sun 10,648,000 times. It is even possible that it may be much greater than this, as Book VI, Chapter 9 will demonstrate.

The sphere of the great orbit of the earth is to be considered as nothing compared with that of the sphere of the stars.

Now even if the radius of the great orbit is of such a size that the radius of the earth is of little significance relative to the distance of the sun, nonetheless, the height of the starry heavens is said by the Copernicans to be so great that even the entire diameter of the great orbit seems to vanish in comparison with the radius of the fixed stars. For they can see no perceptible change in the fixed stars (although the earth is in different parts of the heavens at different times of the year) even though they use accurate and appropriate instruments with the greatest care and attention in their observation.

Nevertheless the Copernicans proposed that the fixed stars may, indeed, have a parallax, not, however, relative to the radius of the earth but only to the whole diameter of its great orbit.

The diameter of the great orbit of the earth is thought to make a parallax in respect to the fixed stars of 30″.

Whereby the distance of the stars from the earth is 8,854,517,430 *German miles.*

Therefore, it has been postulated by Lansberg in his *Commentationes de Motu Terrae* page 6 and by Van den Hove that the fixed stars make a parallax in respect to the great orbit of at least 30". From this hypothesis it follows that the fixed stars are 8,514 radii of the great orbit distant from the earth, or (since, according to the Ptolemaic theory, the radius of the great orbit contains 1,210 radii of the earth) 10,301,940 radii of the earth. Since one radius contains $859\frac{1}{2}$ German miles (ERRATUM, read $859\frac{1}{2}$ for $850\frac{1}{2}$), the total is 8,854,517,430 German miles. Thus the diameter of the stellar sphere contains 17,709,034,860 German miles and following the proportion of the diameter to the circumference, namely 7 to 22, the circumference of the starry heavens is 55,656,966,703 German miles.

If a star follows a daily course, then any star on the equator moves 644,178 *German miles in a second.*

From these calculations it follows (that is, if stars complete a daily course and the circumference of this orbit is divided into twenty-four hours) that any star on the equator within the space of one hour moves 2,319,040,279 German miles, in a minute 38,650,671 German miles and in a second or moment, 664,178 German miles.

Lansberg's measurement of the distance of the fixed stars.

Because, indeed, scholars find that a parallax of fixed stars relative to the great orbit of the earth is either non-existent or at least imperceptible, Lansberg in his *Uranometria*, page 124 later postulated its existence from the orbiting time of the earth, planets, and fixed stars. If, indeed, the earth completes its revolution in the great orbit within the space of a year, the radius of which has 10,000 units, and if the fixed stars complete one revolution according to the Ptolemaic theory in the space of 36,000 years, or according to Lansberg's calculations, 28,000 years, then, Lansberg concludes, according to the Golden Rule that as the earth's time period of one year is to its orbit's radius of 10,000 units, so the fixed stars' time period of 28,000 years is to its sphere's radius of 280,000,000 units. Therefore the stars are distant 28,000 radii of the great orbit of the earth (if one holds that all are contained in one sphere). If these numbers are multiplied by Lansberg's middle distance of the earth from the sun, namely 1,498$\frac{1}{2}$ radii of the earth, the result of 41,958,000 radii of the earth is the distance of the fixed stars. One should also look into Galileo's *In Systemate Cosmico*, Dialogue 3, page 270 in this regard.

The distance of the stars according to various authors.

They assert the distance of the fixed stars to be	Copernicus	indefinite	
	Galileo	13,046,400	
	Kepler (first)	34,077,067	
	(later)	60,000,000	radii of
	Hérigone	1,440,000	the earth

But if they do not allow the parallax of the fixed stars made by the annual motion of the earth to exceed 10″ and propose to maintain the diameter of the annual orbit they have already postulated, the distance of the fixed stars according to Riccioli's calculation, Book VII section 6, at the end, would be:

	Copernicus	47,439,800	
	Hérigone	49,502,400	
	Galileo	49,832,416	
According to	Bouillau	60,227,920	radii of
	Lansberg	61,616,122	the earth
	Kepler	142,746,428	
	Wendelin	604,589,312	

From the distances given by these authors, it follows that a star travels 2,500,000 German miles a second.

From these calculations as to the distance of the fixed stars from the earth, if one takes 40,000,000 radii of the earth for the distance of the fixed stars, it would follow that the diameter of the sphere of stars is 80,000,000 and the circumference (according to the proportion of the diameter to the circumference of 7 to 22) would be 251,428, 571 radii of the earth or 216,102, 856,774 German miles a minute and 2,501,190$\frac{1}{2}$ German miles a second. This discussion will be continued in Book VII, Chapter 1.

CHAPTER 29

THE NUMBER OF FIXED STARS AND THE REVELATIONS OF TELESCOPES IN THIS CENTURY AS TO THEIR GREAT EXTENT

Ptolemy and the astronomers following him observed 1022 stars which were visible from the earth. Even though the actual count of stars is beyond enumeration, all of these cannot be seen by human eyes without help, in view of their great distance and consequent dimness.

Those that are visible, however, were diligently assembled into 48 star clusters or figures or constellations by the early astronomers. In addition to these constellations as time went on, the Spaniards, Portuguese, Dutch, and others who make use of new navigational aids in the eastern and western Indies, came to discover other stars in the southern sky around the Antarctic pole. They supplemented the earlier star groups with these and distributed them in twelve independent constellations, as is evident on the celestial globes.

The number of visible stars does not exceed 1700.

The stars in these more recently discovered constellations number 136 all together, and if these are added to the former, the total is 1158 stars in the heavens which are visible to us. To these, Tycho Brahe, Kepler, and other astronomers add certain others that are difficult to see. Their total sum does not exceed 1700.

The stars which are seen with telescopes are innumerable.

But now at the present time a great many more stars can be seen with the aid of telescopes, and the better these telescopes are, the greater number of stars they reveal to us. Take the constellation of the Hyades, for example (in German, *das Siebengestirn*). With unaided eyes we can make out only six stars in the constellation, but with the help of telescopes, we can distinguish 29, even 30, 40, 50, and more distinctly, depending upon their size. Indeed, with these telescopes, we not only see many more stars in the astral nebulae and in the Milky Way but in fact into whatever part of the heavens we turn our instruments, we become aware of countless stars that are otherwise invisible to the naked eye. One discovers, moreover, many other stars that are below or above one another and which are not contained in one sphere: some of these which are almost directly in our line of vision appear bipartite or tripartite. See also Father Schyrläus in his *Oculus Enoch et Eliae*, page 196.

There are a number of astronomers who only rarely and perhaps never use dioptric instruments, for they assert that this multiplication of stars is suspect and refuse to acknowledge the existence of Jupiter's satellites and the other newly discovered phenomena associated with Saturn, Mars, Venus, and Mercury. They tend rather to ascribe this experimental data to faults and defects in the glass of the instruments. Would they pass the same judgment if a tower which appeared round to them from afar seemed to be square when they used a telescope and would they believe for this reason that the tower was not in fact square, but rather round?

Some astronomers assert that dioptric instruments are suspect, but they err.

I myself learned something about the number of lights seen with a telescope from an experiment of my own. For some reason or other I had two torches and two candles brought out and lit about a half German mile distant from me as I stood at our city gate. They appeared as four lights to my naked eye and the same (but with much more clarity) to my telescope. Thus it is not the part of a rationalist but of one suffering from delusions to accuse a man who sees of blindness without good reason.

Furthermore although most telescopes, as you know, are tube-like instruments with small round apertures of usually no more than two degrees, still with their help, 10, 20, 30, and more stars can be seen, often at the same time. Just imagine how many millions of stars we could see with our naked eyes if they projected from our heads like the tubes of a telescope!

Stars at a very great distance from the earth cannot be seen.

Thus the number of stars throughout the heavens is beyond computation and passes all belief: furthermore, additional stars are always being discovered with better telescopes. Nor should one doubt that better astronomical devices can be made and as these instruments become more improved, a greater and greater number of stars can be observed. The fact that stars at a great distance cannot be seen, even with the most powerful telescopes today, does not prove that they do not exist but only that they cannot be seen at extremely great distances from the earth; for indeed, when objects become increasingly small in Infinite Space (as the study of Optics teaches us), they appear to vanish before our eyes into nothingness. In the same way if our sun were regarded from an increasingly great distance, it would appear smaller and smaller until at last with the immensity of intervening Space, we could no longer see it and it would completely disappear.

CHAPTER 30

THE MAGNITUDE OF THE FIXED STARS

Astronomers have divided fixed stars into six different categories of size. After they calculated their distance from the earth, all of them being on the same plane of the heavens, they proceeded to observe the apparent and true diameter of each star, and from this they computed its magnitude.

The Apparent Diameter of Fixed Stars Observed by Astronomers

Categories of Size	1st	2nd	3rd	4th	5th	6th
Albattani	1′–30″					
Tycho, Longberg and Biancani	2′–0″	1′–30″	1′–5″	45″	30″	20″
Lansberg	1′–0″	40″	30″	20″	10″	5″
Van den Hove	0′–8″	6″	5″	4″	3″	2″
Kepler (earlier)	3′–0″	2′				
Galileo	5″					$\frac{5}{6}$″
Magini	10′–0″	5′–30″	4′–0″	3′–0″	2′–0″	1′–0″
Riccioli	18″	12″	7″	6″	5″	4″
Our computation according to Book VII, Chapter 2	0′–6″ (or 7″)					

The true diameter of a star, hence its magnitude, is calculated from its apparent diameter and its supposed distance from the earth.

A Star is Greater than the Earth – Times

Categories of Size	1st	2nd	3rd	4th	5th	6th
Alfargani, Albattani, Clavius, Maurolico, differ little from one another.	107	90	72	54	35	18
Tycho Brahe / Bayer, Biancani	{45 / 68	28	11	4	1	$\frac{1}{3}$
Riccioli	5355	1728	260	216	92	64
In Book VII, Chapter 2 we compute that Sirius is nearer and somewhat larger than the earth.	681	(as for the others, we are uncertain because of the varying distances)				

The Copernicans, however, find that the stars are at a greater distance by reason of the parallax, as it were, of the great orbit of the earth, whereby Philip Lansberg, in his *Uranometria* compares the fixed stars not with the mass of the earth itself, but with its great orbit. He states:

The assertion of Lansberg that the magnitude of the stars is without foundation when based on the size of the earth.

	of the 1st magnitude are greater	than the mass	67
	of the 2nd magnitude are greater	of the earth,	20 times
Stars	of the 3rd magnitude are greater	that is, the	8
	of the 4th magnitude are smaller	great orbit	$2\frac{1}{2}$
	of the 5th magnitude are smaller	of the earth	3
	of the 6th magnitude are smaller		25

From these calculations it follows that a star of the 1st magnitude is 118,694,587,000 times greater than our earth (because, according to our preceding Chapter 28, the sphere of the great orbit of the earth contains in itself at least 1,771,561,000 spheres of our earth).

Just as this exceedingly enormous size of stars as postulated by Lansberg offends our reason, so too does it prove how preposterous his error and computation are. For in the first place, his later conclusion as to the distance of the firmament is without foundation because the firmament does not move, as we shall see later in Book VI, Chapter 17. Secondly, in the 20th proposition of the aforementioned *Uranometria*, page 130, he assumes the distance to the stars to be 280,000,000 radii of the earth when as a matter of fact, these are merely units of which the radius of the great orbit contains 10,000, as he states expressly on page 125 of his treatise. He should have calculated, however, the distance to the stars as 41,958,000 radii of the earth, of which, according to his hypothesis $1498\frac{1}{2}$ constitute the great orbit of the earth. And thus he calculates the assumed distance of the stars, (indeed without foundation), to be 238,042,000 radii of the earth greater than he should. From these calculations, then, it would follow that the true radius of a star of the first magnitude would be 40,712 radii of the earth which otherwise would be 6,100 radii of the earth. From this there results his exaggerated calculation of the mass of the stars.

Riccioli accuses the Copernicans of error in measuring the magnitude of the stars.

But, since the distance of the stars, as we have said, according to the Copernicans is greater than according to the Ptolemaics, Riccioli set about calculating the magnitude of the greatest and smallest stars, one of the first magnitude with an apparent diameter 18″ and another of the sixth magnitude with a diameter of 4″24‴ He made use of the distances given by the writers in our preceding Chapter 28, as the following table shows.

The Magnitude of the Greatest and Smallest Star, that is, Sirius and Alcor, according to Riccioli's Calculation of the Apparent Diameter Observed as 18″ in Sirius and 4″24‴ in Alcor and the Distance Postulated by the Copernicans, when they Assume that the Parallax of the Fixed Stars Made by the Annual Motion of the Earth Does not Exceed 10″ and Maintain the Diameter of the Annual Orbit as They Have already Postulated.

	Magnitude of Canis Major		Magnitude of Alcor in the Tail of Ursa Major	
According to	Sirius has –diameters of the earth	Its body contains the earth–times	Alcor has –diameters of the earth	Its body contains the earth–times
Copernicus	4170	71,677,713,000	1992	4,378,454,04
Hérigone	4350	82,312,875,000	2068	8,844,058,43
Galileo	4380	8,427,672,000	2092	9,155,562,68
Bouillau	5300	148,877,000,000	2530	15,941,277,00
Lansberg	5424	159,371,956,024	2588	17,333,761,47
Kepler	12550	1,976,656,375,000	6000	216,000,000,00
Wendelin	53200	15,056,882,800,000	25380	1,767,384,872,00

For Riccioli's statement, see Book VII, section 6, Chapter 11, folio 117.

CHAPTER 31

HEAVEN OR THE HEAVENS AS WELL AS THE ATMOSPHERE AND CELESTIAL MATTER

A definition of the sky.

It is commonly said of the sky that it is of extremely wide extent and covers our earth spherically with its very refined and transparent matter. It is a space filled to capacity with the glowing spheres of stars and extends from the earth's and sea's surface to the outermost boundary where it ends. This, indeed, is the common definition.

The many ways the designation "sky" can be taken.

The designation "sky", however, has various interpretations. Frequently it refers to the expanse of air surrounding the earth wherein meteors are born. It is the source of such expressions as "the clouds of the sky", "the birds of the "sky", "manna from the sky", "a calm sky", "rain descends from the sky", "God adorns the sky with his breath", "God covers the sky with dense clouds", etc.

Not infrequently the orbit of one of the planets or its circular periodic course is termed a "sky", for example, the sky of Mercury, Venus, Mars, Jupiter, and Saturn. When, however, people refer to the stars of the sky, they commonly have the higher region of this expanse in mind and call it the starry sky or the Eighth Sphere. It is also termed the ether, as in Ovid the poet, "So many ills have I endured as there are stars shining in the ether".

The Crystalline Heaven.

A fourth sky, the Crystalline or Ninth Heaven, follows these three heavens that have already been mentioned (according to the hypothesis of the Peripatetics). Its existence was deduced from a preconception of the equinoxes.

What the First Mobile is.

Finally, there is the Fifth Heaven or the First Mobile into which they believe all the planets and fixed stars enter in the twenty-four hours from rising to setting. Aristotle in *De Caelo*, Book I, Chapter 9, defines it thus: "The natural body which is the most distant revolving entity of the heavens." This is the same as saying that this heaven is the last self-moving entity of substance in the entire world. See also the preceding Chapter 4.

The Enmpyrean Heaven.

To these heavens that have already been mentioned, theologians, priests, and ministers frequently add another designated as the "Empyrean". This they hold has been ordered by God according to his own nature to receive and house the Blessed. This will be discussed more fully, however, in the following chapter.

What the ether is.

The opinions of the philosophers differ as to what precisely ether is. Anaxagoras, for one, believes that it is the element of fire; Aristotle in discussing the world, claims that this term includes the matter of the heavens and stars, that is, the whole fabric of the stars, sky, and constellations, not only that part which give out flashing light but also that which is always turning rapidly in a circular course ("ἀπὸ τῶ ἀεὶ θεῖν") He holds that this element is different from the four others since it is not only divine but is also indestructible. Basing their theories of this hypothesis, present day philosophers hold that there are two principle parts of the world, the elements and the ether, or the elementary and ethereal world. Under the name, ether, some include whatever is above the so-called element of fire. Others believe that the substance of ether is an infinite expanse of air. Still others accept it as another name for air, as did Vergil, "And the air resounds with shouts"; Apuleius, "The four elements are called Earth, Water, Air, and Fire"; and Lucretius, Book I "He substituted air for Jupiter". These briefly are comments that have been made about air. See Book II, Chapter 5 for our own opinion.

Celestial matter.

In the same way as philosophers clearly differ from one another in this definition, so also do they strongly disagree in the definition of the substance of the heavens or heavenly matter. The Peripatetics together with Aristotle hold that the Eighth or Starry Heaven, as well as the other lower regions which contain the planets are solid and are composed of dense and impenetrable matter. As for the stars, they consist of somewhat denser elements of the sky. The Church Fathers also seem generally to be of this opinion. See Job 37, verse 18: "Hast thou with Him spread out the sky which is strong and as a molten looking glass?"

Is it solid or fluid?

Others, contrariwise, contend that the heavens, or the matter of the heavens, is fluid, being of a substance that easily yields to bodies it encounters like our air which admits passage to all bodies and offers little resistance. For it is non-resistant and yields room to bodies passing through, but like water, its separated parts come together afterwards and coalesce. Many of old were of this opinion, like Pythagoras, Plato in the "Timaeus", Seneca and others. Many of the more modern astronomers have also embraced this theory. Among these is Tycho Brahe who claims that the free passage of comets etc. is proof of the heaven's rarity etc. Certain of the Church Fathers, moreover, decided that this theory was the more tenable one, like Father Arriaga when he says in his discussion, *De Caelo*, page 4, number 53, "The Seventh Heaven is of essentially infinite size. Saturn which is attached to this heaven is only a tiny point in respect to it. Therefore it does not appear to be compatible with wisdom and reason to imagine that the great mass of the Seventh Sphere rotates in order to set in motion so small a star. It is much more reasonable to say that "Intelligence" moves that star alone in the fluid matter, for a huge mechanism of this kind, solid in nature, clearly is not necessary for the movement of that star."

Furthermore there are many other questions, opinions, and discussions regarding these heavens, their substance, matter, nature, form, etc. Aristotle believed that the heavens were animated and endowed with a rational spirit and intelligence. See *De Caelo*, Book II and *De Meteoribus*, Book II. Pythagoras, Plato, and other scholars of antiquity were of a similar opinion. Indeed, they regarded it as a certainty, for Augustine in his *De Civitate*, Book XVIII, Chapter 41, bears witness to the fact that Anaxagoras had been accused and punished by the Athenians because he had said that the sun was not an animate, but only a fiery stone. Philo Judaeus was also of this opinion, as can be seen in his *De Opificio Mundi*, as was Pliny in his *Naturalis Historia*, Book II, Chapter 8 and others.

Aristotle believed that the heavens were animate.

The Athenians declared that the sun was animate.

Furthermore, the philosophers of old, Plato, Anaxagoras, Democritus, and others declared that the heavens are perishable. Aristotle, on the other hand, in *De Caelo*, Book I, Chapter 3 and Book II, Chapter 1, wrote that the heavens are not composed of the same substance as the elements, but rather of a certain fifth essence which is not perishable.

Are the heavens perishable?

The question is then asked whether the heavens were created or not. Ancient disputes apparently arose on this question and Aristotle bears witness to this in his *De Caelo*, Book I. Xenophanes, according to Plutarch, held that it was not created but was imperishable and everlasting. Plato showed himself of two minds, for in *De Legibus*, Book III, he asserts that the universe has existed for an infinite number of years while in the *Timaeus*, he says it was created by God. Furthermore in Definition 6 he is uncertain as to whether the heavens had existed always or only from a past incalculable time. Aristotle believes that the heavens have been in existence forever and that they were not created. He gives the following reason: that which is created and is perishable embodies contrary elements. The heavens do not admit of contrariety and therefore they were not created nor are they perishable. He confirms this with many examples.

Were the heavens created or not?

Modern scholars are divided in their opinions. Some believe that the heavens will perish on the day of the Last Judgment by divine action while others disagree. Those of an affirmative opinion cite various quotations in the Scriptures to support their position: Psalm 102, verses 25 and 26 (ERRATUM, read 102,verses 25 and 26 for 102, verse 27) "And the heavens are the work of thy hands. They shall perish." Luke 21, verse 33 "The heaven and earth shall pass away, with a great noise." Augustine, nevertheless, in *De Civitate Dei*, Book XX, Chapters 14, 16, 18, and 24 interprets these references to the skies to mean the air, in the same sense that birds of the sky are referred to as birds of the air. Suarez was of this opinion also and quotes the aforementioned passage in Peter in verses 5, 6, and 7, where he says "that by the word of God the heavens were of old, and the earth standing out of the water and in the water; whereby the world that then was, being overflowed with water, perished: But the heavens and the earth which are now, by the same word are kept in store, reserved unto fire against the Day of Judgment and Perdition of ungodly men." These words can be interpreted to mean that the heavens which perished in Noah's time during the flood will change into fire on the Day of Judgment, while the celestial spheres which did not perish during the flood, these will be destroyed. Father Thomas in Section 4, subsections 47 and 48, Pereira in his commentary on Genesis, Book I, questions 1 and 2, the scholars of Coimbra in *De Caelo*, Discussion 3, question 1, article 3 and Tanner, Part 1,subsection 6, question 4, uncertainty 5, number 5 were of the same opinion. When, however, the Scripture says in II Peter, Chapter 3, verse 13 "There will be new heavens and a new earth," it is understood that these heavens which already exist are

Will the heavens pass away?

going to be renewed in such a way that they will appear as new. For as the year proceeds in its cycle, the earth appears new with the return of spring and all living things take on a fresh appearance. For a further discussion of this point, see Book V the last chapter.

The heavens are corporeal.

Furthermore, some writers say that the sky is corporeal. Of these, some say it is simple, others, complex. If the body is simple, it is not composed of matter and form; if it is complex, then it is composed of both matter and form.

What is the source of the heaven's movement.

It is also disputed whether the heavens are moved by a natural and intrinsically active Principle or by Intelligence. Inasmuch as essentially all philosophers hold to this point of view, it is generally deemed wrong to contradict it.

The shape of the sky.

It is the common conclusion of astronomers and philosophers alike that the heavens have a perfectly spherical shape, designed as they are for circular motion. This is confirmed by their accepted position, for if the heavens were not spherical, there would be a vacuum beyond them. The subsequent Book, however, will illuminate the discussion of these problems.

CHAPTER 32

THE FIRMAMENT AND THE WATERS ABOVE IT, ACCORDING TO THE SACRED SCRIPTURES

The Expanse is interpreted in the Holy Scriptures as the firmament.

In the first chapter of Genesis, Moses wrote "and God said let there be RAKIAH", that is, "an expanse", (which in certain texts of the Scriptures is translated as "firmament") "in the midst of the waters, and let it divide the waters from the water. And God made the firmament and divided the waters which were beneath the firmament from the waters which were above the firmament: and it was so. And God called the firmament SCHAMAIM" (which numerous scholars say is the same as "ibi aquae" ("There are waters"), but which they translate as "the Heavens." "And the evening and the morning were the second day. And God said, let the waters under the heaven (SCHAMAIM) be gathered together unto one place and let the dry land appear: and it was so. And God called the dry land, the Earth; and the gathering together of the waters, called He the seas, etc", and later God said (verse 14), "Let there be lights in the firmament of the heaven."

From these quotations, it is evident that expanse, firmament, and sky are one and the same thing. Furthermore, in this chapter of Genesis which we have already cited as well as elsewhere in the Scriptures, the heavens are frequently understood to refer to the airy heavens (for birds, indeed, are called "birds of the heavens" and clouds are called, "clouds of the heavens, etc.), it follows that the expanse or firmament or sky is surely that area which begins on the surface of the earth and seas and extends to the outermost boundary of the world. In view of this, one should make the following delineations of the various regions thus: for example, birds of the sky or firmament or expanse cannot be identified with any other region of the firmament than where birds fly. This is also true in the case of the waters above the expanse or firmament. They can only be identified with those watery regions where, as it were, the clouds of the heavens are (this is SCHAMAIM or "ibi aquae"), "for God binds water in the dense clouds" (Job, Chapter 26). Lights, however, are to be linked with that expanse or region of the heavens which pertains to anything luminous. Thus it does not seem to follow from the words of the Scriptures that these waters which are said to be above the firmament or expanse or sky are above the fixed stars; rather they exist in a place suitable for them, that is, in the sky of clouds. The preposition "super" occasionallv signifies "in" or "at" or "among" as in Psalm 137 "Super flumina Babylonic sedent filii Israel", that is, "At the rivers of Babylon sat the sons of Israel". (The Psalm reads, "By the waters of Babylon, there we sat down".)

The firmament or expanse or sky are one and the same thing.

The waters above the firmament are understood to be in the heaven of the clouds.

Furthermore, although the fixed stars are stable and immobile relative to one another, nevertheless, this region of the stars cannot properly be termed "firmament" (if it is assumed that they exist in one region or sphere), as will become evident in due course. Much less, however, should this nomenclature be used by the Peripatetics for if they attribute diverse motions to the sphere of the fixed stars, how can they then term that which they have already said is mobile as stable?

The region of the stars cannot properly be termed firmament.

From this it should be concluded that the word "firmament" cannot suitably be employed here but rather the word "expanse" which is compatible with the Hebraic text. It is also evident that no waters exist above the expanse of fixed stars (since they only are present in the airy sky) and that by the term "waters", clouds and rains are understood here.

Origenes wished to interpret the waters above the heavens as being the angels who enjoy the sight of God, and the lower waters as the devils who have been banished from the heavens for their wrongdoing. This opinion was rejected, however, by the other

Origenes' interpretation of the waters above the heavens.

Church Fathers and Augustine, who at first had believed the waters above the clouds to be angels but changed his opinion later on. Together with Jerome and many others, he came to regard the clouds as the waters above the heavens, and heaven or the heavens as the lower region of air which he then said is known as the firmament which divides the waters from each other. See Compton Carleton in *Philosophia Universa*, page 409. Other scholars and the Peripatetics as well who assert that all heavens are contiguous with one another, locate these waters between the First Mobile and the Empyrean Heaven. Indeed, they could not locate it so conveniently between the other heavens since the moving of one sphere by another would prevent this. For this reason they say that although the heavens move and are contiguous with one another, they are not joined or connected with the Empyrean Heavens. Thus the waters are placed there.

The Peripatetics locate the waters above the First Mobile.

Others declare that these waters pour forth above the outermost bounds of the Empyrean Heavens and use the Holy Scriptures as evidence. Daniel 3, "Bless the Lord, all ye waters which are above the heavens, etc." and Psalm 148, "Praise Him, ye heaven of heavens and ye waters that be above the heavens". Wherefore they hold that these waters are such as to occupy an imaginary space above the Empyrean Heavens.

The opinion of others.

Thomas, Bonaventura, Durand, and others understand the Crystalline Heavens or the putative Ninth Sphere as being the waters above the heavens. They say, indeed, that just as the outermost heaven is called Empyrean by reason of its similarity to fire, so is this heaven called Crystalline because of its similarity to water. Finally others understand the water above the clouds to be some vital matter in the middle of the Empyrean Heaven which nourishes the Blessed who live there. As for us, however, we understand the waters above the heavens to be the clouds and rains of the sky, as we have mentioned above.

CHAPTER 33

THE EMPYREAN HEAVEN

To the five principle skies discussed in the preceding Chapter 31 of this Book, the theologians we have named usually add, and as it were, superimpose, a Sixth and outermost Heaven, the Empyrean, which they say is the home of the Blessed, the dwelling place of the angels, and the majestic throne of God. It is suffused with great light which they believe resembles the light of fire, but without its heat, and for this reason, they call it Empyrean, or fiery. They declare that it shines with no star, is not mobile, square, round, corporeal, nor has it been created, but is a certain celestial something. Furthermore Jacques du Bois, a Leyden theologian, in his *Dialogus Theologicus Astronomicus*, proposition 39, adds that "He who denies the existence of an Empyrean Heaven above the heavens that we see, does not put credence in the Holy Scriptures nor has he learned the fundamentals of Christianity."

Various questions about the Empyrean Heaven.

As is frequently the case in other theories, so also in this do various questions arise. Does the Empyrean Heaven consist of matter and form? Is it a simple body separated from all matter, thus being purely of the spirit, or is it not? Is it mobile? Does it penetrate this lower world of ours or not? Does it suffer from the flame of the stars? Does God reside in the periphery of the Empyrean Heaven or not? Where is the Empyrean Heaven? What form does it have? What size? When did it come into existence, etc?

Some scholars say that the Empyrean Heaven was created at the same instant as the land and sea for they wish the words of Genesis 1 "In the beginning, God created the heavens and the earth" to include the Empyrean Sky. Others say that although God could have created the Empyrean Sky at the same instant as the earth and of the same sort of matter as the earth, there is no reason to believe that he had actually done so.

Is this Heaven solid?

Some declare that the outermost bounds of the Empyrean Heaven are solid while the lower part is veiled by a kind of curtain of thick and dense material which prevents mortals from dwelling here on earth from seeing the light of this heaven, in the same way as once the Holy of Holies was covered by a kind of veil by the Jews.

The shape of this Heaven.

They also encounter difficulty in defining the shape of the Heaven, particularly with respect to its convex portion. As far as the concave part is concerned, they believe that it is most certainly spherical, contiguous as it is to the so-called First Mobile. As for the convex or outermost part however, some want to define it as spherical because,to be sure, this is the most perfect shape. But inasmuch as the Empyrean Heaven is not designed for movement, but for rest, it seems much more suitable to others to regard the convex portion as a square, which accords with the words of Apocalypse 21 where "The kingdom is said to have been founded as a square."

The magnitude of the so-called Empyrean Heaven.

Some wish to give as dimensions for this Heaven a longitude of more than 10,000,000 miles and a latitude of 3,600,000 miles.

This calculation, however, is not compatible with the measurement of the immense distance of the stars. Thus Father Anton Maria

Schyrläus, a Jesuit, in his *Oculus Enoch et Eliae*, Book IV, Chapter 1, section 6, page 195, added another calculation to the hypothetical distance of the fixed stars he had already computed. (See the preceding Chapter 27 of this Book.) He calculated the distance from this region to that of the Empyrean Heavens in the following way (he did not include the waters which are said to be above the Heavens): As the starry heavens are the square of Saturn's distance away namely 1,000,000 solar diameters, so the Empyrean Heavens are the square of the fixed stars distance away, namely 1,000,000,000,000 solar diameters (or 10,000,000,000,000 diameters of the earth, 20,000,000,000,000 radii, or 20,000,000,000,000,000 hours). Since the diameter is to the peripheral surface as 7 is to 22, then the circumference of the Empyrean Heavens will amount to 125,714, 185,714, 285,714 such hours. Thus if one of the Blessed were to measure out this distance in the Empyrean Heaven and endeavor to transverse its vast extent day after day, he could not encompass it in 10,000,000 years.

By means of these calculations Schyrläus wished to exalt the power, great remoteness, and the magnitude of God whom the Holy Scriptures assert is higher than the sky itself.

As to the size of the internal height of the Empyrean Heaven, nothing has been determined other than the calculations of the aforementioned author in another section of his work, Chapter I, page 6. Here he declares that the Empyrean Sky is 8,000,000,000 times greater than the whole firmament. In view of this vastness, he urges the Epicureans to consider the greatness of God's works and bids them to think of the Empyrean Heavens as a million millions times higher than the firmament.

Will the Empyrean Heavens perish or not?

Inasmuch as the sky and earth will one day perish at the time of the Last Judgment, will then this immense and vast expanse of Heaven, of which we have been speaking, also be destroyed at that time? Or will it endure for eternity? Men also dispute this, some asserting that it will perish, others denying it. Some others distinguish between heavens of air, ether, and fire and confine destruction only to the first two. At this point another question arises: if the airy and ethereal heavens are due for destruction, what will remain in that immense space encompassed by the Empyrean Heavens? Will it be empty? Will absolutely nothing remain there? To what end or for what purpose would it exist? What sort of matter would this space contain? For indeed, according to Riccioli's computation (if the starry heavens are 210,000 radii of the earth distant from the earth), it can contain 38,808,000,000,000,000 spheres of earth in itself. Regarding our opinions about all of this, see our subsequent Chapter 11 in Book II.

CHAPTER 34

ARE THERE ANY OTHER WORLDS BEYOND THIS OF OURS, AND IF THERE ARE, CAN THEY BE COUNTED, OR ARE THEY UNCOUNTABLE?

Writers who held that there is a plurality of worlds.

At the beginning of this book we mentioned several scholars who declared that there are many – perhaps even innumerable worlds. In this group were Aristarchus, Anaximenes, Xenophanes, Democritus, Epicurus, Metrodorus of Chios, Leucippus, Diogenes, and others (whom Theodoretus in Book I of his *De Curando Graecorum Affectiones* and D. Thomas, section 1, subsection 68, (ERRATUM read p.q. for p.g.), article 3, had mentioned. They expressed sorrow for Alexander the Great who complained that even though there were many worlds, he was not yet master of one.

Plutarch in his *De Oraculo* states that Plato at one time held that there were many worlds. Furthermore the aforementioned Metrodorus held it absurd to believe that there was only one world in the Infinite on the analogy that we do not see a single spear of grain growing in a large field. It was evident, he said, that there is an infinitely large number of these worlds from the fact that ours consists of infinite causes. It is necessary, therefore, that there be an infinite number of worlds, for where there is cause, there is also effect. These causes, however, may have their origin either in atoms or in elements according to Plutarch in his *De Placitis Philosophorum*, Book I, Chapter 5.

What the universe is.

Empedocles said there was one world but that the universe and the world are not synonymous since a sizable part of the universe, the part that remains after the world, is matter. Seleucus, however, stated that the world is infinite. Diogenes, moreover, declared that the universe is indeed infinite but that the world is finite.

The Stoics maintain that the universe and the Whole are not synonymous, for the universe is infinite in extent and includes the vacuum while the Whole encompasses the world alone, exclusive of the vacuum. Therefore these designations have different meanings. See *De Placitis Philosophorum*, Book II, Chapter 1.

Writers who claim that the universe is a single entity.

Thales and his followers differed from all of the foregoing, asserting that the universe is finite and a single entity.

Plato was of the opinion that the universe is a single entity and based his conjecture on three hypotheses, (1) that it would not be perfect if it did not embrace all things, (2) that it would not resemble its image if it were not unique, and (3) that it would not be incorruptible if there were anything beyond it. Aristotle endeavored to prove that the universe is a single entity by different arguments. (1) He said that it is composed of all matter which is present in nature. His proof of this lies in his statement that perceptible bodies have their being within this universe, and for this reason nothing exists or can exist outside of it. (2) He believes that if there were other worlds in the universe, they must necessarily be composed of identical bodies since their natures and capacities would be identical. If there were many worlds, then the earth of another world would be carried to the middle or center of this world, or vice versa. (3) He contends that these worlds which might exist outside of ours would not be adapted to be here since there is no time to make them grow old nor for any one of them to undergo change: consequently, they could not possibly exist. (4) He believes the motion of one world would hinder the motion of another (at the point where the two worlds come in contact). (5) He denied that the triangular space left between them would be empty. (6) Their number would be uncertain. (7) The mass of matter would be dispersed in many worlds and not assembled in one single world, when, as a matter of fact, nature tends toward such massing-together. (8) He concludes that it is not necessary or right that citizens govern citizens of another world. (9) There would not be a single first Mover. The school of theologians agrees with Aristotle insofar as he declares that the world is a single entity. It disagrees with him, however, in his assertion that it is impossible for many worlds to exist. Theologians, for the most part, believe that many worlds can and could be ordered by God even though He had created one single world of His own will in the course of time. Concerning more questions of this kind, see the last Book of this treatise.

Aristotle's arguments for a single universe.

The opionion of theologians.

CHAPTER 35

IMAGINARY SPACE OUTSIDE THE WORLD

Inasmuch as it was the contention of Aristotle, the first among philosophers, as well as the more common opinion of other philosophers and theologians that many worlds do not exist beyond this of ours, rational thought dictates that this world must also have a terminus. Since, however, it would be contrary to nature and God that anything (other than God) should extend infinitely, the human mind eagerly inquires what there is beyond this world or what bounds it. Is it the so-called First Mobile or the Empyrean Heaven or those waters believed to be above the heavens – or is it anything else? What, indeed, confines or limits it? (ERRATUM read "disquirit. Aristotles" for "disquirit Aristotles".) Aristotle in his *De Caelo*, Book III, Chapters 8 and 9, expressly stated that above the furthermost heaven (which to his way of thinking is the heaven of the stars or the body of nature or the support of the most distant revolution), there is neither place nor time nor vacuity, but nothing at all. Indeed, he is eager to confirm by reasonable argument his belief that no natural body can be there. Some writers declare that there is a kind of imaginary space beyond the world. They call this an "imaginary ubiquitousness." This appears, however, in different forms to the imagination of different scholars, or if they reject that it has this form, at least it serves as a basis for their imagining.

What may be outside the universe.

What is imaginary space?

Aristotle held that there is nothing outside of the world.

The scholars of Coimbra, in a commentary on Aristotle's *Physica*, Book VIII, Chapter 10, question 2, article 4, declare "In the first place, as regards the space which exists inside as well as outside the heavens, this is not an actual quantity endowed with three dimensions: if it were, bodies could not be taken into it. For indeed, the laws of nature do not permit several bodies of such dimensions to exist in the same place at the same time. Furthermore, Space is not a real or substantial entity of another kind since nothing like this has ever been in existence except God. Space, however, has always existed, and in all probability, always will. Secondly, Space is not an entity of reason since bodies can be taken into and placed outside of the world by it alone and without the agency of the intellect. In this imaginary Space, therefore, God can act, not as in some entity of substance but through his own immensity which must exist beyond the heavens in Infinite Space since the whole world cannot accommodate it. Here it should be remarked that in one sense God is present in his own being, in another, in things that are created, and in a third sense, outside the universe. For He is present in Himself insofar as He requires support from nothing, in things that are created, through his own essence, presence, and power, and beyond the heavens because he cannot be excluded from any place, whether real or imaginary. We, therefore, accept that Space beyond the world as it is in fact and not as a real and substantial entity." They confirm this when they write "Bear in mind, however, that although there

The opinion of the scholars of Coimbra.

is no body beyond the heavens, there is, nevertheless, Space which serves as a receptacle for bodies."

Various other opinions.

Others say that imaginary Space is nothing more than nothingness.

Some writers term it a void empty of all reality.

Others declare that it is the negation of all being.

These are answered in the following way: (1) As to Space being nothingness, some philosophers say that since nothingness exists in itself, Space cannot exist therein; for nothingness can have neither capacity, nor breadth, nor length. (2) Secondly they assert that if imaginary Space is Space devoid of all reality and entity, this is the same as saying vacuity; therefore vacuity and imaginary Space are one and the same thing. (3) Thirdly, if imaginary Space is the negation of all entity, then, they reply, it cannot be called Space.

Others, furthermore, wish to interpret imaginary Space as being something purely fictional, whereby they conclude that before the creation of the world,when there was no intellect capable of imagination, there had been no imaginary Space. Moreover, because God cannot be present in that which does not exist (a chimaera, so to speak), but only in that which does exist, and because in their opinion imaginary Space does not exist, God is not present, therefore, in imaginary Space.

Others conceive of imaginary Space as some enormous corporeal mass which has undoubted existence and extends throughout the Infinite; in other words, Space is the "imaginary ubiquitousness" of such a corporeal mass.

Less finds that imaginary Space is God Himself.

Some writers say "Imaginary Space is God Himself, who, in view of his greatness, is everywhere and is necessarily everywhere". Less, in Book II, Chapter 2 of his *De Perfecta Divinitate* writes: "For if one should transcend all heavens and should search for that Space which does not exist, but which he imagines, he most certainly would find God."

There are many scholars who completely reject imaginary Space for this reason: that which our intellect and imagination alone conceives of has no true existence. As an example, one might imagine that he has a thousand gold coins when, as a matter of fact, he does not even have one. Therefore, imaginary Space does not exist.

The opinion of René Descartes.

More recently, however, René Descartes, the philosopher, in part 2 of his *Principia Philosophiae*, number 21, sees in this imaginary Space something of substance, when he says: "Furthermore we perceive that this world or universe of corporeal substance has no limits to its extension. Wherever we imagine the boundaries to be, there is always the possibility, not merely of imagining further Space infinitely extended, but also that such Space actually exists. Consequently the corporeal substance contained in this Space must also be indefinitely extended, for (as he believes he has already shown), the idea of the extension that we conceive any given Space to have is identical with the idea of corporeal substance.

We can also readily come to one and the same conclusion that celestial and terrestrial matter do not differ from one another, for if there were an infinity of worlds, they would consist of one and the same kind of matter. Thus there cannot be a plurality of worlds; there must be only one. Indeed, it is clear that matter whose nature consists merely in being extended substance, already occupies all imaginable Space wherein "other" worlds would have to exist; this is the only concept of matter we can have.

Thus it is one and the same matter that exists throughout the universe. Its distinctive characteristic is extension etc." This is what René Descartes believed and perceived.

What we must finally conclude about imaginary Space.

If one admits the existence of imaginary Space, then it follows that this is boundless, extending infinitely in all directions, in length, breadth, and depth. It likewise follows that it is incorruptible, eternal, static, fixed, and permanent with the result that it cannot be moved for any reason or by any force. Furthermore it can hold within it any body whatever, whether it be great or small. See various references in Book II regarding this theory.

Conclusion of This First Book

This first book was written to present the various theories of the philosophers, both old and new, as to the location of the earth-like bodies, their movement, distance, and size, likewise the so-called matter of which the heavens or that great intermediate Space between these bodies which extends far and wide, is composed. This book likewise presents their views as to what is outside of the world. We have reviewed the opinions and theories of ancient as well as modern philosophers on all these questions and have presented only those that are intelligible and fairly compatible with reason. For indeed, there are many treatises on these subjects which were written more for the sake of a verbose desire to be heard than with the intent to present lucid thought or the truth. Our opinions on these questions will be revealed in our subsequent books.

BOOK II

EMPTY SPACE

CHAPTER 1

WHY THE AUTHOR WAS LED TO INVESTIGATE THE VACUUM

No one will deny that there are indications of omnipotent God in nature, especially in this mighty structure of the world. The validity of this assumption is particularly apparent in Chapters 20 ff. of the preceding Book which describe (1) the incomprehensible distance of the heavenly bodies, namely the sun, moon, and the rest of the planets. It is also evident that some of these bodies are of such great magnitude that they not only exceed our sphere of earth in size but are even many times greater, say ten or a hundred and more times larger. What mortal man can imagine or actually comprehend the extent of this earth of ours, which is 1720 German miles in diameter, 860 miles in radius and 5400 in circumference? When it is compared with the sphere of the outermost planet, Saturn, moreover, the earth is not even of the size of a pea. Thus, although the difference in size of these bodies one to the other and their spheres can be recorded numerically, they cannot be represented or described accurately on a chart in any way or by any means. (2) Saturn bears the same relationship to the sphere of the nearest fixed star that our sphere of earth hears to Saturn. It is indeed the case that fixed stars are not only widely separated from the earth but some of them are removed at so great a distance that they actually stun our senses and the penetrating vision of our intellect. Indeed, the nearest of all the stars (which is probably Sirius or Vega) admits of no parallax at all relative to the radius of the earth. Suppose that this star did have a parallax, however, say of two seconds. Its distance from the center of the earth would be 103,448 radii of the earth. Should its parallax be only one second, moreover, its distance would be 268,965 radii of the

The earth is as nothing when compared with Saturn.

The sphere of Saturn cannot be compared with the circumference of the nearest fixed star.

earth, which is the equivalent of 231,309,900 German miles, as Scipione Chiaramonti computes in his book, *De Universo* and the infallible Tables of Sines, Tangents, and Secants show. Since, however, there is no perceptible parallax of the nearest star relative to the earth, one can easily conclude that the ingenuity of the human intellect is not sufficient to ascertain the extent of so great a distance. (3) As for the number of fixed stars, who can count them? At the present time we can see much more with the telescopes of the astronomers than we once could. Moreover, the longer these tubes are, the sharper they render the blurred images of stars which heretofore could not be clearly seen with the naked eye. (4) As for the magnitude of an individual star, even though a precise calculation cannot be made, nonetheless, there is no doubt that many of them are ten and more times greater than our sun. (5) As for that which is greatest of all, that immense Interval, or Space, or Expanse which encompasses all these spheres, however numerous they may be and however large, in comparison with this, all these bodies taken together are to be regarded as a tiny, atom-like particle.

Stars cannot be counted.

There are many stars ten and more times greater than our sun.

Vast intermediate Space between the world bodies cannot be comprehended by mortal man.

Since I had deliberated on these questions for a long time and had likewise been engaged in active study of the structure of the world, not only did the great mass of these world bodies and their enormous distances lead me to doubt that the human intelligence could comprehend them, but also, and most particularly of all, did vast, intermediate, and endless space fill me with curiosity and infuse me with a desire to investigate it. For indeed, what is it? Inasmuch as it contains all things, it serves as a place for being and existing. Is it by chance some fiery heavenly matter? Is it solid (as the Aristotelians assert)? Is it fluid (as Copernicus and Tycho Brahe believe) ? Is it some sort of tenuous fifth essence? Or is Space completely devoid of all matter, a Vacuum, as it were?

CHAPTER 2

SPACE AND TIME

Because philosophers commonly state that "a vacuum is a place in which there is nothing," it will be necessary to say something about Place. What, indeed, is it?

Aristotles' definition of Place.

Aristotle defines the concept of Place as "a kind of immobile container or receptacle which holds another body" or "Place is the bounds of that which is contained and is in contact with it." "Place is the outermost bounds of that which is contained."

Others say that "Place is the surface of a body situated there," or "Place is the concave surface of an encompassing body and is immobile."

Another definition.

Aristotle, in his *Physica*, Book IV, confesses, nevertheless, that several questions have arisen as to the nature of Place and how things exist there. He concludes, moreover, that one should be uncertain not only as to what Place is but also as to its existence.

The author's description of Place.

As a matter of fact, Place has no existence of and by itself. Rather it can be defined as being a certain determined relationship wherein an object has a position with respect to some other object near at hand; in other words, when it is here, it is not somewhere else.

We usually locate everything in some "πôυ," or "where," that is, in a particular region or within some boundary. This we call Place. Thus, for example, Peter can be described as being in a particular room which is in turn situated in a certain house. The house is on a street, the street in a city, Rome, perhaps. But where is Rome? – in Italy. Where is Italy? – in that fourth part of the earth which is called Europe. Where is Europe? – on the surface of the earth between Asia and America. Then again one asks, where is the earth? – and (according to the Copernican system) one can answer, in its own region or sphere which lies between the spheres of Venus and Mars. Where are the spheres of Venus and Mars? One answers, between those of Mercury and Jupiter. A person pursuing the subject may ask, "Where are the spheres of Mercury and Jupiter," and the forthcoming answer will be, "between the sun and the sphere of Saturn." Where, then is the sphere of Saturn? – and the answer is given "on the outermost bounds of all the planetary spheres which revolve around the sun." Then, where is the sun? Copernicus replies, "in the center of this world of ours." But where is the world? The answer to this will be given in our subsequent Book VII.

Place has no real existence.

And so it is that one can always ask one further question about Place or "whereness." Thus, Place in itself has no real existence. Rather it is a creation of the mind or imagination, an individual and particular relationship, as it were, of one thing with respect to another. When one thing has been taken away, then another takes its place and when this in turn is removed, its place is again filled and vice versa.

Furthermore, just as Place has no real existence, neither does Time.

What is time?

The Peripatetic philosophers define Time as "that which, as an aspect of motion, is closely related to "earlier" and "later." They assert that it is difficult to conceive of Time and more difficult to explain it. It is told of Augustine that he said "If I am not asked, I know what Time is, but if I have to give an explanation of it to someone, I do not know."

In our opinion Time is the measurement of the duration of things relative to their course around the sun. The earth's duration, for example, is measured relative to its periodic or annual course about the sun, a measurement which is called Time and is divided into years.

In matters pertaining to the earth, measurement is made not only by the periodic course of this body, but also by its daily motion, which gives rise to days and hours. Thus Time can be defined as the distinction of duration indicated by years, days, and hours which derive from terrestrial movement, both annual and daily.

Planetary Time.

The duration of the planets and those bodies dependent upon them should be computed relative to the course of each planet, its periodic as well as its individual course. This measurement should be called Time.

CHAPTER 3

EMPTY SPACE

Aristotles' definition of Empty Space.

Aristotle defines Empty Space as "a Place which is not filled by a body but which is capable of being filled." He does not admit of the existence of a Vacuum in nature and argues against it in this way:

"A philosopher should not posit the existence of that in nature which is not the cause of any effect."

Objection to Empty Space.

"A Vacuum is not the cause of any effect in nature."

"Therefore the existence of a vacuum should not be posited in nature."

Explanation.

It is certainly true that when a Place is not filled with a body but is capable of being filled, this is an Empty Place or a Vacuum. Although it does not have real existence, it nonetheless permits relationships to exist and, indeed, can be the cause of a very great effect in nature. For example, a thing which is here can be transferred elsewhere and occupy by its presence a place, (or rather a space) which is not filled by a body, but can be so filled. When Saturn is in Capricorn, for example, its place in Cancer is empty, even though it still can be filled by it. Thus an Empty Place or Vacuum is the cause of a very great effect in nature. To be sure, if Empty Space were not a part of nature and all Space were filled, one body could not transfer its position to the place of another nor would there be movement from one place to another. Experiments in our subsequent Book visually demonstrate an exaggeration of this characteristic when air or water enters evacuated glass vessels, surging in with great force contrary to the usual course in nature.

You may say that this occurs because of an abhorrence of a Vacuum; one answers that if you assume there is an abhorrence of a Vacuum, you also assume that a Vacuum exists. If, indeed, there were no Vacuum, there could be no abhorrence of it, for how can there be abhorrence of anything if one does not admit that such a thing exists? If one wishes to carry on the argument further he can say: "A philosopher should maintain the existence of that which is the cause of any effect in nature." "A vacuum is the cause of an effect of abhorrence as well as of filling (for if vacuums did not exist, nothing could be filled) and of many other phenomena." "Therefore a philosopher should maintain the existence of a vacuum."

Philosophers who agree with Aristotle.

For the most part all philosophers from the time of Thales to Plato were of Aristotle's opinion, as Plutarch records in *De Placitis Philosophorum*, Book I, Chapter 18.

The diverse opinions of other philosophers concerning the vacuum.

Leucippus as well as Democritus, Demetrius, Epicurus, Metrodorus, etc., however, maintained that no vacuum existed in the world, but that beyond, Vacuity was infinite.

Others held that it was the world soul or spirit; that it was incorporeal and unpalpable; that it was formless and could not be formed; that it could not experience or effect anything; and that it was designed only to receive bodies into itself.

René Descartes' opinion.

More recently René Descartes in his *De Principiis Philosophiae*, Part 2, number 16, declared, like Aristotle, that a vacuum does not exist and added as a reason: "The extension of Space or Place is no different from the extension of Body." He continues, "Should someone ask what would happen if God removed everything corporeal contained in a vessel and allowed nothing else to fill its place, one would answer that the sides of the vessel would become contiguous with one another, for when nothing lies between two bodies, they must of necessity touch one another. It is clearly inconsistent with reason that the sides should remain apart or that there be space between them. For indeed, this distance would be nonexistent because all distance is a state of extension and consequently there can be no extension without substance, etc."

The Peripatetics consider the abhorrence of a vacuum to be a fundamental component of the world.

The Peripatetics consider the abhorrence of a vacuum to be one of the fundamental components of the world, because, they say, there is a reciprocal and continuous succession of bodies within it. For this reason there is no corner in the world which does not contain some body or other. Thus the virtue of the heavens, being a topmost body, can pass through intermediate bodies into lower. At this point, one usually asks whether God can make a Vacuum. This leads to a further question: "If no vacuum exists, how can it be said to be the cause of so many different effects?" To the first question Scaliger replies that God cannot make a vacuum because he cannot make anything dissimilar to himself. Since a vacuum is nothingness, it is clearly dissimilar to God, and since creation always has an end in something, God certainly could not create nothing. To the second question, philosophers reply: "It is not a Vacuum but the abhorrence of a Vacuum that is the cause of these things: or to put it another way, it is not that a Vacuum exists, but that it does not."

Experiments which disprove the existence of a vacuum.

Furthermore, daily experience proves that a vacuum cannot exist. For water rises, contrary to its own nature, when air is drawn out of a pipe. Likewise flesh is made to rise, developing into a swelling, with the cupping glasses surgeons are accustomed to use, once they have extinguished the flame beneath them. When the opening of a bellows has been plugged up so that no air can enter anywhere, it cannot be pulled apart. Water does not flow out of a vessel with a very small opening which has been turned upside down, etc. Can a vacuum be made, then, with the help of angels? Oviedo says "no." These, then, are the more common objections to a vacuum that philosophers present. We shall discuss this question further in Book III, Chapters 35 and 36.

We shall demonstrate there is no doubt that a vacuum exists in nature by means of many experiments in the subsequent Book.

What a vacuum is according to the author.

We do not regard a vacuum, in terms of its vacuity, as something of real existence, but rather as a lack or absence as shadow is the absence of light or as blindness is the absence of sight, conditions which have no real existence in themselves. Furthermore, as death is the extinction of life, so death is without reality. In the same way a vacuum is the absence of fullness, but in no way does it follow that a lack or absence or this or that thing (which is commonly called a vacuum) is of real existence.

Inasmuch as the terms 'full' and 'empty' always imply that space or a container or a vessel encompassing something is either full or empty (for simply to say 'full' or 'empty' without consideration or specification of the container or the encompassing vessel is the equivalent of saying nothing), it necessarily follows that there is a containing vessel like Space to hold the various celestial and earth-like bodies of the world. Inasmuch as Space cannot be seen or easily understood, numerous different opinions are brought forth as to whether it exists, and if it does, where, and whether it is empty or full.

Thus the container of the earth-like bodies is nothing other than the Space of the world or a vessel encompassing all bodies in the world. We call it full when there is a body of this kind with its emanations present, and empty when there is no body of this kind.

Therefore we do not speak of a vacuum as such, (which, indeed, is nothing and cannot be mentally comprehended or understood), but rather of empty Space and the use that a good understanding of it can be to the study of all philosophy.

A vacuum as such is nothing.

CHAPTER 4

SPACE

Here we shall not discuss Space in terms of the third dimension, or as it is commonly conceived, in terms of magnitude, amplitude, or capacity for encompassing this or that thing, either a place, a building, a palace, a field, a region, a distance, etc., but rather as a world vessel or container of all things. It is not to be appraised from the point of view of its quantity, longitude, latitude or depth; nor considered with respect to its substance, either in very small or somewhat larger parts, nor partially or separately, nor superficially, internally, nor externally, etc. Nor will it be discussed with regard to its extensions, its enlargement, or its expansion. Rather it will be considered as a container of all things, infinite in extent, wherein all things exist, live and move and one which presents no variation, alteration, or change.

Space, therefore, can be described in two different ways, that is, according to the popular concept or according to its definition as a container.

Space can be understood in two different ways.

When Space is defined as a container of all things, then it is the place where all bodies or substances have or can have their own existences, (either here or there, according to their relative nearness). Space yields to nothing, not even to the body that is to be encompassed. It is eternal, immobile, incapable of division, present everywhere in all things, whether corporeal or unsubstantial, and, in view of these characteristics, causes nothing to be here or there, full or empty, within it.

A description of Space.

In other words, Space as a container of all things in the world is that which is commonly conceived as being nonexistent. It cannot be seen, touched, heard, or perceived by the senses. Rather it is removed from all conditions of natural matter and everything else is completely different from it, with the result that there is, or can be, nothing which approximates it.

Space cannot be compared with anything.

Or, to put it another way, Space is that in which all things are as one and wherein we find all things contained: actually we cannot even find ourselves except in Space. Thus it is a continuum which penetrates and embraces bodies of all sizes. It encompasses all things, whether outside or in, and impartially receives them, extending as it does everywhere, without being moved or disturbed. It is without place (because Space has no given place); it is a part of nothing else and partakes of no form or quality or circumstance. We shall discuss this further in Chapter 6.

How can Space be perceived?

It is clear that all bodies also require internal space, corresponding with their capacity or a third dimension for they cannot exist or have their being without it. Take the following as an example: suppose there is a room in which neither a man or anything else is present other than air (for we know that in this lower part of the earth which we inhabit all space not filled with visible objects is filled with air). Now if someone enters this room filled with air, then he displaces the air to the extent of his own personal substance and that space which had heretofore been filled with air is taken up by his body. Should he depart, then air would again fill his place, or something else could be substituted, a stone statue, for example, which would be chosen to occupy this internal space, but the space itself would remain firm, without action and disturbance.

Air can be removed from the whole room by mechanical means, under proper conditions as our experiments will show, but the space of the room, nonetheless, will remain. Since, therefore, this one and the same space of the room can be filled, now with one body, now with another, and since one body always succeeds to the space of another and all the space within the room can be evacuated of all bodies and air, it follows that there is a great difference between Space and Body. Space, indeed, is capable now of being filled and again of being empty, and no Body in and by itself constitutes Space. Thus the statement of Descartes cited in the previous Chapter, is untrue, namely that "Space or Extension cannot exist without an extended substance." See also Book III, Chapter 9, number 1.

There is a great difference between Space and Body.

Even though Space cannot be seen or touched, it is, nonetheless, perceived by the eyes of the intellect. For example, when we see the intervening distance between two towers or mountains, we can easily perceive that this interval does not create the intervening body of air, but rather that it exists in itself. For indeed, should all air be removed, these mountains or towers would not touch one another. Furthermore, this space can now he empty and again full, as we have said our experiments will adequately demonstrate.

It should be remarked, however, (as we shall see in due course) that although people tend to speak of Space very frequently, it is the scholars who speculate about it. People commonly think of it in terms of the size or capacity of buildings or other such structures, when, in fact, it extends throughout the whole heavens.

People tend to look upon Space as separate, not as extending throughout the world.

CHAPTER 5

THE SPACE EXISTING BETWEEN THE EARTH BODIES IS COMMONLY CALLED SKY

Now the question to be discussed is the great expanse of space which exists between the earth bodies, the sun, the earth, the other planets, and the fixed stars. Should it be regarded as material or immaterial?

In the preceding Chapter I, we have already touched cursorily on how great is the distance, extent, depth, and the incomprehensible height of the interval or intermediate space between these huge bodies. Should we seek to compare its extent with the magnitude of the earth or any other similar body, according to all those learned in such matters, this would not be less comprehensible and credible than if it were compared with something else that is greater.

Is the Space between the earth bodies corporeal or not?

Now this Expanse or Space is commonly called Sky. In Book I, Chapter 31, we have already discussed what philosophers generally consider Sky to be, of what it consists and with what diverse meanings this word is used. Here it is employed in at least five or six different senses.

The Space between earthlike bodies is commonly called Sky.

1. As the airy sky where clouds and meteors are found. From this interpretation the following expressions arise: the clouds of the sky, birds of the sky, manna from the sky, a calm sky, rain descends from the sky, God ornaments the skies with his wind, God cloaks the skies with dense clouds, etc.

Diverse expressions for sky.

2. As the sphere or globe of some one planet.
3. As the sphere (so-called) of the fixed stars.
4. As the Crystalline Sky (imaginary).
5. As the First Mobile (so-called).
6. Finally, as the Empyrean Sky or the Abode of the Blessed, etc.

It should be remarked, however, that people generally are ignorant of such differentiations and designate this whole area as 'sky'.

The author's opinion regarding Sky, philosophically speaking.

In our opinion, however, the term, Sky, is relative and proportional to something, just as we have already said of Place. It is, indeed, not singular and real in itself, but rather is relative, having existence in respect to something else from which relationship it takes its name. If the other body were not in existence, then this would not exist under the same terminology. The following is an example. If there were no earth or any other earthlike body, the space surrounding it or its sky would not exist, although the space itself would nonetheless be there. Were the earth or any other earthlike body destroyed or taken away, the space encompassing it, which is its sky, would also be removed. The space which was there, however, would nonetheless remain.

Thus before the creation of the earth, there was no sky because nothing at all was in existence. When the earth was created, however, its sky was also created, that is, its sky took its origin or being from the proximate space which was there. When the earth will finally be destroyed, however, its sky will also perish although the space will nonetheless remain.

What the Sky of the earth is.

From this it becomes evident that the heavens should be regarded as the Expanse or encompassing Space in which either this earth or another similar earthlike body exists and moves, enveloping both their bodies and their orbits of influence. For example: Our Sky is that space in which the earth, together with its companion, the moon, and its own region of air, revolve about the sun every year. It is believed to extend to the outermost bounds of the earth's influence and to define its limit by Mars' and Venus' orbits of influence.

The sphere of the earth.

Or, to describe it another way, one may say that our Sky (the Sky of our earth, that is) is not only that encompassing space in which the earth happens to be at the moment, but all the space that the earth traverses in the course of its annual motion, and can be considered to be in the sphere of the earth's virtue, being distinct from the space of Mars and Venus which are neighbors to the earth.

The difference between Space and Sky or external and internal space.

How internal and so-called external space can be differentiated.

Now although the Sky of any earthlike body is one and the same thing as its space, nevertheless, because we do not see the internal space which is present in all bodies of this kind as in the case of our earth, while we can observe external space with our eyes, as it were, for this reason they differ from one another. Sky is thought to be external space, while, in fact, Space itself can be both external and internal.

The space, therefore, that is present in an earthlike body proportionate to its quantity or third dimension, is its inner space. External space, however, is the Sky or Expanse extending from the body's surface to the outermost limits of its virtue.

The Sky, or Expanse, namely the external space, of an earthlike body and the Ether are one and the same thing except for the fact that Ether is properly understood to be that space beyond the sphere of air at a very great distance away and widely diffused. Thus it is neither heavy nor light, nor is it forced up by the air. Rather it is that pure space in which an earthlike body has its own existence. In the same way, therefore, as we have established the situation of Saturn in the Ether, so too must we position our earth. Space necessarily exists between all earthlike bodies and this is called the Sky of either this sphere or that. Thus there are as many skies as stars, even though Space is a continuum in itself, and indeed exists as much outside as inside these bodies. As to whether this space is something material or immaterial, however, philosophers differ widely in their opinions. See for example Book I, Chapter 31, to which we have already referred.

The author's opinion as to what Ether is.

Although scholars had imagined for many years that the spheres in the heavens were solid bodies, this theory, nonetheless, has been exploded by more recent philosophers. They now know that, without the penetration of these solid spheres (which is impossible), there can be no free passage for the planets and comets.

Present-day philosophers do not admit that the heavens are solid.

Consequently Tycho Brahe and all his followers, together with Seneca, hold that none of the spheres in the heavens is real or solid in the way that most astronomers have imagined in the past, but that the sky is a very rarefied and tenuous substance like air or fluid matter, or an invisible and very fine essence like air, affording passage for stars to pass through it like birds through the air. Tycho does not believe that the ether is material and Kepler in his *Epitome Astronomicae Copernicae*, page 54, says that all ether can clearly be regarded as a vacuum. We also subscribe to this theory, namely that the sky is pure space and is, in itself, empty of all matter.

The heavens are not material.

The words of Ecclesiastes, Chapter 43, seem to point to this interpretation where it says, "Who would be filled by seeing the glory of God? The lofty height, is pure expanse." As to whether this 'lofty height' or 'expanse' of space is finite or infinite, this is the next question to be discussed.

CHAPTER 6

IS SPACE, THE UNIVERSAL CONTAINER OF ALL THINGS, FINITE OR INFINITE?

If we read the last Chapter of the preceding Book again, which deals with the Space outside the world (also known as Imaginary Space), one must again become aware of the great diversity in the opinions of scholars.

Aristotle declared that there is nothing at all above the starry heavens.

Aristotle declared that neither place nor time nor vacuity exists – in fact, there is absolutely nothing above the starry heavens, which to him is the most distant Sky and the material of which the outermost revolving component consists.

I understand this to mean that if, indeed,there is absolutely nothing beyond the world existing in the Infinite, there will be no Place there either (for Place presupposes the proximity of one thing to another and when there is nothing, then there is no proximity. Accordingly there is no Place). Neither will there be Time (because Time is a requirement of duration and that which does not exist certainly has no duration. Consequently there is no Time.) nor Vacuity (because a Vacuum is empty Space. If, however, there is absolutely nothing existing in the Infinite, then there is neither Vacuity nor empty Space.) The following Chapter will discuss, however, whether absolutely Nothing can exist or be discussed, or whether that which does not exist can be thought about or imagined.

Can absolutely Nothing be discussed or understood?

Philosophers in general say that the Space outside of the world is imaginary.

Philosophers say for the most part that the Space outside the world is a product of the imagination (as is evident in the last Chapter of Book I) and several hold that as it is purely a product of the mind,it is nothing other than Nothingness. Thus they conclude that before the creation of the world when there was no mind to imagine, there was no Imaginary Space.

But since they wish to equate it with Nothingness, why do they contradict Aristotle by positing at the same time that Space exists? For if Nothing is there, as Aristotle contends, neither capacity nor breadth nor depth, then Space cannot exist. This imaginary Nothingness, therefore, becomes a true Nothingness.

Others say that the Space beyond the world is a thing devoid of all reality. In our opinion this is the same as Empty Space.

The scholars of Coimbra hold that Space is as much inside as outside the sky and is eternal. They assert that God functions in it, not as an entity of substance but through his own immensity. See the last Chapter of Book I that has already been mentioned.

From this it necessarily follows: If God exists outside of the world in Imaginary Space into infinity, He has need of nothing else, not even Space (for He is Space and just as He is boundless, so is Space without end, because God Himself is Imaginary Space.) Thus Less declared (as one can see in the aforementioned last Chapter of the preceding Book), "Imaginary Space is God Himself, who, in view of his immensity, is everywhere and necessarily is infinitely diffused, etc." Should someone climb into the heavens and search the Space he has formed in his imagination but which has no reality, he most certainly would find God. If, however, God is Imaginary Space, one can ask whether it is not more likely that true Space exists.

Less declared that Imaginary Space is God.

The school of theologians holds that there is in fact one world, but that many others are possibly in existence. These scholars agree, moreover, that God can create worlds in Space or any other bodies of any magnitude you can imagine there. They hold that before the creation of bodies, however, no intervening Space existed and thus they contend that Space as we know it must have been created at the same time as bodies were.

The opinion of the school of theologians about Imaginary Space.

Some believe that Space was created at the same time bodies were.

When they say, however "He can create bodies there," they presuppose the existence of Place and Space, a containing element, both of which were assumed to have been in existence before the first body was created. As a matter of fact, if Space had been created at the same time as the first body, and had not been diffused in all directions before this time, then a body would not have had an external place in which to move; much less could it participate in the movement of "latio." Thus is would be necessary that God create new Space to enable a body to move from one place to another. It would follow, moreover, that if a body were taken away, Space would likewise be taken away at the same time, something that is contrary to evidence. For indeed, when any body, a tower, for example, is destroyed, the space it occupied still remains and other buildings can be erected in its space. Thus when a man leaves a room, the space he occupied still remains and can be occupied by another person. In the same way when air is withdrawn from a glass vessel, it leaves space behind it in the container and this too can be refilled. Space, therefore, does not exist in substantial form, but rather in itself, being something immobile and present everywhere.

The author's answer to this opinion.

René Descartes held that Imaginary Space "is a thing of substance and that this world or universe of corporeal substance has no limits to its extension, etc." This opinion is quoted in many places, as can be seen in the last Chapter of the preceding Book.

René Descartes opinion about Imaginary Space.

Throughout this treatise of ours, however, we shall give various proofs that this opinion is false and furthermore that it would blaspheme God to declare that something else is Infinite and Immense. For indeed, God alone admits of no boundaries to his extension.

This is false.

We shall continue, therefore, to examine the matter at hand, namely whether Space exists outside of the world and if it does, whether it is finite or infinite.

The author's opinion about Space outside of the world.

Since it is certain that the world is finite, the human mind seeks to discover its ultimate boundary, whatever it is, which contains and envelops everything, (whether indeed, this terminus is the imagined First Mobile or those waters believed to be above the heavens or the Empyrean Heaven or even something else).

If it is posited, in accordance with the more commonly held opinion, that there is Nothing (in the Aristotelian sense) beyond the world, then if one should proceed to its most distant bounds, he could extend his arm from the outermost edge of the farthest removed Heaven, or he could throw a spear or stone beyond this into that Nothingness. If this conclusion is not accepted, then this action would be impossible.

In the first case, if he could do this, then there would be Nothing to stand in his way and for this reason, it would be true Space because Nothing would be there (for if there were no Space, a stone could not be thrown out nor an arm extended. Indeed, nothing can be introduced into that which does not exist, nor can anything exist there, because that which is not, has neither fullness nor breadth and is completely incapable of receiving anything.) Consequently there must be something there.

In the second case, if he could not do this, it necessarily follows that there is something which stands in the way or prevents one body from occupying the place of another. Whatever obstructs this, of necessity will be some hard substance or body. Thus, beyond the outermost bounds of the world something must exist, for Nothing is not able to stand in anything's way. Furthermore, that which obstructs cannot be called "non-ens" (thing that is not). You say, "It is not a body or anything else which will obstruct but Nothingness itself or an absence of Space and Place which repels. Thus Nature repels just as a spear or stone resists a hard body." But who can understand how something without existence can repel and have a resistant nature? Evidence clearly shows us that the absence of Place or Space is nothing other than a hindering body unless one body can occupy the space of another, and that when this has been removed, still another can fill the space. Indeed, since it cannot be denied that the Divine Essence is of infinite immensity outside of the world, it is much more reasonable to regard this immensity as something of fullness, breadth, and depth, that is, as Space or Expanse, and not as infinite and pure Nothingness (which, according to the subsequent Chapter, we are well aware it could not be).

What absence of Space is.

It follows, therefore, that a spear or stone could be projected beyond the world and unless it were hindered by another body, would continue into the Infinite.

Although this throwing out of a spear could extend beyond the world into the Infinite and nothing would stand is its way, nevertheless, such a thrust could only be made in Space.

Thus it follows (1) that that Nothingness beyond the world is one and the same thing as Space, and that so-called Imaginary Space is true Space. For Imaginary Space (as most philosophers believe) is Nothingness and Nothingness is Space. Therefore, they say, Imaginary Space is true Space. The objection is not valid that whatever is imaginary is neither real nor true (although, indeed, this is not always the case), for we must necessarily imagine all that which we do not see and which transcends our experience. To cite an example: when a person does not see Rome, or a spirit, or a strange animal or anything else like this, he must imagine it. In the same way the imagination grasps the Infinite (because no one can perceive it). It does not follow, however, that Rome or the spirit or the Infinite are not true or actual things.

Even if there is Nothingness beyond the world, in the Infinite there is Space.

(2) If the projection of a spear or stone could be continued uninterruptedly into the Infinite and there were no boundary to stop it, it follows that this Space outside of the world must be completely without end as to its extent and therefore is infinitely immense. If it is infinite and immense, it may be asked, what is it? Before we can answer this question, we must first investigate those things which have existence.

Space is immense.

CHAPTER 7

THAT WHICH IS AND THAT WHICH IS SAID NOT TO BE

All that exists is either something that has not been created or that has been created. There is third condition, (or to put it another way, all that exists is either something that has not been created, or something that has been created, or Nothingness.) For Nothingness is the negation of one and the affirmation of the other; consequently, there is no third possibility. Whatever is not something, that has been created, that is something created. There is no third state of existence in thought, concept, or speech.

There are only two things which are: there is no third.

Something that has not been created is the essence from all things that have been created, infinite in extent, immense, eternal, preexistent. and existing in itself, taking its own existence from itself, containing all things and being contained by none.

What the uncreated something is.

What the created something is.

Something that has been created does not take its own existence from itself, but from the uncreated in which it exists and to which it owes its entire being. It is finite and consists of its own matter, (even though this should be spirit, something that we will discuss at the end of this Book), its life and innate virtues. This which has been created has many accompanying characteristics which are not in themselves separate created entities (because they cannot exist independently): rather because they are a part of that which has been created. either internally or externally, in actual fact or relatively. Included in this group of characteristics are quantities, qualities, motions, actions, passions, powers, relations, mental concepts, privations, non-essential qualities, numbers, times, places, names,words, etc.

The accompaniements of that which has been created.

These things which have being can be related to all things which have being.

Because, indeed, there are only two things which have being and no third, what has been created and its accompaniments is Nothing, that is, something that has not been created. Contrariwise, whatever has not been created is Nothing, that is something that has been created. For Nothing cannot be spoken of or understood in these two senses at the same time because in being the negation of one it becomes the affirmation of the other. (It should be remarked that in everyday usage, the word, Nothing, is not understood to mean absolutely nothing because a corollary is always understood in conjunction with it, namely in respect to what it is nothing. Thus a chimaera is said to be nothing insofar as it is regarded as an animal. It is absolutely nothing, however, for it is a concept of the mind attendant upon that which has been created.) Therefore although that Nothingness beyond the world in the Aristotelian sense (which we discussed in the previous Chapter) does not admit of the distinction between that which as not been created and that which has, nevertheless, because no third condition is possible, it is necessary that it be understood to be one or the other. For the existence of the two things that are, cannot be denied simultaneously, and consequently one cannot say here it is one and the same thing that beyond the world there is Nothingness or the Uncreated. Since, indeed, some things, as is believed, do not exist beyond the world, and whatever does not exist cannot be said to be something that has been created, it follows that whatever is uncreated there because it is not something created, this is something uncreated. Therefore it is not precisely Nothing.

The word Nothing cannot be spoken of or understood in these two senses.

Thus if one asks what situation existed before the world was established, and one person replies "something that was not created" and another says "nothing." each response would be correct in its own way.

For the one who says "something that has not been created" replies correctly as does he who says "nothing"; the one makes his answer with respect to something that has been created, which, to the other, is nothing. Nevertheless this was and still is something uncreated.

To say something is created from Nothingness is the same as something uncreated.

Thus we say that the sky and earth were made from Nothingness, that is, created from nothing, or the uncreated. Just as all things have been created from nothing, so too have they been taken into and established in nothing or the uncreated. Where created things now exist and are established, there was nothing here before. They were taken into Nothing or that which was not created because they could not have been received in something that was created, not only because it did not exist, but also because if it had, it would not have been Nothing but something created. But even if it had been taken in, something could not have been received into something: thus it was Nothing into which all things were received, in other words, the Uncreated.

Everything, then, has its place in Nothingness, and should God return the whole structure of the world which he created into Nothingness, nothing would remain in its place but Nothingness (as it was before the world's creation), or the Uncreated. For that is uncreated which has no beginning: we say that Nothingness also has the same characteristic. Nothingness contains all things; it is more precious than gold, free from birth and destruction, more pleasant than the sight of pure light, more noble than the blood of kings, comparable with the sky and higher than the stars, more powerful than a bolt of lightning, perfect and rich in all its parts. Nothingness is always all-knowing. Where it exists, the jurisdiction of all kings ends. It knows no disaster. Above it the earth is suspended, according to Job. It exists outside of the world and is everywhere. It is said that Nothingness is Vacuity, as is Imaginary Space and Space itself.

CHAPTER 8

IS SPACE, THE UNIVERSAL CONTAINER OF ALL THINGS, CREATED OR UNCREATED?

Since Vacuity and Imaginary Space or Space itself is Nothingness, for the latter is the negation of one condition and the affirmation of another (as was shown in the preceding Chapter), it follows that

Vacuity or Imaginary Space or Space itself is one of the two things that exist, either the created or the uncreated. There is no third. The fact that Vacuity or Imaginary Space or Space itself are created from Nothingness is stated by Jacques du Bois, a Leyden churchman who wrote on page 39 of his *Dialogus Theologicus Astronomicus* (in a passage in which he contradicts Galileo, Philip Lansberg, and others who hold that the sun is in the center of the world and that the earth revolves around it); "He who denies the existence of Vacuity above the distant heavens contradicts the infinity of the Divine Essence; for if the skies of the heavens do not encompass God, as Solomon holds in Kings I, Chapter 8, 27, then there is some Place beyond them wherein this Divine Essence abides which cannot be contained by any boundaries. We usually call this Place, Vacuity, because it contains no body. Accordingly before the creation of the world, God was present everywhere and extended infinitely in that boundless abyss which could not be filled by any finite body like the uppermost Heaven, no matter how great it might be. (Note: This writer, like Aristotle, believed that the sky is corporeal.) For, to hold with René Descartes that the corporeal substance of the world has no bounds to its extent, is to hold that creation is infinite. This is abhorrent to God and nature."

Jacques du Bois who regards the universal Vacuum beyond the world as Space and calls this Divine Essence.

From these words, it is manifestly clear that Jacques du Bois feels that God exists and extends infinitely throughout Vacuity, when he says, "There is some Place, therefore, in which this Divine Essence exists," he should have said there is Somewhere or some Place, not "in which," but "which itself" is the Divine Essence. For God can be contained in no Vacuity or Space because he himself is the Somewhere or the Vacuity for all things created or the Space itself which is the universal receptacle of all things.

Furthermore, Father Athanasius Kircher, the Jesuit, a very famous mathematician in Rome, even though he bitterly criticizes Copernicus' world system and admits of no vacant space in nature, writes the following words in his *Itinerarium Ecstaticum*, Dialogue 2, Chapter 9, number 4, where he discusses Imaginary Space:

"Since God is in fact infinite, he necessarily must fill all empty Space with the substance of his presence (which is conceived to be immense and infinite in extent). This in turn banishes all Nothingness or all Vacuity, Emptiness, and Non-existence. Consequently, beyond God there must be no Nothingness or Vacuity left. For were there Nothingness beyond God, one would be compelled to admit that God is not truly infinite because the Divine Substance would then be limited and bounded by Nothingness, and this is impossible. It follows, therefore, that the world is not encompassed by Nothingness but rather by God. When, therefore, you visualize this Imaginary Space beyond the world, you conceive of the fullness of the Divine Substance extended infinitely, but not of Nothingness. For indeed, he who perfected and created all things, receives, conveys, carries, encompasses, holds, utilizes, etc., them all in his own infinity."

The result of Father Kircher's hypothesis.

There can be no other interpretation of these words than that Space, Vacuity and Void, which we regard as the Immensity beyond the world, is not Nothingness or Non-existence, in the Aristotelian sense, but the creation of Nothing, or the Uncreated. Consequently, the Reverend Father speaks inexactly when he says, "God fills all Imaginary Space, Vacuity, or Void,etc, with his own substance and presence." a statement which he himself admits is not in accordance with nature. For how can God fill with his own substance that which is not? Rather he should have said, like the above-mentioned Less, that "Imaginary Space, Vacuity, or the Nothingness beyond the world is God himself," as he finally does when he declares that the Space beyond the world is not Nothingness but the fullness of the Divine Substance.

We should conclude, therefore, that the infinite Essence of God is not contained in Space or Vacuity, for although God is present everywhere, he is not contained either in Space or in Vacuity; rather this Space and the absence of all creation is in himself and for himself. Indeed this Essence is different from all things that have been created for it is confined by no boundary nor enclosed by any barrier (as I shall show), but is infinite in extent, thus being as much outside of all things as inside. Consequently he is present in all our actions and judges us in accordance with them. "He is near at hand, not far, and no one can hide away in the shadows to escape his sight, because he fills the heavens and the earth." Jeremiah, Chapter 23.

Thus in the same way as these aforementioned authors regard the Space outside of the world to be the Divine Essence, so should they consider the nature of the Space within the world. For when they declare that the Space outside of the world is infinite, immense, indeed, the fullness of the Divine Essence, it follows that the Space within the world is the same. Thus this world of ours is contained in infinite Space and cannot be excluded from it (and indeed it is not the size of a point or atom relative to the immensity of Space). For Space yields as much as the world system requires, but in such a way that it is also present in it, penetrates it, and thus receives, conveys, carries, encompasses, holds, utilizes, etc., all existing bodies in this world system of ours; nor does it matter whether the earth be here or located elsewhere, as we shall discuss in the subsequent

Chapter. The scholars of Coimbra confirm all this, as can be seen in the preceding Book, Chapter 35.

CHAPTER 9

INFINITY, IMMENSITY, AND ETERNITY

Philosophers define the Infinite as something beyond which there is always more; an Immensity which is without end, limits, or bounds, extending is all directions, up, down, right, left, before, and after. They describe Eternity through the concept of duration, being without limit, either before or after, that is, without beginning or end. They distinguish an Age from the Infinity of time by reason of its having a beginning.

Inasmuch as Infinity eludes all human understanding, as do also Immensity and Eternity, we must take them for one and the same thing and regard them only as attributes of the Uncreated. This is the only way human understanding allows us to speak of them.

Infinity absolutely confounds all human understanding.

Sane reason, however, declares that Infinity, Immensity, and Eternity is a unity and cannot be a plurality, and although it cannot be defined specifically by numerical measurement, name,or word (unless by that of "being" which indeed indicates its essence but does not, however, reveal what it is), it is that which is, a concept which confounds all human understanding, intelligence, and power of definition. Thus no matter how hard we try to describe this by number, quantity, name, or word, there is always and shall be something wanting. For although it admits of no limit to its size, it also does not recognize any division or separation (for which reason it is said to be one). Moreover, no name can be given that is sufficient to it, for, just as it is immeasurable, so also is it incomprehensible and incapable of definition. At what point, then should we begin to define by a name that which has no bounds to its extent or to give it a beginning, middle, or end?

There is no relativity to Infinite Immensity.

If someone has preceded a great distance into Infinite Immensity, which indeed has no beginning, nor middle, nor end, and had covered countless thousands of miles, he would still be in the same place relative to Immensity. And if he repeated this journey and progressed thousands and thousands of miles farther, Immensity would have remained boundless and he would still be in the same place, not having proceeded a step either towards his objective nor the implementation of his intention. For in Immensity there is no relativity other than that conceived of by our own reason or by that of some other created being.

Furthermore, we are accustomed to regard created things as infinite in number, they being either physical objects or temporal divisions such as years, days, etc., the end of which the human mind cannot conceive because of the great number. Inasmuch as number arises from separation or division (as has been previously remarked), wherein everything is regarded as a unit, with a multitude being collected from units, for this reason number cannot properly and accurately be termed infinite. Whatever has a beginning is not infinite, but finite, for indeed, a number can never be so great that something cannot be added to it. Thus since it is evident that something can always be added to a number, it cannot be infinite but rather must be finite, for a finite number added to a finite number does not produce Infinity. As for what true Infinity is, it is necessary that it be both complete and incomplete at the same time.

Concerning so-called infinite number.

What number is.

Although human understanding can conceive of an infinity of numbers as Infinity, this is not true Infinity: indeed, it can in no way be equated with Infinity which surpasses all human comprehension, (because Infinity, Immensity, and Eternity is equal to nothing else nor can there be anything greater or smaller than it. As for an infinite number, however, it can be either equal or greater or smaller). Nonetheless, this surpasses all human comprehension. Indeed, if God should create an infinite number of men, the infinite number of fingers and strands of their hair would be more than the number of men. What's more, if God should create an infinite number of suns with their ten planets rotating around each of them, there would be more planets than suns. Or, to put it another way: imagine a straight line which extends out infinitely in either direction on which God creates an earthlike sphere every thousand miles. There would be an infinite number of spheres on this straight line arranged in this given distance from one another. Now if every other one were removed and set aside in a similar straight line a thousand miles apart as in the first instance, there would be two infinite parallel lines, each of which would be filled with an infinite number of spheres. Should this process be repeated again and again, each line would still be filled with an infinite number of spheres. Thus many Infinities can be postulated simultaneously. If then, many Infinities can be conceived of, an infinity of Infinities of this kind can also be postulated as well as infinities of Infinities.

An infinite number cannot be defined.

As far as number is concerned, there can be many infinities.

Thus even though we mortals are incapable of understanding Infinity, Immensity, and Eternity, we can nevertheless arrive at a better understanding of the so-called Infinite through ciphers than in any other way. For this reason something should be said about them.

CHAPTER 10

NUMBER

That which is incomprehensible to the inexperienced can only be understood by numerical symbols.

Although no mortal man can imagine Infinity, nonetheless something so huge can be represented by numerical symbols arranged in a series, which seems greater than Infinity, even though this is not so.

To be sure, there were and are many individuals ignorant of the art of stereometry who think that the number of grains of sand on the seashore is so great that words cannot express it nor ciphers represent it.

To Archimedess the number of grains of sand is small.

Archimedes, however, using geometric methods, produced a number which certainly is not only greater than the number of grains of sand on all the seashores, but which is larger than the number of peppercorns that could be encompassed in the solid sphere of our earth. We shall attempt this computation which is greater than that which is commonly undertaken so that an inexperienced person, in following it, can have greater confidence in his concept of numbers:

38 arithmetic ciphers represent more grains of sand than the whole earth contains.

(1) Let us suppose that the whole solid mass of earth is composed of extremely small granules, 10,000 of which correspond to one peppercorn. (As a matter of fact, they will be so small that they cannot be seen by the naked eye.)
(2) Let us take 40 peppercorns and lay them in a straight line. In length they will be equal to a finger.
(3) Let us calculate 10 fingers as one foot.
(4) And 5 such feet as one measured pace.
(5) And 10,000 such paces as one German mile.
(6) Let us set the earth's diameter at 10,000 German miles (which elsewhere is given as 1720 (ERRATUM, read 1720 instead of 1820).

With the problem thus set up, the earth's diameter calculated in paces will be 100,000,000 in feet, 500,000,000; in fingers, 5,000,000,000; and in peppercorns, 200,000,000,000. Furthermore, inasmuch as (according to Euclid, Book 12, theorem 18) the volume of a sphere is the cube of its dimensions, the diameter of the earth cubed will yield a volume of 8,000,000,000,000,000,000,000,000,000,000,000 calculated in peppercorns. If this then is multiplied by 10,000 grains of sand (which are equal to 1 peppercorn), the computation of the earth's volume will be 80,000,000,000,000,000, 000,000,000,000,000,000,000, calculated in terms of grains of sand.

Furthermore a figure of fifty-three ciphers in a series is greater than the whole world, extending up to the firmament (according to the Ptolemaic and Tychonic systems), can encompass in terms of such tiny grains of sand. There have been learned scholars, moreover, who have declared that the number of such sand granules is so great that they cannot be enumerated by many thousand ciphers, even if they were placed one after another at distances of one, three, or even five miles.

The number of 53 ciphers is greater than the whole world can encompass in terms of the smallest grains of sand.

Let us follow Clavius' example and suppose that the Ptolemaic system is correct with its earth-centered world. Then let us set the distance of the Starry Heavens not at 45,225 radii of the earth from the center of the earth as Alfargani calculated (who, of all the Ptolemaics, attributed the greatest distance to the fixed stars), but at 50,000 radii. Assuming this to be the case, then the diameter of the Starry Heavens will be 100,000 radii of the earth or 50,000 whole diameters. The cube, then, of these 50,000 diameters will be 125,000,000,000,000 cubed diameters of which the earth's diameter is one cube. Because, according to Euclid's theorem, which we have already cited, the volume of cubes is the multiplication of their dimensions by three, it follows that the earth is to the Starry Sphere as 1 is to 125,000,000,000,000, that is, the earth is contained in the Sphere of stars 125,000,000,000,000 times. Now continuing these calculations we find that the earth contains the equivalent of 80,000,000,000,000,000,000,000,000,000,000,000,000 grains of sand. This, when multiplied by the aforementioned figure will give the number of grains, according to these computations, that the world as a whole will contain, namely 100,000,000,000,000,000,000, 000,000,000,000,000,000,000,000,000,000,000 grains. (ERRATUM read 100 etc., instead of 10, etc.)

From these numbers it is evident that a figure can be comprehended which is calculated in thousands and tens of thousands, but when it exceeds 10,000 of 10,000 10,000's, then it is incomprehensible, even though it can be described in numerical terms as in the foregoing examples. Because, in fact, an extremely large number like this, consisting of 53 ciphers, can be described, although it exceeds all human imagination, someone may comment that any infinite and completely incomprehensible number can be described by a thousand and more ciphers.

There are few who are aware of what can be understood by arithmetic figures.

I made a bargain with a competent mathematician and by mutual agreement gave him 100 talers in payment for his work if he could compute within a prescribed time, namely three months, a figure which we had agreed upon, the sum of the number 2, multiplied by itself or squared: for example, twice 2 is 4, 4 4's are 16, 16 16's are 256 etc.

He agreed to the bargain and deposited 10 imperials. He wished, however, to complete this calculation, not within three months, but in one, and asked to present the result of his projected work when the time came in the presence of witnesses. He did not realize that this was no work for mortal man in view of the great number of figures that emerged, as the following operation reveals:

$2^2 = 4$
$2^2 \times 2^2 = 16$
$2^3 \times 2^3 = 256$
$2^4 \times 2^4 = 65{,}536$
$2^5 \times 2^5 = 4{,}294{,}967{,}296$
$2^6 \times 2^6 = 18{,}446{,}744{,}073{,}709{,}551{,}616$, etc.

From this operation we see that the number of ciphers is almost always doubled and consequently if the multiplication were to be carried further and two were thus multiplied to the seventh powers there would be 40 ciphers, if to the eighth power, 80, to the ninth power, 160, to the tenth power, 320, to the eleventh power, 640, to the twelfth powers 1280, to the thirteenth power, 2560, to the fourteenth, 5120, to the fifteenth 10,240, to the sixteenth, 20,480, to the seventeenth, 40,960, to the eighteenth, 81,920, to the nineteenth, 163,840, and finally, to the twentieth power, the total number of ciphers would be 327,680.

Who, then, could undertake such a task either of multiplication or addition? In the nineteenth multiplication alone (when 163,840 was doubled), the product would contain more than 26,843,000 figures. Thus there would be as many ciphers written as there are letters in the 1242 volumes of the Corpus Juris (that is, if one computes that the Corpus Juris, together with the Notae Gothofredi contains 1,000 pages, and each page 4 columns, and each column 90 lines and each line 60 letters. According to this computation, the Corpus Juris would contain 21,600 letters, which multiplied by 1262 would yield 26,843,000,000 ciphers). Indeed, where would one find a sheet of paper of such a size that this calculation could be written on its surface?

Thus we see in what an incredible number of ciphers this computation results; but still the overall aggregate itself of arithmetic figures is so great that this number of ciphers, when compared with it, is absolutely infinitesimal.

The number which can be

Now if these aforementioned 53 ciphers contain more in themselves than the whole world can hold of the smallest grains of sand, what an awe-inspiring grand total will the 327,680 ciphers contain when written in a series! For indeed, the addition of one cipher always increases the total ten times (ERRATUM read "decuplo" instead of "decuo"). As an example, one sees that if only six zeros were added to these 53 ciphers, the total would be of such a size as to encompass a thousand thousand of worlds like ours.

represented by ciphers is awe-inspiring and incomprehensible.

0, wretched men that we are, what can our intellects understand of the infinity of the uncreated? If, indeed, there were such a number, how very great it would be! Nonetheless, we are compelled to hold that every number is finite, having, as it does, a beginning by reason of that which comes before, and being without end, by reason of that which comes after.

CHAPTER 11

THE HEAVEN WHICH IS CALLED THE ABODE OF THE BLESSED

It is clear that a certain πο͂υ ("somewhere") will remain for the abodes of the Blessed as well as the Damned after the destruction of this earth (inasmuch as a complete annihilation of the earth lies in the future without the possibility of its changing for the better. (See the last Chapter of Book V.) "For all of us must needs appear in person before the tribunal of God so as to report what each has done in the flesh and whether this was good or evil." There will be a question, therefore, of "where." Will it be above or below the so-called Starry Heavens?

The Last Judgement.

It is understood from what has already been said that Space, the universal container of all things, is infinite and immense. Therefore everything that is created exists in this Space. But just as in Immensity there is no below, no above, no distance, no here, no there, and no Place at all unless it is relative to a second created thing, similarly the place or abode of the Blessed cannot properly be called a Place, unless it is in a relative sense. Especially after what we shall say in Books V and VI about the distances of the stars, we must assume that the abode of the Blessed and the Damned, after the destruction of this earth, perhaps will be that place which the earth had previously occupied and which we commonly call the Sky. For otherwise it would be nowhere in Immense Space unless there is some place known to exist relative to something else. Take this as an example: if a place in the world were not assigned to the earth in respect to other earthly bodies, it could not be said that the earth is here or

In Immensity there is no relativity.

Unless the earth exists relative to something else, it is nowhere.

there in this Immense Space, even if it were before or behind, up or down from here at a distance of inexpressible tens of thousands of miles. Moreover, if there were no other earthly body, (all having perished together with the destruction of the earth), then it would be one and the same that the earth is here or somewhere else at a distance of an infinite number of miles. A tremendous difference of this kind would not be of as much account as one step on a journey to the Indies, for the earth would be in one and the same place, which is nowhere.

The Sky of the earth.

The Sky of our earth, therefore, will be without a doubt (as we said in Chapter 5) nothing other than that expanse of Space which the earth, together with its companion moon, occupies around the sun and as far as their spheres of virtue extend between the skies of Venus and Mars.

Furthermore, just as Space (that universal container of all things) is something uncreated and is always identified with something greater, so the more extensive we conceive Space to be, the more we come to realize that God naturally abides there. Thus it is written in the Psalms: "Who, like our God, dwells in the heavens?" etc. All pagans, moreover, as Aristotle testifies, naturally assigned the uppermost part of the heavens to God and identified them with God when they said "the heavens are witness." In the Lord's Prayer we call upon God "Our Father who art is heaven," that is to say, He Himself is the heavens. (That God exists in the heavens as in some defined region or dwelling-place we know to be otherwise for the Heaven of heavens is not able to hold Him.) Thus He is the dwelling-place of those who are in His grace, while those not so favored belong to Hell and are crushed by eternal shadow.

Why are the upper regions assigned to God?

What Hell is.

After the Last Judgment, in our opinion, the heavens will not be something created, indeed a great palace built of extremely beautiful materials, (for if it were something created and God dwelt in the sky as in some defined or finite region, as is popularly believed, it would follow that God would now be of improved position and richer in possessions, as it were, because He would possess the sky or throne of Majesty which He had not had before). Rather there will exist only God Himself who will embody the grace of God, together with the heavens.

What is the Heaven of the world?

As for the Heaven of the whole world, this is the space existing between all bodies of this world which we suppose was empty before the world's creation but has now been filled in some way with earth-like bodies. Moreover, to each body of this kind, there is now assigned its own heaven or Space, like that region between Mars and Venus which belongs to the earth.

We are of the opinion that this Space can ultimately accommodate as many men as may come into existence. The following calculations will make this clear. From Book VI, Chapter 13, it is evident that Mars is 3739 radii of the earth distant from the sun. Therefore the great sphere of Mars can encompass 52,271,672,419 spheres of the earth, (since, according to Euclid, spheres consist of the cubic ratio of their diameters or radii.) Venus is 2049 radii of the earth distant and therefore its great sphere can contain 8,602,523,649 spheres of the earth. Then if Venus' sphere is subtracted from Mars', sphere, the remaining space belongs to the intermediate sphere of the Heaven of this earth which can contain 43,669 148,770 spheres of the earth. This region appears to be of sufficient magnitude to accommodate mankind, however, populous it becomes. We are not going to discuss here the widely divergent opinion of theologians on this subject.

The magnitude of the earth's Heaven.

CHAPTER 12

THE GREATEST AND THE SMALLEST

Quantity is not anything, in itself without comparison, for it exists only external to and relative to something that has been created. In nature, to be sure, there is nothing that can be said to be great or small, long or short, high or deep, wide or narrow, unless it is in respect to or in comparison with something else.

Quantity is not anything in itself but only relative to something else, e.g. place.

Reason itself dictates that all created things have certain bounds of largeness and smallness, and thus do not admit of being infinitely large or small. This is rather the characteristic of the uncreated. Thus whatever is infinite or immense can never be spoken of in terms of greatness or smallness for Infinity does not admit of comparison. Individual and separate creations, however, are either large or.small relative to one another and cannot be judged to be large or small without comparison.

In the nature of things the greatest creations are the world bodies, such as the sun, or suns, fixed stars, planets, and their satellites. Now inasmuch as variety is a characteristic of things and omnipotence consists in variety,difference also is clearly apparent in all things (especially the homogeneous). Thereby mutual comparisons arise, without which nothing can be said to be large or small. In this way the sun can be compared with other suns and the planets with other planets.

What are the great bodies?

Omnipotence is composed of variety.

Once the largest bodies have been identified, then the smaller and smallest are examined, each body being compared with another of its own kind. For example, on this earth of ours, animals of diverse size are found, whales, elephants, etc., and in contrast mites, lice, and gnats. There are also tiny grains, motes of dust, and other extremely small objects which can be further broken down into smaller and smaller parts. They do not vanish into infinity, however, but only into nothingness. These very minute particles are called atoms and are individual units that do not contain parts of anything else.

What atoms are.

What are the smallest particles?

Although objects can be seen with the help of a microscope,which heretofore were not visible, and we can detect with this optical aid many tiny animals which abound in different substances, things that previously were completely hidden, nonetheless, there are parts of water, air, fire, and other fluids which are so extremely minute that they can in no way be detected.

Even though many of these are very tiny, they cannot, however, penetrate glass or metal. How much smaller, then, must the parts of effluvia or insubstantial matter be which are able to pass through and penetrate such substances? See Book IV on this subject.

Nonetheless whether it be the most minute of all tiny animals with the smallest of entrails, or an invisible mote, or an atom of fire, water, air, or some very rarefied liquid, or the most minute globule, or the essence of some insubstantial effluvium which penetrates all bodies, or even a spirit, still this is contained in Infinite Space and is encompassed by it and has its place therein.

As for the insubstantiality of Space, even rarefied essences are material and corporeal in comparison.

Furthermore, there is no created body of such magnitude that Space is not infinitely and immensely greater. Contrariwise, there is no body so small, insubstantial, or rarefied that the space within it is not infinitely less and more subtle.

From the foregoing discussion, then, it is manifestly apparent that in the nature of things the dust of our bodies however small, and wherever on earth it may be, is in existence in this Space and in it or through it can easily be reassembled and put together. Therefore our body can be resurrected and called upon in the Last Judgment to render account for its actions in this life.

BOOK III

INDIVIDUAL EXPERIMENTS

CHAPTER 1

CONCERNING THE ORIGIN, NATURE, AND CHARACTERISTICS OF AIR

For a better understanding of our experiments, we will set forth certain new and well established facts concerning air which first must be examined and then proved in subsequent chapters.

What air is.

Air is a kind of corporeal virtue flowing out from water as well as from earth and other bodies into surrounding space (See Book IV, Chapter 1). In other words, air is nothing more than an exhalation or vapor or a discharge of waters, soils, and other corporeal substances. This is particularly evident when water or some other fluid is agitated, disturbed, or dispersed (we have discovered, for instance, that wind and organ sounds can he produced through the falling of water). Likewise when a moist body spoils or becomes putrid, then a corporeal vapor or exhalation, which is nothing other than air, flows out even more profusely. Hence air cannot properly be called an element: it is rather an effluence.

Air is not an element.

Air is not perceived by the sense of smell.

Ordinary air, which flows out from water and earth, although it is nothing more than an exhalation of the soil, is not perceived by our sense of smell, since we are accustomed to air from our first moment of life and continue to live in it. We do smell exhalations, however, from other vapor-producing bodies that have mixed with ordinary air and that differ from one another according to their innate condition and our own nature.

Air is something corporeal.

All air vapor is indeed something corporeal, but something very delicate as well, and adapted for extending and expanding itself.

What the sphere of air is.

All the air surrounding the earth partakes of its corporeal virtue: in other words, air is a corporeal exhalation encompassing the earth's sphere in which all animals on the earth's surface are sustained at the same time.

Air is never reduced into water.

Whatever air or vapor has once arisen from water or other things, this never again becomes water but remains air. Even though moist elements evaporate in the lower air through the heat of the sun or fire and are again changed into water through condensation, air, nevertheless, is not changed. One can clearly see this from the experiment in the following Chapter 2 and elsewhere, as for instance in the spark-like particles that glitter, so to speak, in the air in wintertime when it is very cold. This results when fine particles of water that were sprayed and suspended in the air, freeze and become separated from it. The air itself, however, never freezes.

Air is expanded through heat.

Air extends and expands itself more or less according to a greater or lesser intensity of.heat. Warm air, consequently, occupies a larger space and cold a smaller. When philosophers call this change the rarefaction of air, as they are accustomed to do, this terminology is inexact, because air is not expanded in such a way that it penetrates something solid by being more rarefied, finer, and more subtle, but rather that it occupies more space. For this reason, when heat disappears, air is again constricted and occupies less space. Thus air is not increased or lessened; rather it is heat which air absorbs in a greater or lesser degree and gives up when the heat disappears.

Air is condensed through cold.

Just as air is expanded through heat, however, so through a greater and greater remission of heat, that is, through cold, it is condensed, and thereby occupies less space.

Air can be compressed.

Air, moreover, has the characteristic that it can be more and more condensed through violent compression as well as expanded by being given greater space, as will be seen in our experiments. (We see this, for example, with inflated balls and fountains spraying water upwards through the compression of air. Using a syringe or pipe, we forcibly introduce so much air in these as we can without risk of their bursting. After this, when the fountain is opened, the compressed air projects the water that had previously been introduced, on high with very great force.)

The condensation as well as the expansion of air has its limits.

As all things have their own prescribed limits, moreover, so does the condensation as well as the expansion of air have its own bounds. Above and below these limits, extension and compression cannot take place. Thus air can be compressed in a sturdy metal receptacle repeatedly by the exertion of pressure in such a way that at last a solid body, similar to water, is formed. And contrariwise, a very small bubble of air the size of a pin, can expand itself one hundred times over in some empty receptacle and finally disappear into nothingness.

Air has weight and presses upon itself.

The air surrounding the earth presses upon itself because it is corporeal and has a certain weight. Indeed, the upper air presses down increasingly heavily on the lower. Thus it follows that the lower air here, which surrounds us, is much denser than the upper air. And whatever is denser, has more mass, and whatever has more mass, is heavier. Therefore we have more and heavier air here on the earth's surface than that found in towers and mountains; to be sure, the higher the air is, the lighter and more rarefied it is.

Clouds in the air retain their own fixed positions.

In our opinion every kind of cloud, whether heavy or light, retains its own fixed position in air of the same weight. If air were of equal weight everywhere, however, it would be equally heavy above and below and the various kinds of clouds could not be found in the different regions. In this case, they would either descend to the earth or ascend to the uppermost boundary of the air, as with water where an object either seeks the bottom or floats.

Air exerts pressure on everything.

Because of its own weight, air presses not only upon itself but upon all things beneath it with an essentially constant pressure. We do not feel this because we live in this air which surrounds us with equal pressure on all sides and at the same time penetrates us. For as fish have no perception of pressure in the water, so animals perceive it even less in the air.

Air exerts the same pressure as water 20 cubits deep.

The weight of air on the earth's surface is as great as the weight of water about 20 Magdeburg cubits deep. In other words if water should rise 20 cubits above the earth's surface, the pressure it would exert on all things beneath is the same as the pressure of air.

Air surrounding the earth does not always have the same weight.

Air, however, does not always have one and the same weight; when rain falls, for example, it is lighter.

If air did not have some weight, it would fly away and leave the earth as a result of its yearly motion. (We will describe what this weight actually is in Book IV, Chapter 3, that deals with the conserving virtue.)

Water is always present in the lower air.

Some water is always present is these lower regions of the air, varying in amount according to temperature conditions.

Although air presses upon water, it does not press down to such a degree that it penetrates it, since it is lighter.

The function of air.

Air changes bodies and can convey heat and cold, for it is adapted to take in everything, like light, sound, smell, as well as moistures, vapors, and the like. It functions in wondrous ways on animate and inanimate things and by surrounding animals externally, it protects, moistens, and refreshes them by its changes. In order to bring about an equalizing effect, it penetrates their bodies and keeps them alive through breathing.

The ascent of a flame is dependent upon the weight of air.

The ascent of a flame, as well as of vapors and smoke, does not depend upon their own particular and innate lightness, but rather upon the weight of the surrounding air. For this reason a flame, smoke, or vapor, rises upward, as a water bubble rises in water. (Being warmer than air, they are consequently more rarefied.) Thus if there were no air, smoke would not rise.

The air surrounding the earth fills all the empty spaces.

When a space is given up by some body is the atmosphere, no void is left because the space is filled by air. In the same way the space which a fish takes up in water is filled with its body but is again replaced with water when it moves away. (ERRATUM read "in aqua quod piscis" instead of "in aqua, piscis.")

CHAPTER 2

THE FIRST VACUUM EXPERIMENT PERFORMED THROUGH THE EXTRACTION OF WATER

When I came to consider the immensity of Space (with which Chapter 1 of the preceding book deals) and the necessity of its presence everywhere, the following experiment came to mind.

I had a wine or beer keg filled with water and well caulked throughout so that no external air could enter. I introduced a bronze tube, attached to the lower part of the keg by means of which water could be drawn out. Then through its own natural weight, the water would of necessity go down and leave behind in the keg a space empty of air (and accordingly empty of any natural body.)

See Plate V, fig. 1.

So as to achieve results that conformed with my expectations, I prepared a brass pump or a pipe, abc, adapted for admitting and discharging water, of the kind employed in fires, described in German as a Messingen Feuersprutz, with a piston rod, c or f, and piston g. This was carefully tooled so that no air or light could enter or leave around the sides of the cylinder and fitted as well with two leather flaps or valves. One of these, inside of the cover of the pump, a or d, was adapted for admitting water, and the external one, b, for discharging water. When the pump was attached to the lower part of the keg, a, by means of four lugged iron rings, e, I tried to extract the water. The iron loops and screws which fastened the pump to the keg, broke, however, before water could be evacuated.

There was no need for despair, for I remedied the situation by using stronger screws. Finally three hardy men, pulling on the piston of the pump, succeeded in extracting the water that was evacuated through the upper valve, b.

Air passes through wood.

A noise was heard throughout the keg like that of vigorously boiling water and continued until the keg was filled with air which replaced the water that had been drawn out.

This difficulty, then, had to be remedied in some way. I prepared a smaller keg, therefore, and placed it within a larger one. Then renewing my efforts, I saw to it that a longer pump was inserted through the wooden planks of each keg, that the smaller keg was filled with water, its opening sealed, and finally that the larger keg was filled with water. This done, the water was extracted from the smaller keg in whose place a void undoubtedly would remain.

Water passes through wood.

But at the end of the day when work was finished and all murmuring and roaring had ceased, a different sound was heard, occurring intermittently, like the soft twittering of a small song-bird. This continued almost three whole days.

Then when the bunghole of the smaller keg was opened, the barrel was found to be filled in large part with air and water. There was, nonetheless, a part that had been empty because at the time of opening, air rushed in.

Some air is generated from the friction of water.

Now everyone began to wonder where the water came from that had entered this keg, so carefully strengthened and sealed throughout. I finally discovered through diverse and repeated experiments that water passes through wood under great pressure and that the friction produced from the resulting resistance always generates some air from the water in the keg, (a fact which subsequently should be well borne in mind). I remarked, moreover, that a keg cannot be completely filled with air because of the resistance which the wood offers. Thus with the reduction of strong pressure, water and air also cease to enter and the keg is left half-evacuated, as it were.

CHAPTER 3

THE SECOND VACUUM EXPERIMENT PERFORMED THROUGH THE EXTRACTION OF AIR

See Plate V, fig. 2.

Once the porosity of wood had been recognized through visual observation as well as through my own experiments, I finally concluded that a copper globe, whose capacity was about 60 or 70 Magdeburg quarts, was better adapted to this investigation. (Reverend Father Schott in his book on the Magdeburg experiment, calls this a "cacubus" or "copper pot".) I fitted a brass stop-cock, B, in its top; in the lower part, I attached the pump and made sure that it was securely fastened. Then, using the same method as before for extracting water, I now sought to draw out air.

At first it is easy to move the piston rod.

At first it was easy to move the piston, but gradually this action became more and more difficult until finally two strong men were

PLATE V.

scarcely able to draw it out. While they were engaged in plunging the piston rod in and pulling it out and began to feel confident that essentially all the air had been extracted, suddenly, and to the terror of all, with a loud report, the copper globe collapsed like a cloth crushed in one's hand or as if the globe had been hurled from a tall tower and had landed with great force.

The copper globe collapses from external air pressure.

I attributed this to the inexpertness of the craftsman who perhaps had not fashioned the globe in a perfect sphere. For a globe with any indication of a flat surface could not support external air pressure while a perfect sphere could easily have sustained this because of the mutual support of the parts whereby one helps support the other.

A completely spherical globe does not collapse.

Necessity dictated, therefore, that the craftsman fashion a perfect sphere. This being done, the air was extracted as before, with ease at first, but with difficulty later on.

When air ceased to escape from the upper valve of the pump, this served as evidence that the globe had then been evacuated.

Thus a vacuum was produced for a second time.

Air penetrates very easily.

Afterwards, when the stop-cock, B, was opened, air rushed into the copper globe with such force that a man standing in front seemed to be drawn to it. If he moved his face sufficiently close, his breath was taken away. A person, moreover, could not extend his hand over the stop-cock without the risk of its being violently drawn to it.

Although the globe clearly seemed to have been evacuated, nonetheless after it was left in this state for a day or two, investigation revealed that it was filled with air that had entered around the sides of the pump's piston, as well as through the valve and stop-cock. This defect had to be remedied and was done in the following manner.

CHAPTER 4

THE CONSTRUCTION OF A SPECIAL TYPE OF APPARATUS FOR PRODUCING A VACUUM

Because air, being an extremely fine body, in a remarkable way passes into and fills all tiny openings and interstices, no matter how small, and because some air always enters unnoticed around the sides of the piston of a pump, or air pump, as well as its valves (for it is impossible to procure a perfect piston or valve which resists the penetration of air), I built several different types of apparatus in such a way that the air pump could be surrounded by water in the

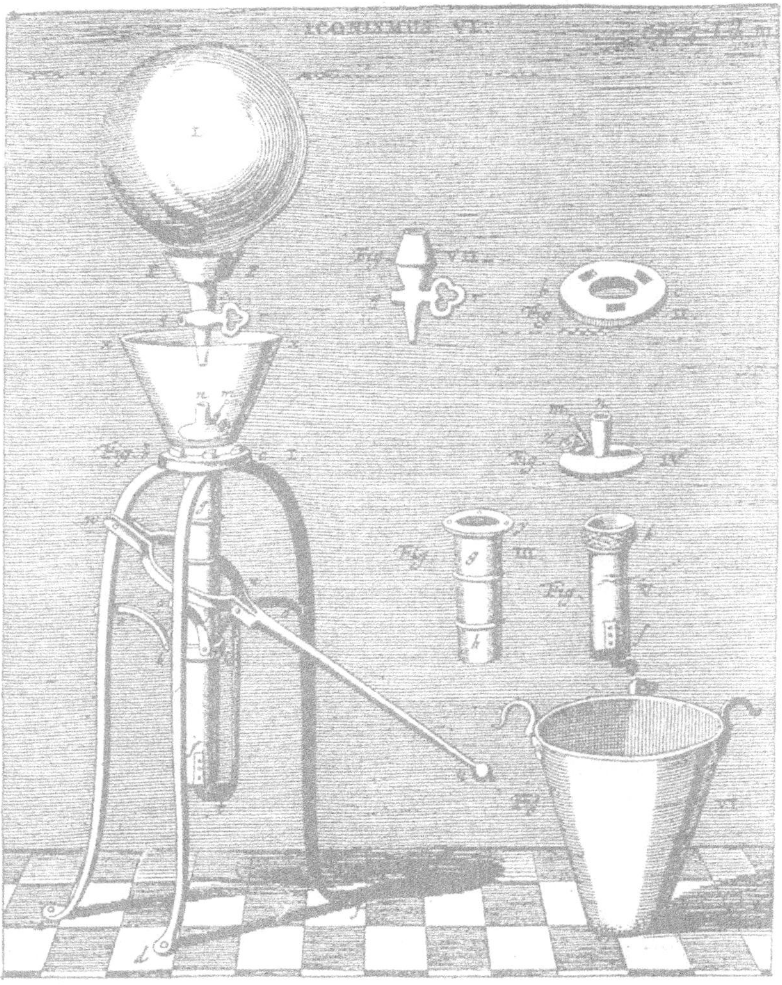

PLATE VI.

lower as well as the upper part. Reverend Father Schott first described these in his *Ars Hydraulico-Pneumatica* and afterwards in Book I of his *Technica Curiosa de Mirabilibus Magdeburgicis*.

Now inasmuch as my Most Serene and Powerful Sovereign, Elector of Brandenburg, my most benevolent Master, graciously wished to see this experiment performed (which the Reverend Father Schott called the Magdeburg experiment), and since these earlier types of apparatus were difficult to transport, I designed the following model.

1. One should have an iron tripod, abcdf, about two cubits in height, fashioned by a smith. *[See Plate VI, figs. 1 and 2.]* Its feet should be joined together at the top by an iron disc, bc, while at the bottom, they should rest on the floor, fastened thereto with iron screws, afd.

2. The air pump, gh, should be of the same type as the brass pump described in Chapter 2 and provided with a lead disc, y, at the top. *[At the same time see figs. 1 and 3.]*

3. One should fasten a brass plate, mn, provided with a small tube, n, *[See figs. 1 and 4.]* (into which the stop-cocks of the vessels to be evacuated can be inserted) to the upper part of the pump, y. Three screws should be used for this purpose and a leather washer should be inserted between the plate and disc.

4. This plate should be provided in the middle of its inner or under side with a leather valve (as is represented in Plate V, figure 1, d) so that when the piston, h, with its piston rod, s, has been lowered, *[figure 5 of this Plate.]* it can draw the air or water from the vessels to be evacuated into the pump, gh. When it has been raised, it can expel the air or water *[fig. 4.]* through the external valve, z.

5. A copper vessel or container, xx, into which water can be *[fig. 1.]* poured, should he soldered to the rim of the pump's lead disc, y.

6. The pump, vgh, together with the attached container, should *[fig. 3.]* be set upon and installed in the tripod through the round opening, e, of the aforementioned iron disc, bc, and with three iron screws *[fig. 2.]* should be fastened at the same time to the outer rim of the lead disc.

7. To prevent the lower part of the pump from moving, moreover, *[fig. 1.]* an iron ring, kk, should be fastened to it by iron screws, with three arms, ooo, attached, which together hold the tripod.

8. An iron handle, wuu, which can be raised and lowered about *[fig. 1.]* a metal pin, w, should likewise be attached to one of the legs of the tripod at point w.

9. Finally, an iron shaft or rod, ut, should be attached to the *[figs. 1 and 5.]* handle. This shaft is connected at point, t, to the aforementioned wooden piston rod, sh, which, in turn, is fitted as before to its aforementioned wooden piston, h. By means of this device, the pump can be made to operate.

10. To prevent air from penetrating into the lower part of the *[figs. 6 and 1.]* pump and around the sides of the piston, h, a deep copper vessel, like a pail, was made to hang by its own three hooks from the three arms, ooo, and was filled with water. In this way the lower opening of the pump within kk, as well as the piston rod and the other

apparatus is always covered by water and thereby air does not penetrate above or below.

The extraction of air depends upon its expanding nature.

11. Now because the extraction of all air as a rule depends upon its elastic nature or its capacity to expand (see Chapter 33 of this book), thus air always expands to some extent, when the pump operates, from the container that is being evacuated into the empty pump from which it is gradually expelled through the pump's motion. Eventually the small amount of air remaining in the vessel being evacuated has no longer sufficient expanding force to open the leather valves (which are usually fitted with brass springs so that they can always close tightly.) For this reason a small pipe can be fashioned in the cover of the pump, zmn, between the valve, z, and the tube, n, and provided with a piston rod, piston, as well as a small knob. Using this, one can skillfully reach in and open and close again the inner leather valve, thus permitting the small residue of air supposedly left to descend more easily and without difficulty into the pump through its own weight. This device, however, is unnecessary and is mentioned only to satisfy those committed to detail.

How a small residue of air can be extracted by means of a special tube.

figs. 1 and 4.

It seems quite evident, therefore, that a vacuum can be created through the use of this apparatus and that the difficulty which heretofore was held to be insurmountable, can be resolved. For if the handle, wuu, is raised, then the piston rod, sh, touches the plate, mn, and the pump is filled up by its own piston rod. But if it is lowered, the internal space of the pump is left empty and the air of the attached vessel enters in. By this procedure, the vessel is finally evacuated.

In a pump, when the piston rod is lowered, the internal space, of necessity, is left empty.

CHAPTER 5

THE THIRD EXPERIMENT DEMONSTRATING A VACUUM

See Plate VI.

For a better proof of a vacuum, I took a glass vessel, designated L, which pharmacists use (usually a $\frac{1}{4}$ or $\frac{1}{2}$ receiver, called in German a $\frac{1}{4}$ or $\frac{1}{2}$ "vorlage") and fastened a brass cap on its neck with a pitch-like substance instead of solder. (Goldsmiths call this "kith" in German.) This cap was a sort of case, pp, furnished with a stop-cock, qr, which I designed to be inserted in the tube of the plate, n.

fig. 7.

Moreover, that even larger objects, like birds, fish, mice, as well as clocks, bells, candles, etc., might be inserted in this glass vessel, I made the stop-cock removable, so that with care it could be inserted into and taken out of the case, pp, as figure 7 demonstrates.

After this glass vessel, L, or a similar one of smaller capacity, had been attached to the tube of the plate, mn, and after the stop-cock, q, had been opened, I had the air pumped out of the glass. Then, when no more air could be expelled by moving the piston rod and after the stop-cock was closed and the glass vessel removed, I submerged the stop-cock in water and opened it. I saw water surge into the glass with great force, seething like a spring gushing from the earth, until it gradually filled the glass to the top, leaving a space, however, the size of a hazelnut, empty. I thought that this empty space was air, a conjecture that was subsequently confirmed.

Water surges into and fills a glass vessel with great force.

Now I readily concluded that the neck of the stop-cock became filled with air from its outer opening to the key (when the glass was removed and brought to the container of water into which it was to be immersed) and that the space that was left (about the size of a hazelnut), perhaps had been this air. Accordingly I thought it was necessary to repeat this evacuation of the glass and to fill the neck of the stop-cock with water before it was immersed. When this was done, a space the size of a pea was left empty of water. And although I repeated this experiment, I could never bring about a complete exclusion of air.

The glass is almost completely filled.

But because I saw that foam and many little bubbles always rise when the water rushes in, I readily conjectured that these tiny bubbles were the cause of this space. I could not perceive, however, where they came from. For although I completely immersed the stop-cock, together with its case, pp, in water up to the glass, nonetheless foam and bubbles of this kind were always produced when the water rushed in. It is worthy of note that although their volume seemed one hundredfold greater during this insurgence of water than the small globule of air contained that remained once the rushing in was at an end, nevertheless there was always some air left.

Through the rubbing and the action of water, little bubbles arise.

CHAPTER 6

THE FOURTH EXPERIMENT: THE PRODUCTION OF A VACUUM THROUGH THE EXTRACTION OF WATER FROM A GLASS VESSEL

Because, as we have just said, some air always seemed to remain above the water in the glass (although, to be sure, it was not even a thousandth part of the evacuated glass), I devised another method of extraction. I procured a glass (commonly called a "viale"), which had a globe or retort at one end and at the other, a long neck (so that the water, through its own weight, might descend more easily from the globe). To this I had a metal cap with a stop-cock attached, as described above. This can be seen in the Plate. I filled the entire glass with water to the opening of the stop-cock and inserted it in the small tube of the apparatus, mn. When the pump operated, I saw that water, indeed, descended into the pump but that soon a sizeable bubble rose up through the glass tube. I perceived that this was air that had been concealed in the key of the stop-cock, (for usually the keys of spigots are cast hollow and are filled inside with gravel and sand instead of metal). Thus it was necessary that a stop-cock be designed in such a way that its key be cast of solid bronze and then perforated.

See Plate V, fig. 3.

Plate VI.

The keys of the stop-cocks should be redesigned.

When I repeated this experiment later on, no bubble of this kind appeared. Some bubbles did rise, however, in part from the pitch-like substance (by which the aforementioned cap was fastened to the glass) and in part also in, or out of, the water itself. And although no more bubbles emerged from the adhesive material when the experiment was repeated frequently in the days that followed, nevertheless I could not in any way prevent some bubbles from always being present in the water itself.

Bubbles also arise from the adhesive substance.

Then, when I had the glass removed from the apparatus and again immersed and opened in a container filled with rather deep water (deep enough, that is, to cover the key of the stop-cock), I observed the same result: water entered with noise and many bubbles, as had happened in the previous chapter when air had been extracted. For again the glass was filled in the same way and perhaps with even more water, but still there was some air left in the glass. Although the volume of this air was small when compared with that in the rising bubbles (for, of these, one hundred or more appeared with the in-rushing water, although ultimately only four or five were seen to remain), nevertheless they served to contradict those who wanted to deny that the glass had been emptied of air.

Through the friction of water, its vapor, which is air, is produced.

These efforts, therefore, were not in vain, for the experiment proved: first, that these small bubbles, which heretofore had not been perceived visually, were hidden in the water. When they reached an empty space, however, they became one hundred times greater than they had been before; second, that many bubbles arise in water when it is forcefully compressed or rubbed: this occurs in the key and neck of a stop-cock; and third, that through the motion of in-rushing water, many small bubbles are produced therein. Had the water remained motionless, they would not have been seen except in a vacuum where they are so greatly expanded and consequently are made visible.

Bubbles are greater in a vacuum than they are otherwise.

CHAPTER 7

A FIFTH AND MORE ACCURATE METHOD OF PRODUCING A VACUUM

After the foregoing results had been confirmed by experiment and were duly explained, I had the glassmakers make another glass vessel, A. It was rather long (a Magdeburg cubit in length), hermetically sealed on one end, and almost as thick as a man's arm. On the other end, which was open, I had a case of thin metal, B, together with a stop-cock, C, soldered on as before. This stop-cock was attached in turn to still another case, D, provided with two other stop-cocks, E and F, (one of these,E, was on the side for operating the inner stop-cock, C; the other, F, was designed to be inserted in the small tube, n, of the apparatus). After I had filled the glass, A, as well as the case, D, to the top, G, with water, I laid it on its side for several days, with the stop-cocks closed. Thereafter, when I set it upright, I saw an air bubble rise. I perceived that this had been caused by the motion of in-rushing water and that after an interval of time, many air bubbles that had been present, came together as one. Accordingly I made this bubble run out by inverting the glass and with a drop of water filled its space. I repeated this for several days.

See Plate VII, fig. 1.

It is necessary that the stop-cock be surrounded with water. See fig. 5.

Because of motion of inrushing water, bubbles arise.

Then when the glass was set on the pump in the customary way and essentially half of the water was extracted from the glass by turning the keys of the stop-cocks, no bubbles appeared as in the earlier experiments; instead, the water descended peacefully. After the stop-cocks were closed, however, and the glass was removed, behold! not only did some bubbles arise in the water because of the motion arising from moving and tipping it, but at the same time the noisy water so dashed against the interior of the glass (as though it were made of some strong material, like stone), that the glass suddenly was broken, I know not why. It was necessary, therefore, to make another similar glass tube, one especially thick, and to carry out the experiment as before. When this was done, and the glass was vigorously shaken up and down, I saw that the water expanded and split itself in two, that an empty space appeared in the same, and that straightway the water rushed together with a crash sounding like two sticks being struck one against the other. In this collision, a small bubble was always produced in the middle of the water.

Water enters a vacuum with noise.

From this it was manifestly evident that there was nothing in this hollow space produced by the separation of the water, and that the space was empty of all air. As for the small bubble that was produced

Water in a vacuum expands itself and produces an empty space.

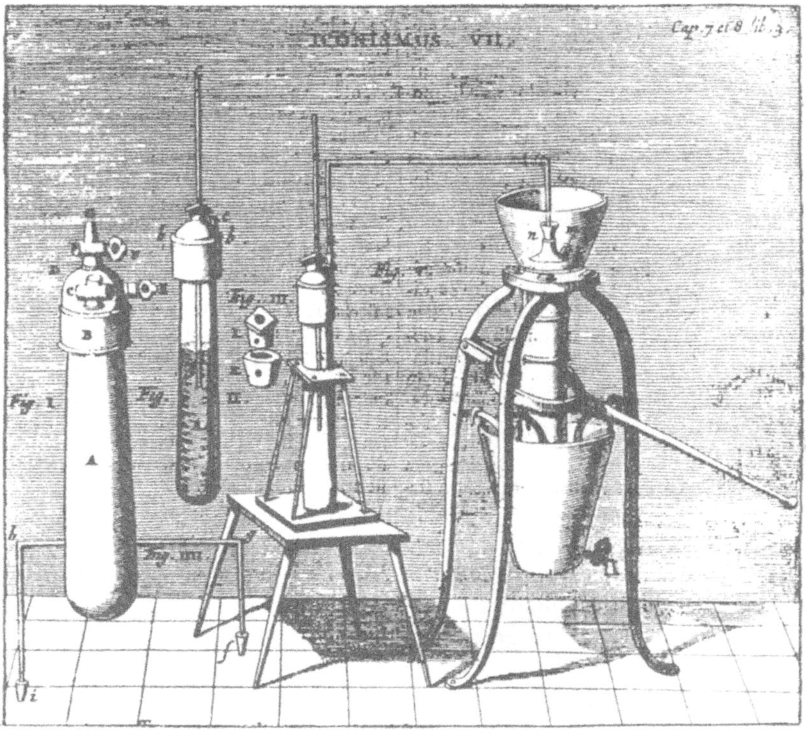

PLATE VII.

Some air is always produced through the collision of water.

through this parting in the middle of the water, it was clear that this had its origin in the vigorous friction or violent collision of the water and that it was so tiny that, unless it were in an evacuated glass, it could scarcely be perceived. It was evident, I repeat, that the space was empty because the glass had been completely sealed so that all external air was wholly excluded. Moreover, no internal air was present (had there been any present above the water in the upper part of the glass), for air could not have penetrated the water, which by nature is always under rather than above it, whereas this separation takes place either in the bottom of the glass or in the middle of the water. As a matter of fact, the strong collision accompanied with such a loud noise indicates that no air was present in the space produced by the separated water. Furthermore, no one could stimulate or produce this collision except in an evacuated glass of this kind.

Demonstration of empty space.

CHAPTER 8

THE SIXTH EXPERIMENT TO OBTAIN THE BEST POSSIBLE VACUUM

In the foregoing experiment, a bubble was always produced because of the separation of the water, which was undoubtedly caused by the vigorous rushing together or collision of the water when the tube was shaken (that is, when the keys of the stop-cocks were closed.) This bubble rose into the empty space in the upper part of the tube and gave those who did not want to admit there was a vacuum, a reason for suspicion. (For even though Doctor Deusing in his treatise *De Vacuo* together with others, does not believe that air has an elastic quality or a particular ability to expand itself unless there is an external force to drive it into a confined place, and that therefore a small air bubble of this kind cannot fill the tube because there is no external force to cause it to expand and thus fill the tube, they raise an objection, nevertheless, to this experiment by saying that a very small particle of air can so extend itself in an evacuated glass or tube as to fill it completely. Therefore, they would only concede that the tube was empty when they saw (1) all the water leave the tube, (2) absolutely no air was able to enter through any passage to replace the water that left, (3) no bubble, even though it be very small, develop in the water itself.) For this reason, I designed another piece of apparatus with two tubes. One of these (the narrower and longer) was always protected from the penetration of air or the formation of bubbles, and in consequence preserved the best possible vacuum.

Some scholars do not agree that air expands, but nevertheless, they say that a very small bubble of air can fill a vessel.

This is selfcontradictory. Another piece of apparatus to produce the best possible vacuum.

For this purpose I took a glass tube, A, almost a cubit long, similar to the one in the preceding Chapter. I enclosed the open end in a case or metal covering, bb, as I had done before with other similar vessels.

The construction of the apparatus.

I inserted a narrower glass tube, cd, almost a cubit and a half long, through the cover of this case. It extended to the middle of the larger tube at point d, as the diagram indicates. In the side of this case (at point, e) I attached a valve, e, (or K) which was fashioned in a special way to that it would serve in place of a stop-cock. Finally, I fastened this case, bb, to the glass, A, with the same material as that used by goldsmiths.

See Plate VII, fig. 2.

Description of the valve, e:

I had two small hollow receptacles, K and L, resembling thimbles, made of brass. The smaller of these, L, is closed on the bottom. The inside of the larger one, K, has been carefully rubbed by a metal-worker so that the former fits into its cavity with precision. Both

See fig. 3 of this Plate.

have a small opening in their sides and when L is inserted in K, the one hole corresponds with the other perfectly. Then when L is turned within K, this valve can be opened and closed.

After the apparatus had been constructed in this way, I filled the aforementioned tube, A, with water so that the opening of the narrower tube, d, should always remain immersed in water no matter how the apparatus was turned, whether it be up or down. Then I tilted it so that the narrower tube, cd, was completely filled with water. When this was done and the apparatus set upright again, water remained suspended within the tube. This did not result from an aversion to a vacuum, as is commonly stated, but because air, pressing upon the water within the tube, A, prevents the descent of the water in the tube, cd, because of its weight, as we shall see.

The use of the apparatus.

There is no aversion to a vacuum.

After this I placed this apparatus at the side of the equipment we described in the preceding Chapter 4 and connected the two with a specially designed thin metal pipe, fghi. (See Plate VII, figure 4.) This was done in such a way that one end, f, was attached with precision to the small tube, n, of the large piece of apparatus and the other end, i, to the valve of the smaller at point, e, as can be seen in figure 5.

After all these preparations had been duly carried out, I evacuated the air (using the small tube, m, which was discussed in Chapter 4, number 11. This allowed the leather of the inner valve of the pump to be opened and all remaining air, if there were any, to be given a free means of escape.) When this was done, I clearly saw that the water immediately ran down from the glass tube, cd, and that many bubbles arose in all parts of the lower as well as in the upper tube. At first I had believed that these bubbles came from the water alone, but later discovered that they arose as much from the glasses themselves and even the soldering material, as from the water. I particularly noticed that many of them clung to the glasses and the soldering material and that, increasing little by little, they finally broke through the water and rose.

See Plate VI, fig. 3.

The bubbles which cling to the glasses are an exhalation of the same.

Then we continued pumping until the air had been completely drawn out of the larger tube, A, at which time the water of tube, cd, sank and reached a level equal with the water of tube, A, as we confirmed visually. The reason for this is that the pressure of ordinary air is not only excluded through the insertion of siphon, fghi, in the valve, e, but also the dispersing or expanding force is removed from the air, which was enclosed in tube, A, through pumping. (This expanding force, before exhaustion, held the water suspended in tube, cd, inasmuch as no external air could enter in tube, A, through the closed valve.) For this reason, water necessarily must descend from the tube, cd, because of its innate weight, and seek a lower place. It is not a question of aversion to a vacuum, since this is non-existent in nature. For if the water of tube, cd, descended by reason of aversion to a vacuum, why should it leave tube, cd, empty only to fill the other tube, A, since it would be natural for it to satisfy or take care of its own needs first before another's?

The so-called aversion to a vacuum is caused by the weight of external air.

After this had been done and all air had been pumped out so that the water of tube, cd, was level with that of tube, A, the valve was closed and the apparatus was ready to produce a vacuum. Now when you want to demonstrate a vacuum, tip the apparatus so that the tube, cd, is filled with water. Set it upright in such a way that the opening, d, always remains submerged in water, and you will see the water descend from the tube, cd, immediately and the space within this tube become empty. Now since the opening, d, has remained submerged in water and no air can enter, the space inside this tube must necessarily be empty. Next when one quickly moves or shakes this apparatus up and down, as we did in the previous chapter, one will see the water in tube, A, divide, thus creating a completely empty space in the middle of the water which then straightway rushes together with a crash. This could not happen in any way, unless that space were empty, as we have said in the foregoing chapter.

There is no aversion to a vacuum.

How an empty space can easily be demonstrated.

Water in a vacuum rushes together with force, like stones striking one another.

It should be remarked, furthermore, that in the first six or eight days when the water, as well as the glass and soldering material are still fresh, the aforementioned little bubbles always appear in great abundance. For this reason, I thought that air must surely penetrate the pores of the glass. I was strengthened even more in this opinion because, when the tube, cd, was tipped and again filled with water, there was always some air, about the size of a pea, retained in the tube, cd. Although the air again burst into the space of the larger tube, A, (when it was set upright, that is), nevertheless this same condition always prevailed. This moved me to enclose tube, cd, in another glass tube, to fill the latter with water, and to seal its top opening in such a way that the tube inside, cd, was completely surrounded with water. I thought that in this way I could prevent external air from penetrating the tube, cd, (since air surely could not permeate both glasses as well as water). Nevertheless, these small bubbles continued to appear in the tube, cd, until I finally discovered after six or eight days of repeated experimentation, that these bubbles had their source even in freshly washed glass and always disappeared gradually within several days until they were no longer perceptible. In the course of time, numerous repeated experiments confirmed this observation. For when the apparatus had

Air does not penetrate glass.

There is visual proof that air does not penetrate glass.

stood open and unused for several days and I repeated the procedure, bubbles of the same kind appeared as before, in the water and the glass tubes.

From these observations I clearly saw that these bubbles originated in the effluences not only of water, but also of glass, and that corporeal effluences of things (we shall make mention of these in Book IV, Chapter 1) are nothing other than air, and that air is nothing other than the corporeal effluence of all bodies. We shall demonstrate more fully (in our subsequent Chapter 10) that everything has its corporeal effluence or exhalation, and in consequence that not only water, but also glass, metal, and all other things emit their own exhalations into the space surrounding them.

All bodies have effluences which are their exhalation or air.

Useless effort was expended, therefore, in completely surrounding the glass tube, cd, by another tube filled with water, since clearly no perfect glass permits the passage of air through it.

Now to proceed further, the following should be remarked. Even though no more bubbles appear after the aforementioned six to eight days, and the water descends quietly from the tube, cd, without any frothing up of bubbles, and no air at all enters in any way, (just as the above-mentioned scholars wished), still, one small bubble, resembling a pea in size which was mentioned before, appears in this tube. (This occurs, to be sure, after the tube has been pumped out and tilted with its end, c, down, so that it can be filled again with water.) And although this bubble is again removed when the tube is filled and the water once again is forced to leave the elevated tube, nevertheless, after repeatedly filling it with water, I always found that a bubble of this kind remained, having been produced from the water itself in a completely evacuated tube.

Water emits an exhalation or effluence in a totally evacuated tube.

Once the apparatus has been opened, this bubble is so compressed by the weight of in-rushing outer air that it escapes almost without being seen. Still it should be understood from this that water in a vacuum always imparts some air or exhalation into the empty space surrounding it. From often repeated experiments I also noticed that when this bubble had appeared several times in the tube, cd, and passed over into the space of tube, A, through the turning of the apparatus, a perceptible amount of air was introduced into the larger tube. This was evident because the water in tube, cd, did not maintain the same level as before with the water of tube, A, but began to rise, by reason of the natural tendency of air to expand.

And although I frequently pumped out this air from the larger tube, A, still, as often as I tried, a bubble of this kind remained in the tube, cd. One must deduce, therefore, that water continually discharges its exhalation, which is nothing other than air, into adjacent space (particularly when it is empty), as long as there is any water left. For this reason nature gave the earth the conserving virtue, (Chapter 5 of the following book discusses this question), whereby air maintains a certain force and weight, so that all water does not vanish or become dissipated (by freely discharging its exhalation into empty surrounding space in which it is carried around by the sun's rays in a great and never ending circuit. (In Book V, Chapter 16, this movement is termed "annual".)

The weight of air prevents water from discharging its exhalation.

From the foregoing, it is evident that we can produce a vacuum only so far and cannot proceed beyond this point.

No better vacuum than this can be produced.

What can be said, however, about a vacuum produced through the sinking of mercury is to be seen in our subsequent Chapter 34.

Finally one should not neglect to mention here: (1) that if the apparatus is partially filled with beer (halfway, that is, as had been done before with water) and the air is pumped out, then the beer is completely reduced to foam and rises through the valve so that it partially enters into the stop-cock. (2) It should be noted that when the apparatus has been evacuated and the valve is opened too quickly, then the external air rushes into the larger tube, A. It presses upon the water so forcefully that the water not only rises in the tube, cd, immediately, but even breaks the upper portion of the vessel and dashes it against the ceiling of the room, thereby lifting up the whole piece of apparatus in the process. This apparatus should always be tipped slowly so that water does not rush into the tube, cd, too quickly and shatter it, inasmuch as there is no air left in the tube to withstand the in-rushing water. For the water flows in with such force and noise that, in its precipitous flood, it breaks these glass tubes with its rock-like impact. In this event, external air would again rush through this tube into the larger tube, A, and would shatter it, thus endangering the eyes of those standing around. (4) One should also take care not to shake the apparatus too much, since it breaks (just as happened to me in the preceding chapter). (5) It is worthwhile to mention here that water would not descend from the smaller tube, cd, even if the tube were a hundred cubits long (when the apparatus has stood inverted for some time, enabling the tube to be filled with water, and later set upright). I feel this is the case because the water is so cohesive itself that in no place will it begin to separate or break apart unless the apparatus is struck against a table or floor. Then as a result of the blow, it noisily breaks apart in some place or other with a great risk to the tubes. Well-cleaned mercury acts in the same way and this reaction is independent of the weight of the cylinder of air. We shall deal with this in Chapter 19 and the following chapter as well. Although these fluids

Remarks
1. concerning beer.

2. The glass can be broken in three ways.

Liquids that cohere well because of the affinity of their parts are separated in a tube without force.

Liquids always emit some exhalation or air.

in time, however, become so cohesive that they only separate as a result of a blow, nevertheless, we find this assumption in error, for when the tube remains standing for a day or two full of water, (for all fluids, to be sure, especially when they hang suspended, like the water in the tube here, give out a kind of air-like discharge which in time accumulates in the top of the tube and causes a parting or separation of the elements, namely the glass and the fluid.) Then, contrary to expectation, the water usually is found to descend.

CHAPTER 9

DOES A VACUUM EXIST IN NATURE, OR NOT?

Corporeal exhalations of things are air.

Since the corporeal effluences of things are air and since nothing exists without an effluence of this kind (for even though the exhalation or corporeal effluence of certain things like metals, precious stones, glasses, etc., is very small and imperceptible to a person, nonetheless, in an evacuated glass container where this exhalation has room to expand freely, it can be perceived in the form of a very small air bubble. We shall deal more fully with this experiment in the following chapter.) It follows that in these lower regions, or in other words, near our earth, there is no space so completely empty of all effluences and exhalations of things that nothing at all is present there.

A vacuum taken in a rigorously exact or ideal sense is not present in these lower regions.

Inasmuch as it is certainly true that all air, in a rigorously exact sense can never be completely excluded (for the actions of mortals do not take place ideally or mathematically, but rather in a physical way, following an ideal concept, for just as it is impossible to describe, delineate, construct, measure, or weigh any line, surface, or body with perfect accuracy, since this can only be done physically, in the same way as here on earth where there are things everywhere discharging a bodily effluence or air, it is impossible to make or render space empty in an ideal sense), nevertheless, it does not follow from this that a vacuum does not exist in nature even though no one on earth can produce such a space that is completely empty of all matter. We rather seem to demonstrate the opposite to be the case in our foregoing as well as our subsequent experiments wherein we show:

There is empty space in a mathematical sense beyond the limit of the earth's virtue.

1. that air is an exhalation or corporeal effluence of the earth.
2. that the earth retains this air with a certain weight for its own preservation.

3. that, as a result of this preserving virtue,the air, (indeed the lower more than the upper) is greatly compressed.
4. that the weight of air undergoes frequent change.
5. that its weight varies sharply on different mountains.
6. that air, by reason of its own weight, rushes into all space which is not otherwise occupied by a body.
7. that aversion for a vacuum does not exist in nature; this only appears to be so and rather is dependent upon the weight of the surrounding air.
8. that all air can be artificially withdrawn from glass vessels or tubes whereby a vacuum can be produced, provided only that new air, taking its origin from the water or even the vessel itself, does not interfere with this. (The amount of this air, however, is very slight and should be regarded as nothing.)

Certain facts which demonstrate that the space beyond the limit of the earth's virtue is empty.

From these and many other experiments it must be concluded:
1. that the corporeal virtue of the earth, being nothing other than its exhalation, or air, or effluvium, flows out into proximate space and wherever else there is no matter, just as the odour of a rose, if we can compare the small with the large, flows out into adjacent, not infinite, space within prescribed limits of its virtue, until it gradually becomes weaker and weaker and finally disappears into nothingness. Similarly air, which is an exhalation of the whole earth, does not extend out to infinity from around this sphere but up to a defined limit where it finally ends. This matter will be investigated in Chapters 7 and 9 of Book V. (ERRATUM, read Chapters 7 and 9 instead of 7.) Where the air ends, there, of necessity, pure space begins, empty of everything corporeal. For it can reasonably be said that no matter, whether of the nature of a heavenly fire or other things solid or fluid, should be imagined to be there, nor would there be any purpose in there being so extensive an area of corporeal matter between the stars (whereby their effluences as well as their movements, virtues, and influences, would surely be rendered ineffective). Those scholars are certainly dreamers, then, who think that space is a corporeal thing whose nature is solely dependent upon its expansive quality, and accordingly that space without body is illusory and purely imaginary (like the line of the equator, for example, that the astronomer posits). They should bear in mind, nevertheless, that that small amount of newly generated air cannot fill the void in evacuated vessels in any way, much less is it able to produce or maintain a condition such that vessels of this kind could not sustain a vacuum since their sides would collapse against one another, unless this small amount of air were present or regenerated as René Descartes affirms and is cited in Book II, Chapter 3. This would not be

The limit of the earth's virtue does not extend infinitely.

Where this ends, there is empty space.

Space and corporeal things are completely different from one another.

expected, however, with round vessels, as we shall show in the following Chapter 26.

2. that when water, air, or some other substance is so completely evacuated from vessels that not even a hundredth or a thousandth part remains, and nothing else replaces it, this would be sufficient proof of the existence of a vacuum in nature. For even though a vacuum in a strictly scientific sense cannot be produced here on earth, nonetheless we perceive the reason, for clearly, the effluences of matter prevent this. Where substances are no longer found, however, far above and beyond the earth, there will be no effluence of things. With the cause lacking, the effect is absent. Consequently space must be empty.

Where there is no substance, there is no effluence.

3. that because air pressure undergoes frequent changes and sharp fluctuations on different mountains, one can easily conclude (if he considers the great height of stars) that the vast expanse of surrounding space is not filled with air as far as the moon, much less the sun, etc. Nor has one any grounds to infer that air makes, creates, or constitutes this space, so that if there were no air,elemental fire, or other corporeal substances, no space could exist either and in this case, the earth, moon, sun, and the other stars would be contiguous to one another. Who can imagine that the exhalation of the earth, namely air, or some kind of hypothetical fire, or anything else, whether solid or fluid, can produce the vast space between the stars, and that unless there were a thin substance of this kind between these earthlike bodies, there could be no distance between them, for they would touch one another? Would it not be more rational to hold that whether there is an intermediate substance or not, that Space, nonetheless, exists and endures, independent of motion and rest and that it makes no difference whether anything corporeal is present in it, or not?

Because air pressure fluctuates sharply on a mountain, one can conclude that it does not extend half way to the moon.

4. that those are led into error who assume that not only the thousandth part of air remaining in the glass, but even the smallest particle possible, can nonetheless fill a receptacle. In truth, can anyone (apart from those scholars referred to in the beginning of the previous chapter, who hold that air does not have a tendency to expand itself), persuade himself, unless prejudiced by the opinions of others, that a thousandth part or the smallest possible part is of as great a size as its whole? Reason supports our theory that the more that air is drawn out and rarefied, the greater the vacuum that is created in the space previously occupied by the air, and that finally the air completely disappears, leaving behind an empty space.

If nature admits to an almost perfect vacuum, one that is perfect is also psssible.

5. that others, moreover, have different grounds to confuse the question of a vacuum when they postulate the existence of a kind of

finer air which they believe can penetrate a glass after the coarser air has been evacuated. But surely it is evident from previous experiments as well as from Book V, Chapter 9 that there is no such difference in air and that it cannot pass through glass. Should it be assumed, however, but not admitted, that a finer air of this kind does exist which is capable of penetrating glass, one may ask why it does not penetrate glass before the extraction of the supposed coarser air, inasmuch as there is always some space in every vessel wherein more air can be blown or pressed. Because air does not enter, however, it must be assumed that all air is of equal fineness or coarseness. The difference in air lies rather in its rarity or density (when it is more expanded or compressed) or in its dryness or humidity which arises, in fact, not from the air itself, but by accident, and in no way contributes to its penetrability. And even if someone wishes to grant the existence of a kind of air which can penetrate glass and other hard substances so that when water has been extracted, it enters the space left by the water, and when the water rushes in again, it passes out through the glass; this, however, is not compatible with reason because air of the kind which could penetrate glass could also penetrate water and would not yield place to it as it rushes in again. Rather this air would be present in the glass before the water was pumped out and would remain there afterwards. This is the same as saying that air of this kind is without body and penetrates everything, just as we maintain space penetrates everything, yet at the same time is motionless and present everywhere. Nor would it make any difference if this space were here or there or filled with a body or not. This statement can be made in no way about air.

All air is equally fine.

How space penetrates everything.

It must he concluded from the foregoing, therefore, that above or around the earth as far as air extends, there is no space empty of air, unless it has been produced artificially in vessels made of a solid material. Beyond or above the region of air, that is, where air ends, there Space begins, pure and empty of all bodies. This subject is discussed in more detail in Book V, Chapters 7 and 9. Now as for the incorporeal earthly virtues (which we shall discuss in Book IV) should one say that these also fill up Space? We shall have to distinguish between those virtues which proceed from the sun and those which flow from the earth and the other planets. The latter, although they are dispersed at greater or lesser distances around these earth bodies in empty encompassing Space according to the individual characteristics and nature of each virtue, nonetheless, they do not fill Space because they are without body. Nor do they remain for a long time in one place but follow the body in its periodic orbit, thereby leaving the space behind it empty. The former, moreover,

Pure Space is found outside of the atmosphere.

Are virtues without body capable of filling?

exists only as long as some object presents itself within the limit of its virtue which is susceptible to its influence. For indeed, that which does not exist, cannot fill, and therefore the immense Space between the earthly bodies is empty of all bodies.

CHAPTER 10

EXPERIMENTS CONCERNED WITH EXHALATION AND FERMENTATION

That all things of substance emit a corporeal exhalation from, or out of themselves was demonstrated in the preceding Chapter 8 and can be seen even more clearly when these substances are immersed in water or some other fluid. Then many bubbles form around it and (when this experiment is performed in a vacuum), they boil up and ferment, as it were.

All things have an exhalation, which is air.

This effluence of theirs is an exhalation, or air. For example, if you immerse a bit of gold, silver, quicksilver, or any other similar metal in a glass vessel partially filled with water, you will see that countless bubbles appear a short time later. They cling to the glass on all sides and rise to the surface of the water little by little, where they burst and mix with the air.

But that which is difficult to perceive when such a method is employed, is much easier to observe in a vacuum. For just as air expands more and more in a vacuum, so do these bubbles expand,which are nothing other than air or an exhalation.

We do not smell ordinary air.

Even if we do not smell ordinary air which is nothing other than the exhalation of the whole earth (the reason for this is that we live in it, by means of it, and are accustomed to it from our first moment of life), nevertheless we perceive the effluence, or air, of separate things by our sense of smell to a greater or lesser degree, according to our own nature, as well as the individual properties of a substance which emits air in more or less abundance.

And although we distinguish an exhalation as being good or bad, this is based upon our own opinion and not, in fact, upon the smell itself. For an unbearable stench can be a most pleasant odour to another living creature.

An exhalation does not penetrate hard bodies.

Just as air, a thing of weight and substance, does not penetrate hard bodies like glass and metal, so it is with all exhalations.

Now moist substances in particular, which have a tendency to change or decay because they are more subject to fermentation than others, emit a stronger odour or exhalation. Thus all decayed earth under the water of ponds and swamps always emits many bubbles. This can be seen when a rod or pole or some other implement is stuck in the bottom and many bubbles immediately rise to the surface. I often wondered where bubbles of this kind came from on the bottom and under the water since air, being a lighter body, floats over the water and cannot descend into it. Many bubbles are also found in ice and have the same origin as those in water. When they rise to the surface these bubbles freeze at once, thereby making the ice lighter than water.

Decaying things emit a greater exhalation.

What is the origin of the bubbles in ice?

For the same reason, the bodies or corpses of men that have been submerged under water also come to the surface: for when they begin to decay after several days, a new exhalation or air is generated, swelling the bodies and making them lighter than before. Thus they are compelled to float. When I had perceived this, I performed the following experiment. I placed a small dead carp in a bowl-shaped glass vessel filled with water. I covered this with another glass in the shape of a cup in such a way that all air was excluded, having completely filled the cup with water so that the little fish was totally immersed in it. After several days had passed, many bubbles arose from the dead body causing it finally to float. Since these bubbles could not mix with ordinary air because of the covering glass, they gathered in its upper part and produced newly generated air.

Why cadavers float.

Experiment involving the generation of air.

Now this fermentation or generation of new air, as we have said, can be seen in particular when these experiments are performed in evacuated glasses. For then these bubbles appear much more numerous and larger for a period of several days until they finally are seen to develop no longer.

For a further discussion of fermentation, see the last chapter of Book IV.

CHAPTER 11

THE EXPERIMENT IN WHICH CLOUDS AND WIND AND THE COLORS OF THE RAINBOW CAN BE PRODUCED IN GLASSES

It is a known fact that air expands through heat and contracts through cold. We have learned this in particular from thermoscopes or thermometers which were invented about thirty years ago. From the rarefaction and condensation of air enclosed within them, the variations in heat and cold can be examined from day to day and hour

The explanation and use of thermometers.

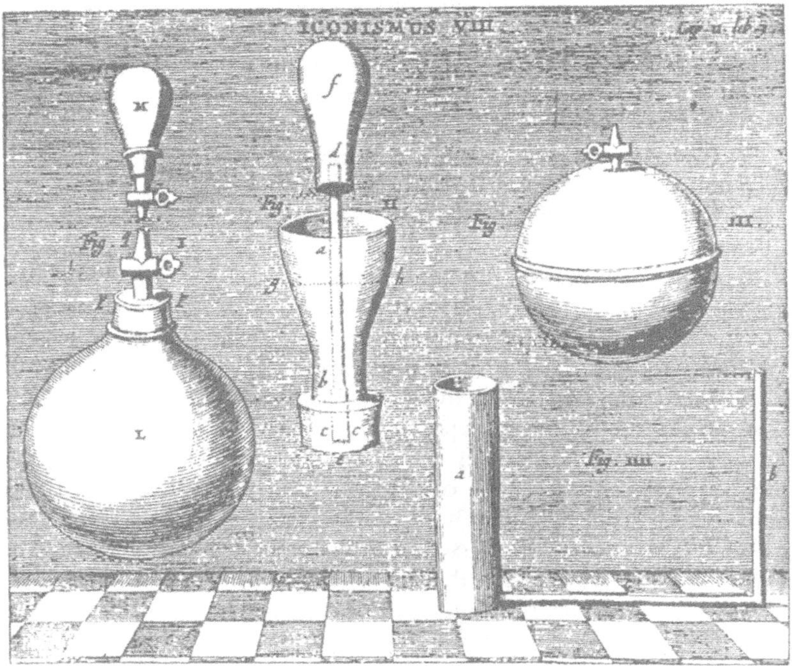

PLATE VIII.

to hour. Because many others have written about this, so as to save the reader time and for the sake of brevity, we shall refer him to the Reverent Father Kaspar Schott's *Mechanica Hydraulico-Pheumatica* page 229 and other writers mentioned therein. Likewise see our invention below in Chapter 37.

See Plate VIII, fig. 1.

To carry out the experiment on which our discovery was based, one must have at hand the receiver, L, that was described in Chapter 5, and provide himself with an additional transparent glass, M, which chemists commonly call a cucurbit or retort, called in German a "kolbe". The latter, like the receiver, is provided with a stop-cock and cap that has been cemented on to fit into the stop-cock of the receiver.

Wind is produced in the lower glass, L.

When the receiver has been evacuated and these two glasses joined together, the smaller being superimposed on the larger, the stop-cocks of both should be opened. Then one will see the air of the upper glass descend into the lower evacuated glass with a rush, seething noisily and at the same time several objects inserted in the bottom glass, be they only pebbles or hazelnuts, scattered and thrown about.

Because of this sudden expansion of air in the upper glass and its descent into the lower, the residual air is very much changed and diminished. A large amount of air, however, can contain more humidity than a small amount and for this reason, the air leaves its superfluous humidity in the upper glass. This can be perceived visually in the form of the small drops of water which little by little drip down to the bottom.

The water concealed in air is made evident.

The more air is compressed, the more water it contains.

This observation becomes even clearer the more the inside of the glass is filled with humidity. For then numerous bubbles rise in greater profusion (especially when the stop-cock of a cucurbit evacuated in this way is immersed in water and opened, for then it draws up water which is violently dispersed throughout the whole glass and makes the air enclosed within it much more humid) so that they produce a vapor which can be dispersed into clouds through the introduction of a small amount of air. For when the stop-cocks of both vessels are closed, 'the glasses being separated from one another and the stopcock of the cucurbit opened a little so that a small amount of air is admitted, then this vapor is changed into clouds.

How clouds can be produced.

From these experiments it is evident that through the contraction or reduction of air, the water contained in it separates itself from the air and collects in clouds. But then, if the stop-cock is completely opened and air is freely admitted, the clouds and vapors immediately vanish because they are absorbed by the incoming air.

A large amount of air can contain a large amount of water.

Furthermore, when this cucurbit has held vapor for a short time, the descent of the vapor and its separation from the clearer air can very well be seen. For such movement and undulation can be produced in the glass as perhaps the mists in the upper air have. This experiment leads to a better understanding of the motion of air.

There is no doubt that the vapors in air undulate.

From all these observations, moreover, the cause of clouds and winds can be learned. For when vapors arise from the base of mountains or out of subterranean caves or some other new kind of air is generated (just as we find the air in goldmines that blows out of the passages at different times), this movement of internal air, as it were, causes it to contract and change because it possesses a different quality from external air and makes the moist substance which is present to separate and condense, thereby causing clouds to be produced. Now, for the same reason it happens that animals exhale breath similar to vapor from their mouths in the winter. Thereupon this warmer air becomes condensed in the colder air, and being condensed, it occupies less room. Consequently, it cannot hold as much water as the expanded air. It leaves its humidity, therefore,

The cause of clouds.

Why and when the breaths of animals are visible.

for us to see in the form of many tiny cohering particles of water. And in the same way we see glasses and wine jars that have been carried from cold store-rooms in the summertime or in warm places, sweat, as it were. The reason is that the air surrounding the jar is chilled by it. Thus this air contracts, leaving its moisture which clings to the sides of the jar.

Why glass jars sweat.

In conclusion, the following should not be passed over in silence. When the aforementioned glasses are exposed to the sun during the experiment, the air in the upper glass at first begins to grow exceedingly bright and afterwards the colors of the rainbow become very clear.

The colors of the rainbow are produced.

CHAPTER 12

FIRE IN A VACUUM

See Plate VIII, fig. 1.

I lit a candle made of white wax and inserted it in the aforementioned receiver, L, (see Chapter 5 as well as the preceding Chapter) by removing the stop-cock, qr. Then I placed the receiver near to, or more specifically, at the side of the apparatus (Chapter 4) and by moving the pump vigorously I extracted the air. Thereupon I saw the flame gradually grow less bright and become contracted as the air was being evacuated. When a blue flame appeared at the end of the wick, the candle went out, although the wick glowed for one or two minutes after the flame had been extinguished, it emitted smoke immediately, but this also soon disappeared.

From these experiments that were frequently repeated, it was observed that a fire cannot live without air and that it is extinguished at the same time that air is removed. It was also evident that a flame is elongated in shape through the weight of air which lifts it and to a degree forces it upwards (like water, naturally heavier than air, that makes a bubble rise) and that a flame continues to adhere to the burning end of a wick because the surrounding air does not exert more force than the cohesion of the flame with the wick. For this reason when a lighted candle is set in low-lying places where the air is very heavy and impure, we see that the flame flares up and is consumed within a short time. Were there no weight in air, however, all flames would not be elongated but rather round and sphere-shaped, just as the flame of the sun is in the form of a sphere.

Why the flame is elongated.

Why is it extinguished through impure air?

At another time I inserted a lighted candle in this receiver in the same way as before and sealed it so that no air could leave or enter. I found that the candle remained lighted for three or four minutes

A flame does not exist for a long time in sealed containers.

before it became extinguished. It went out, moreover, in almost the same way as it had previously except that the flame was not blue and did not die at the end of the wick but rather in its middle.

I could not very well understand the reason for the candle's extinction except that the fire took some sustenance from the air and thereby used it up. Once it was gone, the candle could no longer burn. I therefore carried out an experiment which is described in the following Chapter.

The reason could not be ascertained in this way.

Now many things have been said about the lamps used underground by the Romans that burned for centuries. All accounts of their existence are based upon the testimony of others, for no one actually saw them. In the writings of early antiquity there is complete silence regarding the kind of oil used in them that was not consumed. Since experiment shows that fire cannot endure without air, however, no one can prove conclusively that these lamps continued to burn for many years.

The lamps of the Romans.

CHAPTER 13

AN EXPERIMENT IN WHICH AIR IS CONSUMED BY FIRE

I discovered that fire consumes air in the following way. I took a glass vessel which is commonly called a cucurbit or retort. Its top was cut off, as figure, ahbg, indicates. I fastened the narrower part, b, into a thin metal case, cc, with the soldering material I had used before, and inserted a thin metal tube, de, at the same time through the middle of the glass and case in such a way that its opening, e, could be attached to or fitted into the stop-cock, q, of the aforementioned receiver, L, (See the preceding Chapter 11.)

Plate VIII, fig. 2.

Plate VIII, fig. 1.

Then I had the glass filled with water beyond the halfway mark up to point, gh, and inverted over it another smaller cucurbit, f, which almost reached the bottom, b.

I attached this apparatus to the removable stop-cock, q, of the receiver, L. In its bottom I placed a lighted candle which I had inserted at the same time the other device was attached. In this way no air could leave or enter the container except through the tube, ed, leading into the glass, f, that was inverted over it.

When this was done, I immediately saw the glass, f, rise because the heat from the candle's flame instantly expanded the air enclosed in the receiver and sent it out through the tube, ed, into the glass,

The air in the receiver at first expands itself.

f. Since the lower part of the latter remained submerged in the water, no air could escape.

This continued for one or two minutes, at least, whereupon the glass, f, began to descend again and to settle gradually to the bottom. But then, indeed, I saw all the water rise into the glass, f, and at the same time to draw up many bubbles. This was a clear indication that some of the air contained in the receiver had been consumed. Indeed, at least a tenth part of the air was consumed by the flame of the candle and perhaps all might have been consumed had not the flame gone out so quickly.

It is partially consumed.

What causes the flame to go out?

Now the lack of air in this closed receiver can certainly be a partial reason for the flame's going out, but it also seems to me that its extinction was caused by the air's impurity resulting from the scattering of wax or tallow.

How the impurity of air can be seen.

For when the air that still remains is extracted from the receiver, it immediately becomes darkened because a residual black smoke is left which previously had not been visible in the denser air.

Does fire reduce air into nothingness?

Finally the question arises whether fire consumes air in such a way as to reduce it to nothingness or changes it into a substance of some kind. I believe the latter to be true although the resulting substance may be so nebulous that it is imperceptible.

CHAPTER 14

LIGHT IN A VACUUM

There are those who assert that our evacuated glass vessels cannot be empty of all matter since the things placed in them are visible.

Now if these things cannot be seen because of a state of vacuity, it follows that empty space is something which hinders the passage of the light-producing virtue and that whatever can do this is only a dense body. Therefore a vacuum must be a dense body. But since empty space is not a corporeal thing, it must therefore be invisible; but if it is invisible, then it cannot appear dark when it is daytime. For where would the darkness in the glasses come from when they are surrounded on all sides with daylight? The latter is present in the air and at the same time, so are all other objects that are there. Daylight is also present in the glass jar itself since it passes through it. The situation would be otherwise if we did not exist in the air nor on this earth where in daylight hours light is reflected which also illuminates the empty space in the glass as well as everywhere around it.

Now if we existed far from here in the sky where there is no resistance to nor reflection of sunlight but the sun's virtue (light, that is), freely passes through the air, then we would see nothing else but darkness (if our face were turned away from the sun) and, at the same time, that small space in the evacuated glass would be dark and shadowy. (See Chapter 12, Book IV and the end of Chapter 9 of this book.) We have a classic example of this phenomenon at night, for at this time, we have the earth behind us and our faces turned away from the sun. And since there is no lighted body before us (unless by chance on a clear night certain planets are visible, or even comets lighted by our sun, or fixed stars showing their own light as our sun does), we also see nothing. Thus that small empty space in the receiver is also invisible or dark in the same way as it is lighted by day.

The virtue of the sun in empty space does not manifest itself.

The virtue of the sun, or light, works only on corporeal things.

CHAPTER 15

SOUND IN A VACUUM

In an evacuated vessel, bells, stringed musical instruments, and other objects which resonate in different ways, give forth very weak sounds. To carry out my experiment, therefore, I inserted in the glass receiver (which has been mentioned very frequently) a sharp striking mechanical clock, suspended by a string from the vessel's opening. I so adjusted the clock beforehand that it produced a clear tone at definite intervals during a half hour period by the striking of its hammer on a bell. Then when the receiver had been sealed, I began drawing out the air and noticed that a very weak sound was emitted when the air was partially evacuated. When the receiver was completely evacuated, I found that I could hear nothing at all. Then I moved my ear near the glass and perceived a weak sound when the hammer struck. It was the same sound as when someone strikes a bell of this kind, which is completely encompassed in his hand, with a hammer. The bell, then, produces a rather dull sound or noise but no real ringing tone. On the other hand, the receiver was no sooner opened for the air to return than the sound of ringing recurred.

Bells do not ring in a vacuum.

They nevertheless produce a sound.

The presence of air causes the ringing sound of a bell to return.

After this, I had a carpenter replace the bell with a box, a kind of square rattle constructed from four shingles. It was of a size, moreover, that could fit into the opening of the receiver. Having first fastened the clapper of a bell in it, I suspended this rattle in the receiver and inserted a stopper. When I turned the receiver this way and that in my hands, the clapper struck the sides of the box. We heard the sound which was produced by the rattle's clapper just as clearly before air had been extracted as we perceived it after the air had been evacuated. No difference in sound could be detected.

The clapper produced a sound which is the same in an evacuated as a non-evacuated vessel.

After that we found that sound-producing bodies, like bells, cymbals, glasses, and stringed musical instruments and other things of this kind bring forth their own tones with the assistance of air, depending upon the force of the agitation or shaking with which they strike the air. On the other hand, we learned that a sound or noise which is produced solely by the friction or rubbing of objects, one against the other, results from the sound-producing virtue itself and does not depend on the presence of air. (See Book IV, Chapter 10 on this subject.)

Ringing is produced with the assistance of air.

The ringing sound has its source in the sound-producing virtue.

Although the receiving vessel had been sealed with extreme care so that air was not permitted to enter or leave freely, nonetheless, not only sound but also noise penetrated the glass. When there was no air, however, then no ringing sound was produced but noise and din persisted without interruption whether air was admitted or extracted.

Indeed, one should bear this distinction in mind, namely that the sound or noise which exists or is produced by air also takes its origin through the assistance of air, as we shall see in Chapter 26. Here we find that glass vessels or bottles with four sides are not only broken but also produce a loud noise when surrounding air violently streams into that space which has been evacuated in the vessels. For the same reason the noise of thunder and guns is produced in the air, for when fire appears very suddenly, expands the air and is quickly extinguished, it leaves behind it an empty space into which air forcefully rushes and fills, thereby causing noise.

Noise is produced from the concussion of air in the same way as through the rubbing together of hard objects.

The sound of guns results from the concussion of air.

It is known with certainty, moreover, that if a mechanical clock were carefully placed.in a glass or copper globe so that no air could escape, and if this, in turn, were suspended in the oft-mentioned receiver with all air being evacuated from the latter, the sound of ringing would clearly be heard. Since there is little doubt, however, that a glass globe like this would be broken by the pressure of interior air (as will be seen in Chapter 33 in the experiment with the "vesica" or bladder), I did not wish to engage in this experiment.

A clock enclosed in a sealed vessel containing air and suspended in an evacuated receiver continues to ring.

CHAPTER 16

EXPERIMENTS PERFORMED WITH ANIMALS IN A VACUUM

The glass vessel, L, that has often been referred to and is called a receiver by chemists, was constructed with a removable stop-cock, qr, (see figure 7, Chapter 5), so that certain small animals and other creatures could be inserted in it.

Plate VI.

I first placed a sparrow in the receiver. The bird beat its wings and flew about from this side to that without stopping. Then I joined this glass, by means of a connecting tube, fghi, (described in Chapter 7) to the apparatus which extracts air, at which time the bird's flying came to an abrupt halt. At first (when the air was being vigorously pumped out) no change was noticed except that the bird drew its breath with difficulty through an open beak. Finally it stood completely still without panting, with open mouth, and remained in this position until it fell forward and breathed out its life. One could very clearly see from this that all bodily movement ceased in the animal with the loss of air and that its life-force, residing in the heart, had been extinguished (like a flame of wine spirits).

Plate VII, fig. 4.

Experiment with a bird in a vacuum.

At another time I poured water into this glass and introduced various species of fish, like pike, perch of the larger and smaller variety, suckers, barbels, and other more common types. These reacted in different ways to the extraction of air according to their own natures. The pike opened their mouths and their bodies became more and more distended. They disgorged the smaller fish they had fed upon and at length their bodies became so swollen that it seemed possible they might burst, until finally they lay at the bottom of the receiver, belly up, as though dead. Then I observed that their bladders were constricted and that they had no other outlets for discharging air, which, because of their bodies' capacity for expansion, made them swell up. On the other hand, suckers and other common table fish discharged air from their bladders.

Fish in a vacuum.

Now a great difference was seen in both varieties of perch. Their jaws, stretched open in a remarkable way, gaped, and their eyes bulged out from their sockets. They died in the same way as the pike and did not escape this end even though air was admitted when the glass was unsealed. The others that had not been left for too long a time in a vacuum revived and came to.

Finally, it should be mentioned here that when grapes are inserted in a glass receptacle of this kind and it is evacuated and set in a

Grapes are preserved in a vacuum for a year.

rather cold place for a half year, the grapes are not changed as far as their appearance is concerned, but they lose all flavor.

The reason for this is that the essence of the grapes flows out into empty space. Otherwise this essence is retained by the weight of the surrounding air and is conserved in the body of the grape.

CHAPTER 17

THE CONSTRUCTION OF A KIND OF HYDRAULIC-PNEUMATIC APPARATUS WHICH CAN BE USED NOT ONLY FOR CARRYING OUT NUMEROUS EXPERIMENTS BUT ALSO USED AS SOURCE OF MENTAL RECREATION AND STUDY

One should set up an iron tripod, abchi, whose feet at the top are inserted in an iron ring, bc, and are fastened with screws. At this point it should have two moveable pieces of iron so curved that the glass receptacle, eqr, which is usually called a $\frac{1}{4}$ receiver and is of the same design as that described in Chapter 5 (the latter being fitted with a brass cap or case so that air has no way of entering or leaving) can be suspended by means of these supports between the legs of the tripod as well as removed again.

Plate IX.

Another glass receptacle of the same kind, L, with a stop-cock, d, should be set up and fitted to the lower glass, eqr, as the diagram indicates. Both ends of this stop-cock are so designed as to fit with precision into the two cases of these receptacles. In addition four other smaller stop-cocks (of which only three are indicated, fxq) should also be mounted around the stop-cock, d, in the case of the lower receptacle. With care a tube, pp, can be inserted in the mouths or openings of these stop-cocks, fg, for drawing in and expelling water. The stop-cock, g, should have a thin brass pipe, q, attached to its lower end, which extends to the bottom of the glass.

The two other stop-cocks, however, of which a fourth cannot be illustrated, do not have mouths but tubes, which also extend to the bottom of the glass and are so pointed on their ends that only a pin can be inserted in their openings.

(Note: All these stop-cocks and other similar devices should be carefully designed and finished so as to fit into the openings in the case with precision and to prevent air from entering it. To insure the exclusion of air they are surrounded by a small ring filled with water.)

The middle tube, r, furthermore, should be made of the same

material as the pipes, but with greater capacity and of about a finger's width in thickness. It should pass from the larger stop-cock, d, (at the point where it is removable) through the middle of the lower glass and should extend almost to its bottom. By means of this, water can be transferred from the lower to the upper glass.

Finally, the aforementioned tube, pp, should be made of a thin sheet of tin plated iron of such length that it extends from the openings or mouths of the two stopcocks, fg, almost to the floor of the room. It should be provided in the upper part with a brass hook-like attachment, m, precisely fitted to the mouths of these two stop-cocks and at the other end with still another stopcock, n, as the Plate indicates. I leave the rest to be filled in and pursued by the curious reader.

CHAPTER 18

THE USE AND OPERATION OF THE AFOREMENTIONED APPARATUS

Method employed in evacuating these glasses.

1. There should always be some water in the lower receptacle up to point, st. Then both vessels should be evacuated by means of a copper sphere (as illustrated in Plate VIII, figure 3) whose capacity is essentially the same as that of the two receivers. This sphere should be evacuated and inserted into the stop-cock, f, and in this way the air contained in the receptacles will enter into the sphere. This process should be repeated several times until almost all air is extracted (a state that can be recognized when little air or none at all rises out of the water from the upper glass through the middle tube, r. Then one should close the stop-cocks, d and f, simultaneously; the other two stop-cocks, however, which have pointed tubes, should be opened a little (if there are three or four of these, so much the better). Then one will see and hear at the same time air breaking through the water with a hissing sound, more or less loud according to the position of the stop-cock keys. This can continue day and night for one, two, or three weeks and can serve as a way to pass one's time. One should bear in mind, before all else, however, that the danger of air's seeping in unnoticed should be prevented: and that if pins are inserted in the mouths of the tubes a more pleasant sound can be produced. For the smaller these holes are, the more the air is constricted and is forced to hiss.

The first application for the machine's use is to produce a lasting sound.

2. Introduce the pipe, pp, with its stop-cock, n; open into a pail filled with water and at the same time insert its hook-like attachment,

PLATE IX.

m, into the mouth of the first stop-cock, f. Then open this stop-cock and you will see the water from the pail rise into the lower glass, eqr, with a rush. When the glass is about half full of water, shut the stop-cocks, the lower, n, as well as the upper, f, detach the pipe, pp, now full of water, and lay it aside.

Water rises contrary to its nature.

3. If the third little stop-cock, g, (connected with the tube, gq, ERRATUM, read "gq" for "xr") is opened, air rushes in so forcefully, again breaking through the water in the glass, that the glass seems scarcely able to withstand this stress. This condition prevails until the glass has been filled with air.

Air rushes into the evacuated glass with force.

4. When the first little stop-cock, f, (which has no tube,) is turned so that air can freely enter into the glass, the stop-cock, d, of the upper glass, L, can also be opened. Then you will see water suddenly rush into the upper receiver noisily and with such force as to break through its top or shatter the glass into many small pieces. One should take care (after the water has been drawn up) that the air which follows is not admitted without restraint, a situation the key of the stop-cock, d, should regulate. Otherwise the whole apparatus could be endangered by the passing of air through the water with too much force.

The water of the lower glass rushes with great force into the upper glass.

5. As has been said at the end of Chapters 2 and 4, many bubbles generated in the water from the strong pressure it exerts as well as its friction in the stop-cocks are later reduced to air. Nevertheless, they are invisible to the eye unless they are present in a vacuum. We have an observable demonstration of this fact in our experiment. When the upper stop-cock, d, is opened only a little so that the water in the lower receiver is forced to enter its narrow opening, the water is dispersed into very small air bubbles because of the resultant vigorous friction, but when these bubbles appear in a vacuum where they have free space, they resemble smoke on the surface of the water. Otherwise, outside of a vacuum, they are not visible. The reason for this is that inasmuch as the pressure of air over the whole earth as well as all water is great, these bubbles are greatly compressed in water so that they are invisible to the human eye. In a vacuum, however, where the pressure of air is absent (through the glass's being completely sealed), the bubbles have room for expansion, as is the nature of air. The remaining experiments will make this fact increasingly clear.

Water in the lower container can be changed to ressemble smoke.

Invisible bubbles are rendered visible in a vacuum.

6. This apparatus clearly demonstrates that the air in the lower regions where we live is greatly compressed because of its own weight, but that it has the natural ability of expanding itself. For when air is extracted from these two glass vessels in the manner described above, then the air which is in the upper glass descends

Air is greatly compressed.

through tube, xr, by reason of its expansive nature, passes through the water, expands in the lower glass, and finally enters into the copper sphere. It should be remarked here that if only the lower glass is evacuated (with the stop-cock, d, closed) and then if this stop-cock is opened quickly, the air of the upper glass rushes in with such force through the water of the lower vessel that it immediately breaks the glass. This accident happened to me.

The so-called aversion of a vacuum has its origin in the weight of air.

7. By means of this apparatus one also clearly sees that the so-called aversion of a vacuum is nothing other than the pressure air exerts upon itself and all things beneath it. For when these two glasses are evacuated at the same time, the water in the lower makes no effort to rise or fill the evacuated upper glass. At the precise moment, however, that a little air is-introduced into the lower glass through the stop-cock, f, the water rises with great force into the upper one and continues to rise until the stopcock, f, is closed. When this is done, the air which has been pressed down in the lower glass by the weight of external air holds the water, suspended at the height it has risen, through its capacity for expansion. Once again, when this stop-cock, f, is opened and remains open until the upper glass is filled with water and finally with air, then this water is held in the upper glass in the same way as before through the presence of the air (which is pressed down in the lower glass) and not through an aversion of a vacuum. And even though this stop-cock, f, should be closed, the water, nonetheless, cannot descend because were it to do so, it would press down upon the air enclosed in the lower glass, something which cannot occur because the air and the water of the upper glass are in equilibrium with the air of the lower. On the other hand, when some air is again extracted through the agency of the copper globe, all the water descends and the air finally follows until the glass is rendered empty.

How the descent of water is brought about.

8. Now if you wish to return this apparatus to its original state, that is, bring the water that has risen back into the lower glass, it is necessary that a free passage be given for air to enter into the upper glass, L, from somewhere else, preferably through some tiny opening. This can be done by means of a small tube, u, passing through the brass case that has been cemented on, from below into this upper glass. The tube is provided with a stopper, o, and when this is removed, air can fill the place left empty by the descending water.

Finally all water is extracted through a syphon.

9. When the water has again returned to the lower glass, the pipe, pp, should be connected to the mouth of the third stop-cock, g, by its hook-like attachment and the stop-cock of the pipe, n, should be opened. The the water which heretofore had been retained in the

pipe, pp, will descend and all the rest will follow as though through a siphon. In this way the apparatus which drew up water from one side through the first stop-cock, f, will expel it again from the other through the third stop-cock, g. There are other additional and different experiments which one can devise through the actual use of this apparatus rather than through verbal descriptions of its uses.

CHAPTER 19

A NEW DISCOVERY THROUGH THE USE OF THIS APPARATUS WHICH INDICATES THE WEIGHT OF THE ATMOSPHERE

When I had set up the apparatus, which has already been described, on a table in my study for the sake of mental recreation and performed the experiments mentioned in the preceding chapter for my friends and others who were investigating these discoveries, several thoughtful onlookers, seeing the water rise with such force and quantity from the floor of my room into the evacuated glass through the pipe, pp, asked to what height water could actually be raised in this manner. Since I was as yet ignorant of this information, but nonetheless did not suppose that water rose in the glass to an infinite height, I could not neglect to perform an experiment that pertained to this question. Therefore I had the pipe, pp, lengthened (so that it extended from the window of the middle floor of my house to the ground outside) and placed a container full of water under it in the my same way as was described in the preceding Chapter, section 2. Again I saw the same phenomenon occur, namely that the water, contrary to its own nature, rose into the evacuated glass.

See Plate IX, Chapter 18.

Water rises from the ground to the middle floor of my house.

It was absolutely essential, therefore, that the apparatus not only be carried to the third floor of my house but also that it be provided with a larger pipe or tube, and the experiment be repeated. When this was done, the results were the same as before, that is, the water rose to the third floor as well, but in relatively less abundance than before. Then I betook myself to the fourth floor of my house, carried out the required preparations and resumed the previous experiment. At this time I perceived that no water rose in the evacuated glass but rather was suspended in the tube.

It rises still higher.

Because the precise place to which the water rose was uncertain, however, it was necessary that a glass pipe be firmly embedded in a pitch-like substance in the usual way at the point where the water seemed to reach and stop, and that the experiment be repeated a fourth time. Then when this was done and the key of the stop-cock was turned, I saw the water begin to rise immediately and after this, at different times to rise and fall in the glass tube and finally to come to a stop and remain quiet. In this way, the place where the water had risen before was evident. I recorded this point and dropped a line from here to the ground outside. This distance was measured in Magdeburg cubits and I found these to number essentially nineteen.

Water rises to 19 Magdeburg cubits.

And although I did not cease repeating and demonstrating this proof in the same way at different times, nevertheless I saw that the water always maintained the same height. Afterwards, however, when this experiment had been set up for several days, I saw some variation from one day to the next. Sometimes the water was the width of one hand or two or three higher than before, and sometimes that amount lower. I could not explain the reason for this.

The height of water varies.

Although I suspected that the cause lay in the glass which perhaps had permitted some air to enter in the course of time, thereby lowering the water accordingly, nevertheless, when on another day I saw that the water had risen almost a half cubit above the eighteen and a quarter previously recorded, I could not ascribe the cause to the apparatus being insufficiently sturdy in construction. To verify this observation, I had the glass evacuated again and the experiment repeated. Thereupon the water reached and remained at the same height it had attained when this experiment was begun. This lasted for several days until it again underwent a change.

From this experiment whose results I had not anticipated, I could only conclude that the aversion of a vacuum is nothing other than the weight or pressure of air around the earth, which, when it finds a space empty or free of anything corporeal, exerts pressure on water to enter into or occupy this space. This penetration is as great as the weight of the air.

The aversion of a vacuum is the weight of air.

For if the aversion of a vacuum, as the Peripatetics (ERRATUM read "Peripatetics" for "Paripatetics") aver, is a theory which precludes the existence of a vacuum in nature, then water would necessarily follow a vacuum to any height one suggests and fill it. This theory, nevertheless is counter to our experimental and visual observation, since water does not seek out a vacuum above nineteen or nineteen and a half cubits. That the pressure of external air causes this ascent of water is manifestly evident from the fact that water does not always maintain one and the same height. If this ascent were dependent upon the aversion of a vacuum, however, water would either rise without limit to any height at all or would always remain at one constant height. The fact that the height varies is a

The theory is refutable.

very clear indication that not only does this ascent of water have its origin, but even its variation, in an external cause.

Water is not elevated in a tube, therefore, by reason of nature's shunning a vacuum but rather from the existence of a state of equilibrium between a cylinder of water and one of air.

CHAPTER 20

OTHER EXPERIMENTS OF THIS KIND WHICH DEMONSTRATE THE WEIGHT OF THE ATMOSPHERE AS WELL AS THE LIMIT TO WHICH AVERSION OF A VACUUM EXTENDS

Since it was visually perceptible in the foregoing experiments that nature does admit of a vacuum and that the phenomenon which is commonly called an aversion of a vacuum has its origin, at least in these lower regions, in the weight of air (when the atmosphere, through its own weight, not only forces water into a place where it finds empty space, but even presses it upwards until it is in equilibrium with air itself), two other shorter methods will be presented for measuring the weight of air and for examining the aversion of a vacuum.

First Method

Have four tubes or pipes, ab, cd, ef, gh, made of thin brass and the size of the small finger in thickness, essentially five cubits in length,(which can be joined together in various ways, or separated, and consequently lengthened or shortened). Their openings should be so constructed that one can be inserted into the other with precision, as for example, d into a, f into c, and h into e and at the same time be surrounded with water through the device of small bowl-like vessels, a, c, e, g, so as to prevent the penetration of air. Then three of these tubes should be joined together and set up near the wall of the house. The end of the lowest tube, ab, should have at point, b, a stop-cock that can be closed and opened. The top part of the third or even a fourth tube should have a mouth or opening into which another stop-cock, l, attached to a glass vessel, ikl, can also be fitted precisely so that air cannot enter. (The latter is two or more cubits in length and is of the type represented in Plate V, Chapter 2, figure 3.) *See Plate X, fig. 1.*

Then join these three pipes, ab, cd, ef, and set them up near the wall of the house (once they have been tested under water to see that they do not allow the smallest air bubbles to pass through), at first, for example, having them extend to a height of 15 or 16 cubits with the lower stop-cock, b, closed and immersed in a container, mn, filled with water. Next fill these tubes as well as the glass with its stop-cock, l, which has been mentioned before, with water and insert it in the mouth, e, of the tube, ef. Turn the keys of the two stop-cocks and then you will see that no water descends from the glass. The reason is that the total height of the tube together with the glass is only 17 or 18 cubits, while the atmosphere exerts more pressure than the weight of water at a height of 17 or 18 cubits. The water, therefore, cannot descend. But if you repeat this experiment and add another tube, gh, 2, 3, or 4 cubits long so that the total height of the tube, together with the glass vessel, extends to 20, 22, or even 30 or more cubits, you will see the water descend from the glass and then even fluctuate in the glass tube, but finally come to rest at its set limit of 18 or 19 cubits.

Water remains in equilibrium with the pressure of a cylinder of air.

Remark

It makes no difference whether the tubes are of a large or small diameter for water does not exert more pressure in a wide than in a narrow tube, other things being equal. For example, let two tubes be set up perpendicularly, making them of the same height but of different diameters, their bottom ends being joined with a horizontal tube, c, so that water can pass from one to the other. Then let one of the two vertical tubes, a, for example, (whichever it may be), be filled with water. You will see water rise in both tubes simultaneously to an equal height. The large quantity of water in the wider tube cannot make the smaller amount of water in the narrower tube rise the least bit higher; rather it remains at the same level in both a and b. The reason is that air and water, being fluid materials, press upon one another in accordance with the height of the perpendicular tubes. In these experiments, therefore, it is not quantity of water but the height of the perpendicular tubes that is the decisive factor.

See Plate VIII, fig. 4.

The height of water, not the quantity, drives it upwards.

Second Method

Procure for yourself the tubes or pipes that have been mentioned, as well as the glass container which chemists commonly term a cupping glass, cucurbit or retort ("eine Kolbe", or in other words a $\frac{1}{4}$ receiver, for the larger such a container, the more suitable it is). This glass, (provided with a thin metal case and stop-cock, as was described in the preceding Chapter 5) should be evacuated of all air and inserted at point, g, in the pipe set up near the wall. (This pipe, with its lower stop-cock open, must be immersed in a pail filled with water.) Then, when you open the stop-cock of the glass, you will discover that the water rushes with force into the tube up to the aforementioned 19 cubits (or thereabouts). The reason is that external air presses upon the water in the pail, mn, while the water finds an empty space in the evacuated glass into which to pass. For this reason the water rises to as great a height as it can be forced upwards by the weight of the surrounding air.

Remark

It should follow from these experiments:

Water cannot be conveyed higher than 19 cubits by means of a syphon.

(1) that water cannot be conveyed with the help of a syphon over a mountain or other similar elevations more than 18 cubits in height. (2) that in the pipes of wells no water can be raised by means of suction or hydraulic equipment whose valve and suction pump are located higher than 18 cubits. When the valve and the pump are placed lower, however, water can be raised not only to this height, but even higher than the plunger which is moved up and down in the aforementioned pipes.

The height to which water can be raised by means of a suction pump.

Many other things that were not understood until now, moreover, were clarified through these experiments, details of which I reported to Rev. Father Kaspar Schott in a "Reply" dated Dec. 30, 1661. He had this reprinted in his *Technica Curiosa*, Book I, "De Mirabilibus Magdeburgicis", Chapter 21. The words of the letter on page 52 of the same read as follows:

The little man which serves as an indicator of different kinds of weather.

See Plate X, fig. 4.

"In order to ascertain the weight of air at a specific time, I devised an original mechanism wherein a small figure made of wood in the form of a little man moved up and down in the upper part of a glass tube and pointed with his finger at certain marks indicating the air pressure at a given time. The internal mechanism, housed in the lower part of the glass, however, is not visible, being completely surrounded and well protected by a thin metal sheath. Thus onlookers do not have access to the secret and the entire device is rendered unaffected by change or breakage. I call this small invention "Always Alert". It can also be reasonably called "Perpetual Motion" even though it does not move continually, but only upon the change of air in the entire region surrounding it. It is not a thermoscope which is changed by heat and cold.

"Would that your Reverence and all friends of true philosophy take some pleasure in this experiment and ponder the following. Since it

PLATE X.

is clear that air does not always maintain the same weight, it follows that if the sphere, L, once it has been completely evacuated of air, is hung from a true and accurate balance, the sphere is rendered lighter when the air is heavier. Contrariwise, when the air is lighter, the sphere is heavier. For the heavier the medium is, the lighter the object contained therein. Take this as an example: if water is heavier than air, then the sphere will be lighter in water than in air. If these facts are considered. they will open a new area for investigation. For example, one can ask whether or not it is raining widely in the surrounding region; indeed, whether it has begun or stopped, etc.

See Plate X, fig. 3.

A completely evacuated vessel that is suspended from a balance, varies as in weight.

"When in a previous year (1660) a mighty windstorm occurred, I perceived a singular and extraordinary change of air by means of the device I have just mentioned. The air was rendered so much lighter than usual that the finger of the little man was lowered below the lowest point marked on the glass tube. When I saw this, I said openly to those who were present that there was undoubtedly a great storm somewhere. Scarcely two hours had elapsed before a gusty wind roared into our parts; it was somewhat less violent, however, than it had been at sea."

An indication of a storm by means of the little man.

CHAPTER 21

CONCERNING THE WEIGHT OF AIR

We perceive the weight of air in two ways; it can either be weighed collectively or in parts. We term the collective weight of air the pressure it exerts in a perpendicular direction as in a cylinder. For example, a person can imagine to himself a perpendicular cylinder of air (even when its height is not specifically determined), the base of which in its circumference or roundness, if you will, is similar to that of an groschen, a gulden, or an imperial thaler, or to that of a man's head or the brim of a hat. I say that the weight of a cylinder of air, whose circumference or roundness is that of a groschen is the same as that of water in a tube (which is like a groschen in roundness) 19 Magdeburg cubits high. This is to say that the weight of air presses down as much as the weight of water 19 Magdeburg cubits high. And again I repeat: the weight of a cylinder of air whose diameter is that of a Rheinish gulden is the same as the weight of water in a tube of a like diameter 19 cubits high. And so it continues proportionately, as will be seen in the following Chapter.

The twofold aspect of air.

The weight of air around the earth.

The weighing of air in parts can be performed by the following method (as can be seen in the illustration). Since it is known that

See Plate X, fig. 3.

air has weight, it follows that when a glass vessel or receiver, L, such as that mentioned in Chapters 4, 8, and 11, is evacuated, its weight is not the same after this has been done as it was before. For example, hang a glass of this kind from a balance and weigh it, (with the stop-cocks open, however, so that you may be certain that the glass is filled with air). Then evacuate all air from it and you will discover that after this has been done, it weighs one or two ounces less than its capacity. Thus after my receiver has been exhausted and is surrounded by the pressure of the atmosphere, it it will be found, when examined on the balance, to be two ounces lighter than before, being equal in weight to two imperial thalers. Then prepare to let air enter once again (only a little at a time, however, so that it does not rush in so forcefully that it breaks the glass) and not only will you hear air enter with a hissing sound but you will even see the glass gradually taking on its former weight. This is a very clear demonstration of the weight of air. It also follows from these experiments that air presses upon itself; in other words because of its weight, air is greatly compressed and condensed here in these lower regions where we live. The higher or the more distant it is from the earth, however, the less it weighs. In consequence, it is not as compressed in the upper reaches of the atmosphere but is more rarefied, more expanded, and lighter.

The weighing of air in parts.

Air presses on itself.

It likewise follows that condensed air contains a greater mass of air than expanded air, but that in its composition, there is no difference. Rather this difference lies in a greater or lesser amount of air, which always varies with the greater or lesser amount of space the air occupies. In a higher place where it is not pressed down by so much weight, it expands itself more and more and occupies a greater space until it finally disappears into nothingness. The great expanse of Space which air cannot fill, moreover, is of necessity left empty.

Condensed air contains a greater mass of air.

There is no difference in air itself.

These facts, along with the others, should be carefully noted for they will produce a better understanding of subsequent experiments.

Furthermore, it should be remarked here that a glass vessel of this kind which is attached to a balance with its stop-cock open, always fluctuates in weight. For in warmer weather, when the air is expanded, the receiver is lighter. When it is colder, however, the vessel is heavier because it holds more air in it. (See the end of the last Chapter of this book.)

The sphere attached to the balance fluctuated in weight.

Finally, some very learned scholars have tried to establish the comparative weights of air, water, and mercury. But since, in fact, the lower the air is, the more compressed and consequently the heavier it is, and contrarily the higher it is, the more expanded and hence the lighter, all endeavors in this direction are disappointing and generally unsuccessful.

The true difference in weight of air and water cannot be established.

CHAPTER 22

HOW TO DETERMINE THE PRESSURE THAT A CYLINDER OF AIR OF ANY GIVEN CIRCUMFERENCE EXERTS

How to ascertain the weight of a cylinder of any given capacity.

Because a cylinder of air can exert as much pressure as a cylinder of water 18, 19, or at the most 20 Magdeburg cubits high, the weight of any given cylinder of air is calculable.

For this purpose, then, the weight of a cylinder of water should first be determined; one 20 Magdeburg cubits high, for example. This calculation can be made as follows.

A Magdeburg cubit is divided into 100 parts.

(1) Divide a Magdeburg cubit into 100 equal parts. Using these parts as units, measure the base of a tankard or Magdeburg quart (commonly called "ein halb Stübichen Maas"). You will find that this measures 19 hundredths of a cubit in diameter and 38 hundredths in height.

The area of the base of a Magdeburg quart.

(2) To obtain the surface area of the quart's base, employ the ratio of the diameter to the circumference, 7:22. You will find that it contains 285 squared parts.

Calculation

7–22–19	$418:7 = 59\frac{5}{7}$ The circumference of the base of the Magdeburg quart.
19	
198	
22	
418	

(3) Multiply the radius with half the circumference of the circle and you will find the surface of the base of the Magdeburg measure.

Calculation

$59\frac{1}{2}$	circumference
30	half the circumference
$9\frac{1}{2}$	half the diameter
270	
15	
285	squared parts, or the surface of the base of the Magdeburg quart.

(4) Furthermore, if these figures are multiplied by 38, the height of

this measure, the result will be 10,830 cubic parts for the capacity of the Magdeburg quart.

The weight of the Magdeburg quart.

(5) When this measure is filled with water and weighed, the water weighs $4\frac{1}{8}$ pounds (or 4.125 pounds to three places after the period in decimal numbers); the weight of a pound equals sixteen imperial thalers.

The weight of a cylinder whose base is that of a Magdeburg quart and whose height is 20 cubits.

(6) First the capacity and then the weight of a Magdeburg quart 20 cubits high should be calculated. For as the height of a Magdeburg quart is 38 parts of a cubit and its weight is $4\frac{1}{8}$ or 4.125 pounds, so water 20 cubits, or 2,000 parts, in height weighs essentially 217 pounds.

CALCULATION

The height of a Magdeburg quart reckoned in parts of a cubit	It weighs in pounds	How much does a cylinder 20 cubits or 2,000 parts high weigh?
38	4.125	Result: $217\frac{4}{38}$ pounds

A cylinder of water which is 20 cubits high, the surface of whose base contains 285 square parts of a cubit,weighs this much.

ANOTHER METHOD

The base of a Magdeburg quart contains 285 square parts. If these are multiplied by 20 cubits or 2,000 parts, this vessel will contain 570,000 cubic parts. Since 10,830 cubic parts equal 4.125 pounds, then 570,000 parts equal 217 pounds.

CALCULATION

40830 (ERRATUM read "40830" for "80830"	4.125 pounds	570,000 equal $217\frac{1130}{10830}$ pounds

In this way the weight of a cylinder of air can be calculated, provided that the base of the cylinder in question is known. For just as the surface of the base of a Magdeburg quart contains 285 square parts relative to its weight of 217 pounds, so the base of any other given cylinder is related to its own weight.

Calculation

If the diameter of the base of a cylinder is 67 hundredth parts of a cubit, then its circumference will be 210.6 (according to the ratio of 7 to 22) and the surface area of the base will be $3527\frac{1}{2}$ square parts.

If the diameter of a cylinder's base is 67 parts of a cubit.

Now since the surface of a cylinder's base of 285 square parts weighs 217 pounds, a cylinder whose base is $3527\frac{1}{2}$ square parts will weigh 2686 pounds.

Calculation

The surface of the base of a Magdeburg quart	Weight of a cylinder 20 cubits high	Surface of a base whose diameter is 67 parts
285	217 pounds	3527.5
		217.0
		246925
		35275
		70550
		765467.5

765467.5 divided by 285 equals 2685.8 pounds

Then the cylinder of air weighs 2686 pounds.

A cylinder of air whose base has a diameter of 67/100 of a Magdeburg cubit has this weight.

Another Method

As the aforementioned 10830 cubic parts are to its weight, 4.125 pounds, so the cubic parts of this cylinder whose base is 3527.5 square parts, is to the weight being sought. These cubic parts are easily calculated when the square parts are multiplied by the height of 20 cubits or 2000 parts.

Calculation

3527.5	square parts
2000.0	parts, or 20 Magdeburg cubits
7055000.0	cubic parts

Cubic parts	weigh	What will cubic parts weigh?
10830	4.125 pounds	7055000
	7055000	
	20625000	
	20625	
	288750	
	29101875.000	

29101875 divided by 10830 equals 2687.1

Thus a cylinder of air whose base has a diameter of 67 parts of a Magdeburg cubit has this weight.

CHAPTER 23

THE EXPERIMENT DEMONSTRATING THAT AS A RESULT OF AIR PRESSURE, TWO HEMISPHERES CAN BE JOINED TOGETHER IN SUCH A WAY THAT THEY CANNOT BE SEPARATED BY SIXTEEN HORSES

See Plate XI.

The construction of the apparatus.

I had two hemispheres or bowls, A and B, made of copper about 3/4 of a Magdeburg cubit in diameter, or, to put it more exactly, 67 hundredths of a cubit, inasmuch as craftsmen cannot usually make vessels of as exact proportions as one demands. These fitted together with precision, provided with a stop-cock or rather a valve-like device, H, soldered to one of the hemispheres. By means of this the internal air could be extracted and the external air excluded. We described a valve of this type in the preceding Chapter 8 and illustrated it in figure 3 (ERRATUM read "3" instead of "4") of Plate VII found there. Iron rings, N, N, N, N, moreover, were attached to these hemispheres so that horses could be hitched to them, as can be seen in the Plate. I also had a band or ring of leather made, which was stitched together and smeared with a wax and turpentine mixture so that no air could pass through.

I set these hemispheres one upon the other with the band inserted between them and then had the air pumped out quickly (by means of a special tube, F, G, H, I, which has also been described in Chapter 8 and illustrated in Plate VII, figure 4). At this time I saw the two hemispheres come together on the band with great force. Thus by the pressure of external air they were held one against the other so firmly that sixteen horses could not, or only with the greatest difficulty, were they able to pull them apart. When a very great force is finally applied, however, these hemispheres are sometimes pulled apart, at which time they produce a noise similar to the explosion of an air gun.

When air is admitted, the hemispheres can easily be pulled apart.

As soon as the valve or stop-cock, H, is opened and air is permitted to enter, they can easily be separated or pulled apart from one another by one's bare hands. If you want to know the strength of the force which holds these hemispheres together, refer to the preceding chapter for the weight of a cylinder of air whose diameter is 67/100 (ERRATUM read 67/000 for 79/000) of a Magdeburg cubit. We have cited this example there for the purpose of clarifying the experiment.

One discovers here that the weight of this cylinder is 2686 or 2687 pounds. From this it follows that the weight of air pressing one hemisphere down upon the other is 2686 pounds. Thus one hemisphere has the same pressure exerted upon it that eight horses put forth or produce in drawing one hemisphere apart from the other, namely a weight of 2686 pounds. On the other side the eight horses exert a like pressure to draw apart the hemispheres.

The weight which presses the hemisphere together.

Now, although eight horses can pull a wagon loaded with a weight of 2686 pounds rather easily, they have greater difficulty, however, in pulling apart the hemispheres, for this effort is exerted against a whole cylinder or air, an action which is more opposed to nature, so to speak, than the pulling of a weight in the middle of a wagon. It follows from these facts, then, that if these two hemispheres are joined together and are suspended, and then if a weight of 2686 pounds is fastened to the lower hemisphere, this hemisphere can be separated or drawn apart from the upper one by the application of this weight. It should be remarked, however, that the weight necessary to accomplish this fluctuates according to the quality of the air (because the weight of air is sometimes greater and sometimes less). The weight of the whole expanse of the sky, moreover, can be calculated in the same way (that is to say, according to the dimensions or rather the base of any cylinder you wish). Thus if one wishes to know the weight of the whole expanse of the sky, that is, the weight of all the air around the earth, he first should calculate the extent of the earth's surface in square miles and reduce these miles into square cubits and then should proceed according to the Golden Rule set forth in the preceding Chapter to find the weight he seeks.

How they are separated by means of a weight.

The weight of the expanse of sky.

Finally one can deduce from the foregoing that the earth (since the surrounding air has its own separate and particular weight) is an individual and separate body of the world, existing freely by itself and contained in and surrounded by no other body or other celestial matter.

The earth, together with the air, constitutes one separate celestial body.

PLATE XI.

CHAPTER 24

A FURTHER EXPERIMENT WHEREIN THE HEMISPHERES WHICH COULD NOT BE DRAWN APART BY TWENTY-FOUR HORSES ARE SEPARATED BY THE ADMISSION OF AIR

Because some damage almost always occurs when these hemispheres are separated and especially because they tend to lose their perfect symmetry as a result of their falling or striking the ground when they are being pulled apart by the horses, I had two other larger ones made, a cubit in % diameter. (But because craftsmen can rarely make vessels of this kind according to prescribed specifications, I found these hemispheres had diameters of only 95/100 parts of a Magdeburg cubit.) I was of the opinion that twenty-four horses could not pull these hemispheres apart but that a person could separate them by admitting air.

The larger hemispheres cannot be separated by 24 or 30 horses.

The calculation proceeds according to the method set forth in Chapter 22, namely:

7–22–95	2090:7 = 298.5 The circumference
22	
190	
190	
2090	

298.5	circumference
149.3	half of the circumference
47.5	radius
7465	
10451	
5972	
7091.75	area of the base of this cylinder

The area of the base of a Magdeburg quart	The weight of this cylinder	The area of the given cylinder
285	217 pounds	7092
		217
		49644
		7092
		14184
		1538964

1538964:285 = 5399 pounds
The cylinder in question has this weight.

Another Method

Cubic parts of a Magdeburg quart	They weigh
570000	217 pounds

3077928000:570000 = 5399 pounds
The cylinder whose diameter is $\frac{95}{100}$ parts of a cubit has this weight.

What will the cubic parts of a cylinder weigh whose area is 7092 (that is, a cylinder whose base is multiplied by the height of 20 cubits or 2,000 parts)?

```
  14184000
       217
  99288000
    4184
28368
3077928000
```

Because (according to the preceding chapter) sixteen horses are required to draw apart a weight of 2686 pounds, one asks how many are needed to draw apart one of 54000 pounds. According to the Golden Rule, the number is thirty-four horses.

How the hemispheres are separated by the admission of air.

See Plate XI, fig. 4.

Larger hemispheres cannot be drawn apart by 100 horses.

When only twenty-four horses are hitched up, therefore, they certainly will not be able to pull apart these hemispheres and they will remain joined together as a sphere. In order to separate them in some way, therefore, air must be drawn into the sphere by means of a pump through the stop-cock, with a valve provided on the under side of one of our hemispheres which does not permit it to escape. Thus at first the stop-cock is opened so that air can enter and fill the sphere. Then when air has been drawn in three or four times with the help of the pump, the hemispheres by themselves fall apart from one another. And thus we can say in truth that that which twenty-four horses, nay even one hundred or more horses could not draw apart (when the hemispheres are larger), can be separated in this way by one man alone through the admission of air.

PLATE XII.

CHAPTER 25

ANOTHER EXPERIMENT WHEREIN IT IS SHOWN THAT THE AFOREMENTIONED HEMISPHERES CAN BE SEPARATED BY A WEIGHT

In addition to the questions that have been treated thus far, I was further concerned with the following: whether one should assume that these hemispheres, AB, would emit a crashing noise like a cannon shot when a ball is ejected by gunpowder if they were broken apart by a falling weight. For this reason I had these hemispheres fashioned in the usual way and the air extracted from them. When they were joined to form a sphere and were suspended at the top, I

See Plate XII.

The experiment whereby these hemispheres are separated by a falling weight and produce a loud crash.

had a hundred pound weight suddenly released which was attached to the lower hemisphere. What, then, was the result? The hemispheres were not broken apart. Moreover, because there was danger that one of the onlookers might be injured or even killed from this suspended weight falling unexpectedly and suddenly and since this experiment had been attempted without success a third time, I completely gave up the undertaking. For, indeed, as often as I released the weight, the hemispheres were not broken apart or separated.

See Plate XII.

The method of separating these hemispheres by means of a weight.

That those smaller hemispheres mentioned in Chapter 23 might not seem to have been made without purpose (inasmuch as I could not provide the requisite number of horses at that time, I had an upright post, which was set in the ground, erected on my estate near the corner of a wooden fence enclosing my garden. On its top part a transverse beam or arm was attached. Into this an iron hook, K, was inserted from which the two hemispheres which had already been united and evacuated could be hung by their own iron ring, L. Then a good sized square plank, M, (ERRATUM read "libripendens" for "libripendes") was attached (of the kind used by merchants in the markets to weigh out merchandise), with four chains fitted on its corners. These, in turn, were inserted in the four rings, NN of the lower hemisphere, (provided with the valve, H). On this plank a weight was laid which was sufficiently heavy to break apart the hemispheres.

To find the weight which can separate the hemispheres.

They give forth a crash.

Now since we have accurate information as to the weight of air about the earth from Chapter 22, and since the weight of air pressing upon one of the hemispheres has been demonstrated to be 2686 pounds, according to Chapter 23, it can be concluded from these data that these hemispheres, or the sphere resulting from their being united, can be separated if 2686 pounds are placed on the plank. When the hemispheres are broken apart there is a loud crash, but duller nonetheless than that which would result from a falling weight. The reason is that the weights are placed on the plank, not all at once, but gradually, and that the hemispheres are not separated equally all around, but rather on one side more than another. Thus they are not broken apart evenly, for the air enters them more from one side than another and does not pour in all around at one time.

PLATE XIII.

CHAPTER 26

ANOTHER EXPERIMENT DEMONSTRATING THAT ALL VESSELS CAN BE COMPRESSED AND BROKEN BY AIR PRESSURE

Because the air surrounding us exerts pressure upon entities on the earth's surface, it follows that it enters into all bodies that are not completely solid and fills all their small internal spaces (as we find is the case in wood, stones, and other materials of this kind that are not absolutely solid). What is more, all vessels, even though they have very small orifices or pores, are filled with air. If it were otherwise and they were not filled with air, then such vessels could not exist because of the great pressure of the surrounding air which

Air fills all small space.

This is true for all vessels.

would compress them, unless they were perfectly spherical. Take as an example the glass receiver, L, that has been mentioned before (see Chapter 21, Plate X, figure 3). When this has been completely evacuated, attach to it a foursided glass flagon which can be connected to the stop-cock of the receiver in such a way that no external air can enter. (This can better be seen in the Plate XIII.)

Evacuated bodies, unless they are absolutely spherical, are compressed by the external air.

Then place a small howl, B, under the flagon (so that glass fragments are not scattered throughout the room) and open the stop-cock gradually. Then you will hear the hiss of the air contained in the flagon as it rises into the upper evacuated receiver. Because the flagon is not spherical, it cannot withstand the pressure of the external air, (that is, the air surrounding it). Thus of necessity it is compressed and is broken into a thousand or more pieces. At the same time, the external air, entering that same space in the flagon left empty of air, produces a loud crash.

But when vessels are spherical, then they are not broken, for because of their roundness which does not permit of their collapse (as is true of vaulted arches which we see become stronger the more they are pressed upon), they remain intact. When they are not absolutely round, however, they are subject to damage, as I discovered at my expense. This occurred when I contracted with a coppersmith to make a copper sphere, (which I mentioned in Chapter 3) and because this was not perfectly round, it was compressed by the surrounding external air, as I described there, and was destroyed.

If a person breathes into an evacuated vessel, he suddenly dies.

It follows from these facts, then that if a man breathes into such an evacuated receiver, he will suddenly die. For the external air because of its weight not only squeezes the breath of life from the body of a man or any living animal into the evacuated glass, but also brings about severe internal injury to the body itself.

CHAPTER 27

A GLASS VESSEL WHICH CAN FORCIBLY PULL MORE THAN TWENTY, INDEED FIFTY OR MORE STRONG MEN

Certain experiments which the author performed for our Sovereign Majesty in the Diet held at Magdeburg in the year 1554.

When, engaged in state business, I attended the Imperial Diet which was held at Regensburg in 1654, I presented examples of certain of my experiments (among which was the new one mentioned in the preceding Chapter) at the bidding of his Sovereign Majesty, Ferdinand Third, as well as his Roman Sovereign Majesty, Ferdinand Fourth, both of most glorious memory. In my opening remarks I

PLATE XIV.

had stated, in regard to these experiments, that if a man should breath into the aforementioned receiver, he would breathe out his life at the same time. The Prince of Auerberg, Major domo of the Court of his Sovereign Majesty, could not be persuaded to accept my oral account until he had seen with his own eyes what had taken place with the glass receiver. But since at that time something occurred causing these reliable results of mine to seem unreliable and called into doubt, I thereupon proceeded to undertake a certain new experiment to prove and confirm my theory. In an another way, then, I succeeded in proving its validity, namely that this glass vessel can pull along more than 20 or 30, nay, if you prefer, 50 or 100 men and (if they do not cease this activity), lay them out on the ground.

That 20, 30, even 50 or 100 men are laid out by the glass vessel.

First of all, a coppersmith made a copper vessel, a. It was in the form of a basin except that its sides were exactly parallel and vertical

Plate XIV.

like those of a cylinder. It was more than a cubit in height and its circumference was equal to the smaller hemisphere described in Chapter 23. A tin handle, s, was attached to its lower side.

Construction of the apparatus.

Next a piston and piston rod, PQR, was designed to fit into this vessel with precision: P was made of iron, Q of wood, and the piston itself, R, of an oak plank. Its circumference was provided with a groove around its edge as in a well pump so that rope or oakum could be inserted and packed all around. This device, then, was introduced into the vessel, a, so designed as to prevent air and water from entering or escaping, (just as the piston rod with its piston is inserted in the pumps used in fires).

Thirdly, a rather large basin-like container, a, was brought over to the heavy post, I, and by means of a metal projection or handle, S, was fastened to an iron arm, BW, (which fitted quite firmly into the post). Then the piston and piston rod, PQR, was lowered into the container and the iron rod PQ, was inserted in such a way through a hole, V, into another iron arm, OV, that the piston rod could be raised and lowered in a straight line without any deviation from the perpendicular. It could not, however, be raised higher than the projecting section, Q, of the piston.

Operation.

Fourthly, when the container was attached to the post in this way and the piston rod, together with the piston was lowered almost to the bottom, a rope, to which an iron hook, F, had been attached, was lowered through a block, T. When the hook was inserted into a grommet at the top of the piston rod, and the stop-cock, x, was closed, twenty or more men pulling manually on the rope strove to raise it. They could only draw the rod about halfway up and were not able to pull it completely to the top.

Fifthly, a receiver, to which I have often referred, was attached to the stop-cock, x, in such a way that the surrounding air could not enter or escape. Meanwhile, although the aforementioned men exerted as much force as possible, still they could not succeed in raising the piston rod completely. When the two stop-cocks were opened (that of the receiver and that of the basin-like container), however, the piston PQR, suddenly fell to the bottom of the container and all effort on the part of the men was rendered useless. For this reason I would venture to assert that this glass could draw twenty, thirty, and more men to itself (and even a hundred, if the apparatus were larger). And if this is the case, why can it not draw one man to it and cause him to rupture internally?

The reason for this result.

Air pressure should be regarded as the cause of the foregoing result. For when air is extracted from the inside of the container, the piston immediately falls. And since the container is of equal

PLATE XV.

circumference to the hemispheres of Chapter 23, the air pressure equals 2686 pounds, a weight which twenty, thirty, or even fifty men cannot sustain when it falls so quickly.

CHAPTER 28

AN EXPERIMENT TO RAISE A GREAT WEIGHT

A boy twelve to fifteen years old can lift a very heavy weight by means of the apparatus which has just been described.

Leave the container, a, attached to the post, I, and when the piston, R, has been raised to its rim, let down a chain through the *See Plate XV.*

How a boy 15 years old can lift a great weight.

block, T, and a second one, H, which is attached to hook, K. Then fasten the chain at one end by the hook, F, to the piston rod, QO, at point, P. On the hook at the other end, N, attach the plank supporting the aforementioned weight of 2686 pounds.

Then prepare a pump similar to that described in Chapter 2 (one which is not too large, however, for a boy to handle). Attach it to the stop-cock, x, and then bid the boy to put the piston rod in motion. In this way air is evacuated from the container and the piston, R, slips down slowly and gradually into the container. At this time the plank supporting the weight of 2686 pounds is raised.

Thirdly, if it should be necessary to raise a heavier mass, the diameter of the container also must be greater, for the diameter stands in direct proportion to the weight which is to be raised. This computation can easily be made from the data given in Chapter 22.

The weight of a mass varies according to the fluctuations of air pressure.

Fourthly, you should note that since air pressure is not constant at all times but from time to time fluctuates up to a twentieth part of its weight (as we shall show in. Chapter 31), it follows that it does not always lift a mass of the same weight. You can measure this fluctuation by the fingers of the little man hanging in the glass and can calculate it very accurately. In this way you can establish with precision the air pressure for any given time and if this is not possible, at least ascertain its approximate deviation.

CHAPTER 29

AN EXPERIMENT EMPLOYING A NEW AND HERETOFORE UNTESTED AIRGUN

A method was found several years ago whereby air was so compressed within a pipe or tube of iron or copper that a lead ball could be shot from it with essentially the same result as from an ordinary gun using gunpowder. What others have attempted to do by means of a great amount of air in these airguns, however, we bring about through a deficiency of air in an inverse or contrary way, as it were. The procedure is as follows:

The construction of the gun.

See Plate XVI.

Have a pipe, abd, made of a sheet of brass with a circumference such as to accommodate the lead balls which one wants to shoot. The brass disc, a, functioning as a valve and similar to that pictured in Plate V, figure 1, letter b, should be soldered on one end of this pipe into which a leather disc, K, can be inserted to cover and close the opening, a. On the under side of the pipe at point, g, there should be a slit a half finger in length and no more than the back

PLATE XVI.

edge of a table knife in width. A small pipe shaped like the tile of a rain gutter and extending from a or g to point b or e, should be attached to the aforementioned pipe, ab, which terminates in A funnel or cone, e. This cone should be made of a piece of metal with a thickness such that the stop-cock, c, of the receiver, I. already referred to, fits it exactly. Should this not be the case, then one should wrap the stop-cock in linen or oakum so as to prevent the passage of air in or out.

The operation of this airgun.

So much for the construction of the airgun. It operates as follows: (1) Once the receiver has been completely evacuated, it should be placed on an open-ended stool, a basket without a bottom, H, or any other similar support. (2) The cone-shaped tube, e, of the pipe, abd, should be inserted in the stop-cock, c, of the receiver. (3) The leather disc, K, (which is made of cordovan leather and is stamped out by a special round piece of iron designed for this purpose and of the kind used by shoemakers), should be inserted in the brass valve in the opening, a. (4) Then a lead ball which fits the cavity of the pipe precisely, abd, should be inserted in the opening, d, and the airgun should be directed at a target.(5) The key of the stopcock, c, should be turned once very quickly. When this has been done, there is a sudden explosion. Thereupon, if another leather disc, K, is again inserted in the valve at point, a, and another lead ball is placed in the opening, d, and then if the key of the stop-cock is once again turned quickly, there is a second explosion. This operation can be repeated until the receiver gradually fills with air.

In this way, moreover, an explosion can also be produced by the use of a glass. This instrument, however, still requires improvement, for repeated testing indicates many changes that still should be made. Experience shows, for example, that the length of the pipe, abc, should be proportionate to the size of the lead ball which is to be ejected. I have observed, moreover, when the said pipe is $4\frac{1}{2}$. cubits long that the ball is ejected with a much greater force than when it has been shortened to a length of 3 cubits.

CHAPTER 30

EXPERIMENTS WHICH SHOW HOW AIR PRESSURE VARIES AT DIFFERENT ALTITUDES

Lower air is more compressed and therefore heavier.

Because that effect which was attributed elsewhere to an aversion for a vacuum has its origin in the pressure of air, it follows that the lower parts of the atmosphere, being pressed down upon by the upper strata, are compressed by this weight. Thus they are heavier or lighter according to the difference in altitude.

Air pressure is weaker on mountains.

As proof, it follows that on towers and mountains the so-called aversion for a vacuum is weaker, or, to put it another way, that air does not exert as much pressure on high places as on the lower parts of the earth. To perform our experiment, we should first place the glass container (see Plate VIII) in some lower place, a church, for example, or a cemetery, with its stop-cock open. Then it should be carried with its stop-cock closed, to the top of the steeple. If the stop-cock is opened there, air will flow out. Now if the stop-cock is again closed and the vessel is removed once more to a lower place, air will flow in when the stop-cock is opened.

It is apparent from these results that in high places air is not subject to as much pressure as in the lower regions and that consequently in the former, there is some decrease in the height and weight of a cylinder of air. It should be noted that when this experiment is undertaken, the temperature of the air should be the same within the church as without. For when it is warmer in the church than at the top of the steeple, the results of the experiment are not accurate. Thus it is a better technique to close the stop-cock of the receiver outside of the church in a place where the temperature approximates that at the top of the steeple, and when the stop-cock has been closed, to transport the receiver to the top of the steeple, or rather, to draw it up by a rope and then to carry out the experiment there.

An experiment revealing the variation of air pressure with the difference of altitude.

Others have tried to corroborate the validity of this experiment by recording the fluctuation of mercury in a glass tube on different heights of a mountain (we shall describe this in Chapter 34). In Paris M. Pascal discovered that at the foot of a mountain,mercury in a glass tube fell to about 27 inches. At a height of about 156 feet on the mountainside, the cylinder of mercury in the tube remained steady at 25 inches. Finally, at the mountain's peak, 206 feet above its base, the same cylinder of mercury registered about 24 inches.

Another experiment with mercury.

Now I undertook an investigation of this problem four years ago on the Brocksberg. I had good results at first with the experiment performed at the base of the mountain, but when the glass tube which was enclosed in a thin metal sheath was given to a servant to carry, he broke the glass in a fall and the whole undertaking was brought to nought.

No matter what these results may be, it is clear that their variations are caused by air. The air on the top of towers and mountain ridges is under less pressure from higher air than that in the middle or at the bottom, and for this reason it is lighter and makes the mercury less high in the cylinder.

It is truly remarkable, therefore, that the Hon. Doctor Deusing in his treatise *De Vacuo*, sought to discredit M. Pécquet who shared in our belief, by making the following objection. "If the strata of lower air are weighted down by upper layers lying upon them whereby they are rendered more compact and heavier, does it not follow that we mortals who live in the lowest region of the air and sustain on our heads the whole weight of the upper air, should perceive the weight of the air which lies above us and should be pressed down to the earth, etc.?" M. Deusing should bear in mind, however, that air does not only press down upon our heads but also circulates all around us. Thus, just as it presses from above upon our heads so it presses from below on the soles or bottoms of our

Dr. Deusing objects to the experiment.

Because air surrounds our bodies, we do not perceive it.

feet as it circulates around all parts of our bodies. As a matter of fact, the body itself is completely enveloped by air and all its interstices which would otherwise by empty, are filled with air. And, I might add, just as fishes in water do not perceive its pressure, we who live in air, perceive its weight even less.

CHAPTER 31

EXPERIMENT SHOWING HOW AIR PRESSURE FLUCTUATES ACCORDING TO THE WEATHER

The weight of a completely evacuated vessel hanging from a balance fluctuates.

I discovered from the experiment cited in Chapter 19 that air does not always have the same weight. I perceived the same result in the experiment with the receiver in which air is weighed (Chapter 21).For when the receiver has been exhausted of all air and has remained hanging in equilibrium for several days in this state, its weight fluctuates and we see that this glass globe is raised or lowered from time to time, something that never occurs with the counter-weight suspended on the other end of the balance. We have discussed this observation above in Chapter 20. On this account, I had a sphere the size of a $\frac{1}{2}$ recipient made of a sheet of copper and provided it with a valve similar to that described in Chapter 8 in place of a stop-cock.

Air becomes lighter.

The heavier the encompassing material is, the lighter the encompassed object.

I hung this, when it had been evacuated of all air, with a counter-weight made of lead, on a balance designed for this express purpose. This indicated with a marker the fluctuation in the weight of the sphere on a semicircular brass plate which was divided into intervals and was attached to the side of the balance. This fluctuated slightly on one day and from time to time on the second, third, fourth, and fifth days. When it rains, much water falls from the air, thus rendering it lighter. When this occurs, the evacuated sphere becomes heavy and sinks, On the other hand, it rises (or becomes lighter) when the air is heavier. One must conclude that the reason for this is that the heavier the encompassing material is, the lighter the object encompassed. Take wine vapor, for example, which is heavier than air. If the sphere is suspended with a counter-weight of wine vapor, it is lighter than it was in air. It is lighter, moreover, if it is submerged in pure water, and lightest of all if immersed in salt water, inasmuch as this is heavier than pure water. And thus it continues through many degrees according to the quality of the encompassing material. In this way it has been discovered that this globe becomes heavier or lighter depending upon whether the surrounding air is lighter or heavier.

Since in fact it is very time-consuming and extremely difficult to construct a sphere in such a way that no air can enter or penetrate it (for when this occurs with the passage of time, the increase in air becomes noticeable and leads to unlooked for errors) and to fabricate a balance provided with such delicate parts that it can indicate the slightest fluctuations more correctly and accurately in another way. (One should also consult the end of Chapter 20 on this subject.) As you know, I placed the small figure of a man carved carefully and in detail from wood in a glass in such a way that it was supported from below by air pressure and thus ascended and descended according to the variation in the weight of air. It indicated this fluctuation at certain points with its finger. Is it not possible for one to learn and forecast from this whether or not rain is going to fall on the earth and whether it is beginning or ending?

The wooden figure supported by air pressure.

See Plate X, fig. 4.

At this juncture I shall pass over the storms and violent winds which can be forecast by this device even if they occur one hundred and more German miles away. The defect in this apparatus lies in the fact that it does not indicate the locale of a storm and cannot be easily transported from one place to another.

CHAPTER 32

THE CAUSE OF SUCTION

Since the air enveloping the earth presses upon it so heavily, it follows that air penetrates and fills all cracks and interstices of bodies (unless the bodies are perfectly solid.) Now if this is the case, then air also fills the bodies and all the cavities of animals and in so doing it renders them unaware of its pressure. Once a body is filled, however, it takes in nothing more, but rejects it. However, a container outside a body which is not filled, but is empty, or at least partially empty, is attached or connected to it and this body is subjected to pressure (just as our bodies and all things existing in these lower regions are pressed by the weight of air), of necessity whatever is fluid in it passes into the attached empty container so as to fill it. For example, the human body is pressed on all sides by surrounding air, but once it has been completely filled,'the pressure of the encompassing air is imperceptible. When, however, an infant is placed upon such a body at the breast of its mother, for instance, and creates an empty space in its mouth by retracting its tongue,

Air penetrates all bodies unless they are solids or metals.

That which is pressed, yields when it finds empty space.

The cause of suction.

that is, when it sucks the fluid, or milk, which is near the surface, air is necessarily directed into the infant's empty mouth through the pressure on the body and breast.

The cause of suction with cupping glasses.

The result is similar with cupping glasses. The larger the opening in a vessel of this kind, the greater the pressure, which.we call suction. Therefore if an evacuated receiver, for example, having an opening in its stop-cock larger than that of a cupping glass, is applied to a body, the flesh is gradually drawn into its neck, until the skin which has been drawn up finally breaks.

The operation of suction pumps which we use in wells.

The hydraulic pump designed to raise water from wells is constructed on the same principle. (The suction mechanisms, or air pumps, in these wells draw up the water through pipes.) For when the piston which precisely fits the smooth inner surface of the pump is raised, it leaves an empty space. And through the piston's frequent and repeated movement, water rises into the pipe from the well up to the pump. This occurs not by reason of an aversion for a vacuum but because of the pressure of the air lying on the water in the well. If this resulted from an aversion for a vacuum, water would rise to any height whatever in the pump. This is not the case, however, (for as soon as this mechanism exceeds nineteen or twenty Magdeburg cubits in height, that is, when this suction pump, together with its valve is not less, but more than this height, no water can be raised and the theory of an aversion for a vacuum is rendered completely untenable). The ascent of water in the pipes, therefore, occurs by reason of the weight of air which forces the water up to nineteen or twenty cubits.

CHAPTER 33

EXPERIMENTS CONCERNED WITH THE EXPANSION AND CONDENSATION OR COMPRESSION OF AIR

The operation of thermometers.

Thermometers clearly demonstrate that air has the characteristic and quality of condensing or compressing, as well as becoming thin or expanding itself. If these instruments are cold, the air is contracted and occupies less space, but if they are warm, the air is expanded and fills or occupies a greater space. Additional support for this thesis can be seen in the pneumatic fountains which are set on banquet tables. By using a pump, we can blow so much air into them through an opening that they finally are broken. Furthermore, if no new air is introduced and only that is left which is usually present in them and fills all spaces, and if, (after all openings have been closed), a fire is set under them, these mechanisms are broken in the same way. Experiment similarly shows that in an airgun, air can be compressed into one fifth less space and this is done solely through the manual use of a pump.

Pneumatic fountains.

Air is greatly compressed in an airgun.

Since the envelope of air around the earth is pressed upon itself by reason of the latter's conserving virtue (see Book IV, Chapter 5), it necessarily follows that air, thus contained, takes on a certain weight and heaviness. For this reason it becomes compressed and condensed and the nearer the earth it is, the more it is compressed. For just as the more bodies that are placed upon one another and the greater the heap that is piled up, the more the lower bodies are pressed down. In such a way the lower strata of air nearest the earth are compressed by a heavier weight than the upper, a reasonable assumption that is confirmed in the foregoing Chapter 30.

Air is greatly compressed in these lower regions.

It is clear from the numerous experiments already performed that because of this attraction the entire mass of surrounding air exerts pressure upon itself and that it has a certain innate tendency to expand itself. (Some philosophers call this characteristic "elasticity", when the lower parts of the atmosphere, which in particular are under pressure from the upper parts, strive to extend themselves by reason of a natural tendency to expand, provided, of course, there is opportunity and space). This is especially evident in the experiment in Chapter 11 where air flows with violent force into a glass which has previously been evacuated. Furthermore the same result pertains with the machine described in Chapter 18, number 6 when I happened to forget to evacuate the upper glass and thoughtlessly opened the middle stop-cock that was fastened to both glasses. This resulted in the air's rushing into the lower glass (which was evacuated of air and partially filled with water) and struck the water with such force that I was frankly fearful that the whole apparatus would be fragmented.

Air has a natural tendency to expand itself.

Now if a person does not wish to believe that air has this capacity for expanding itself (for there are many who reject this hypothesis), he should place a bladder full of air inside of two copper hemispheres of the kind mentioned in Chapter 23.). Then if he pumps the air out from these attached hemispheres, he will find that the bladder rup-

Another experiment with a bladder.

tures very soon after the start of the operation. Again if one takes a bladder of this kind, fills it half full of air, and suspends it in a receiver (as was done in the experiment with the watch in Chapter 15), he will discover that the more air that is taken from the receiver, the more the bladder swells until it finally is destroyed completely.

Air can be extracted from a container although the evacuating apparatus is placed under the roof of a house.

I have also discovered that if an air pump is placed under the roof of a house with a tube connected to it which extends down to the ground level and is inserted into a receiver or some other vessel that is meant to be evacuated, air rises into said air pump because of its expanding nature. This observation is clearly apparent to a person who is occupied in investigating these things.

From all the foregoing examples, it is clear that the lower strata of the atmosphere are compressed more than the upper. One must conclude, therefore, that the more space air is given, the more it expands itself.

What is the origin of air pressure?

One must further investigate for what reason this compression of air occurs. Some believe that it has its origin in the pressure of its weight toward the center of the earth (to which they assume all things gravitate). Others, however, lay the cause to the radiating light of the stars which streams from all sides and exerts external pressure.

But now why do they attribute to individual things a weight or pressure and a natural tendency to direct themselves towards the center rather than accord this property to the conserving virtue of the earth? Indeed, anyone can see from Book IV, Chapter 5 that the capacity for this kind of attraction exists in the nature of things and that is why this inclination downwards should not be attributed to things in themselves, but to a certain common virtue. Furthermore, this force which we assume compresses air, does not originate in the astral bodies above us, for if this were the case, the earth would be subjected to this force as a body standing in the way, and would sustain the pressure of the stars. When, however, two bodies press against one another, a body placed between the two is subjected to equal pressure on both sides. For this reason the lower strata of air would necessarily receive the same pressure as the upper, something that experiment contradicts.

The height of the atmosphere is insignificant compared to the distance of the stars.

Finally, since more pressure is exerted on lower than higher air, and since we are aware of this not only on very high mountains but also on church steeples, it follows that air does not extend upwards from the surface of the earth very far and its height is insignificant compared to the great distance of the stars. The boundary or limits of the atmosphere is nothing other than pure and empty space, a subject to be further dealt with in Book V, Chapters 6, 7, 8, and 9.

CHAPTER 34

AN EXPERIMENT DEMONSTRATING PROOF OF A VACUUM THROUGH THE DESCENT OF MERCURY IN A GLASS TUBE SEALED AT THE TOP

When I attended the Imperial Diet.in Regensburg (as we mentioned above in Chapter 27,) among other things, I demonstrated certain experiments of mine to the electors, various princes and legates, and on this occasion I became acquainted with the Right Reverend Capucchine Father, Valerianus Magnus. He performed an experiment he had devised, so he said, for demonstrating a vacuum. The procedure was as follows: (1) He took a glass tube which was longer than 6/4 of a Magdeburg cubit and completely sealed one end. (2) He filled this tube with mercury and inverted it (the opening was tightly blocked by his finger so that nothing could escape) and submerged it, with his finger still in place, in a small container filled with mercury. (3) He withdrew his finger gradually, and behold! the mercury descended little by little in the tube until it reached a certain height, actually about 5/4 of the height of a cubit. Provided only that the finger was not withdrawn suddenly, it remained unmoved at this height and descended no lower in the tube. If the finger was quickly removed, however, then the mercury fluctuated at different levels, now ascending and now descending before it came to rest at the same point.

The experiment of Father Valerianus which demonstrates a vacuum through the descent of mercury.

The Reverend Father Valerianus stated that the space in the upper part of the tube which contained no mercury was properly, and in a real sense, a vacuum since no other body in the meantime could have filled the place of the mercury that had descended. He gave me his book entitled *Demonstratio Ocularis*, (1) of a space without something contained in it, (2) of a body moved gradually in a vacuum, (3) of light which adheres to no body. Now although I gathered as much information from this book as I did afterwards from other writers, I saw that this experiment had first been undertaken by the renowned Evangelista Torricelli, mathematician of the Grand Duke of Tuscany.

As for my assessment of this very significant experiment: I have no doubt that the small space in the upper part of the tube which contained no mercury was empty of essentially all air, although, in fact, it contained some air because a few bubbles always remained visible clinging to the inside of the tube after the mercury had been poured in. (It should be noted here that it does not seem possible to introduce mercury into the tube without them.)

The opinion of the author concerning this experiment which produces a vacuum by means of mercury.

Later on, when I prepared glass tubes at home in the same way and repeated the experiment several times in order to produce a satisfactory vacuum, I set up the apparatus in such a way that I could join one of these tubes to the evacuating pump that has been described in the preceding Chapter 4.

See Plate VII, fig. 5.

First of all I filled the tube a third full of mercury and then connected it to this pump by means of a small pipe (which we mentioned above in Chapter 8). Now I could see clearly that bubbles rose into the upper empty space and that at the same time, the mercury in the tube rose perceptibly with these. The mercury that was poured in, however, did not always remain at the same height but sometimes sank the thickness of a stalk of straw. And when I poured in some more mercury in the same way as before and extracted the air in the manner already described, a similar result occurred. It was clear after this procedure had been repeated for a third time that all the mercury contained in the whole tube fell about a half finger's width. From this it was quite evident that so much air is always left between the mercury and the sides of the tube as is essentially the width of a half finger. Therefore a vessel can be evacuated of air more completely by means of our method of pumping it out than can be done simply with mercury alone.

Now these bubbles of air which cling to the inside of the tube are in no way without importance, nor should they be compared with those greatly expanded bubbles which arise from the water in evacuated vessels and increase ten, nay, one hundred times by reason of the empty space in such vessels. In the first place these bubbles in the mercury are not only under considerable pressure from the outside air whose weight is equivalent to twenty cubits of water, but in addition are under even greater pressure from the mercury itself.

Thus when these enclosed bubbles are drawn out or are evacuated through the agency of an air pump, no one can assert that a more perfect vacuum can be made or produced in this glass tube than by an evacuation of the kind described.

But indeed, this very small space, empty of air, ought not be used for other purposes.

The experiment performed in longer tubes with water.

Finally, it should be remarked that the Reverend Father Kaspar Schott asserts in his *Mechanica Hydraulico-Pneumatica* on page 308, "that some men (when they heard that various scholars had sought to prove that a vacuum was produced by the aforementioned experiment), tried to accomplish this same thing by using water instead of mercury in much longer tubes. They found in the upper part of the tube that did not contain water, however, that the space was not completely empty. For this reason they attached a glass vessel to the upper end of this same tube, so joined to the neck of the tube that no air could possibly enter, and within this container they cleverly fastened a bell with an iron clapper to the sides of the vessel in such a way that the little clapper, attracted by a magnet outside of the glass, was able to strike a very clear sound that was heard by all who witnessed the experiment. This certainly could not have happened in a vacuum, for a sound such as this could not have been propagated through an empty space. Father Melchior Cornaeus confirmed this result in his discussion of the earlier experiment which he included in his *Cursus Philosophicus*.

I subscribe to this explanation so far,for if they heard a sound of ringing, then there was no vacuum present since ringing cannot be reproduced nor travel through empty space. On the other hand, another kind of noise or sound can be propagated through a vacuum, as the experiments in the above Chapter 15 prove.

CHAPTER 35

COMMON OBJECTIONS TO THE EXISTENCE OF A VACUUM AND THEIR REFUTATION

The Reverend Father Kaspar Schott in his *Mechanica Hydraulico-Pneumatica* on page 25 describes certain experiments by which scholars endeavor to prove that there is no vacuum.

The first objection arising from the drawing apart of the sides of a bellows.

1. "The sides of a bellows cannot be drawn apart by any force, even through superhuman power, if the opening is sealed through which air is to enter and fill the space between the sides to be separated. Why should this be the case unless it is that a vacuum cannot be produced? Nonetheless this procedure would be possible if air or some other body were excluded when the opening was sealed and yet the sides were drawn apart."

The reply.

Reply: If the sides of the bellows are sufficiently strong and the leather thick enough, there is nothing to prevent these bellows from being separated, nor would superhuman strength be required. For when the lower side alone is firmly attached to the ground, the upper side can be raised with the hands, or, if the bellows happen to be rather large, with the help of a lever. As a matter of fact, through the computation given in Chapter 22, a clear-cut answer can be given as to how much force is required to draw apart any bellows, since the force is proportional to its size. And if scholars concede at this juncture that a vacuum is produced (if air is excluded when the opening is sealed and yet the sides are separated), then everything is in order. Actually a better and clearer experiment is seen with the air pump, gh, pictured in Plate VI (ERRATUM read "Plate VI" instead of "Plate IV"), figure 1. When its opening, w, has been closed, the piston rod, nonetheless, with its piston must be set in motion under the water and drawn down. No air can enter this space since not only is the opening closed, but also the operation is performed under water and anyone can see with his own eyes that neither air nor water is able to enter. In this way a vacuum is produced in the pump.

2. Father Niccolo Zucchi in his discussion *De Vacuo* says, "A brass plate was once placed before a young man who took great pride in the strength of his muscles. He was instructed to raise the plate by means of a handle projecting from its middle from a marble table to which it conformed perfectly. At first the youth scorned this as a ludicrous task for a young man to do, but when, in the presence of his friends, he strove for a long time to lift it up with both hands and was unsuccessful, he attributed his lack of success to the presence of a foreign and very strong glue endowed with such strong cohesive powers that he could not pull it away. But then he saw that another youth solved this problem very easily by bringing the plate to the edge of the table, tipping it, and thus removing it.

Reply: Had we been apprised of the size of the surface of this plate, we could easily have found how much force would be required to lift it from the calculation given in Chapter 22. In this way we could have refuted the supposed aversion of a vacuum.

3. "If you immerse a glass container in water, fill it, and then invert it so that the open end is turned downwards in the water, and then raise it in its inverted position and try to draw it out, you will see that the water contained in the glass is raised at the same time. And if you hold this container in your hand, partially drawn out and partially immersed, you will see the water, enclosed in the glass, remain raised above the surface of the surrounding water in equilibrium and even exert pressure, because of its weight, upon the hand holding the glass. Why is this? This occurs because air cannot succeed in entering the space between the descending water and the top part of the container which is made empty by the descending water."

Reply: If that glass container were longer then twenty cubits, you would see the water enclosed in it descend with so much space remaining as the height or length of the glass exceeds twenty cubits. And because air then cannot succeed in occupying the space between the descending water and the top part of the container which is left empty by this descending water (since the open end of the container remains immersed in water), for this reason, then, a vacuum is left there.

4. "When those vessels that are commonly used for watering gardens are filled with water and are lifted up, the water does not flow down (although their bottoms are perforated and pierced with many holes) as long as a finger pressed.on the opening in their necks prevents air from entering and filling the space which, if the water flowed down, would be left between the upper surface of the vessel and the descending water. Under these conditions the water remaining in the vessel weighs as much and exerts as much downward pressure as it would if it were contained in a vessel with a non-perforated bottom and open neck. When the finger is removed, however, and air is permitted to enter through the top opening, the water immediately flows out of the holes in the bottom. But if a finger again obstructs the opening when the water has half flowed out, once more the water remains suspended, despite the natural pull of its weight which even then exerts pressure on the hand, instead of issuing out into the open air."

The fourth objection arising from an experiment with vessels used for watering gardens.

Reply: Now air with its often mentioned weight also surrounds this vessel made for watering gardens. When the opening in the top of the neck is blocked up, water cannot flow out through the holes in the bottom because it is checked by atmospheric pressure (unless, of course, the water happens to be twenty cubits high, in which case it exerts more pressure than the external air). Thus it is not the aversion for a vacuum that holds the water in the vessel, but rather the weight of the surrounding air. When the finger is removed, however, air can enter through the opening in the top and then water flows out from the holes in the bottom because of its own natural weight.

The reply.

5. "If you insert one end of a tube in water and suck out air with your mouth from the other, the water immediately is raised, in spite of its weight, and follows the air that has been drawn out to the mouth of the person who is sucking. Therefore no vacuum is present in the pipe when the air has been extracted."

The fifth objection arising from suction.

Reply: You should conclude that this elevation of the water in the tube results from the pressure of external air upon the water which forces it into the empty space of the tube, provided only that the height of the water is not more than twenty cubits. Were this the result of an aversion to a vacuum, however, the water would rise infinitely high and fill the pipe.

The reply.

6. "If you move a copper globe with strong sides, so that it does not easily collapse, near a fire and heat it, the air contained within it is greatly expanded, and then if you immerse it in cold water, the air in the globe that had been greatly expanded, again contracts and occupies less space. Then water is drawn into the container because

The sixth objection arising from an experiment with a globe being moved to a fire and with cupping glasses.

no other air can take up the space that has been left empty lest a vacuum be present in the sphere. In the same way if you suck air vigorously and without interruption from a pipe whose one opening has been closed and the other open, or from a well-made hollow stop-cock and then when you have stopped, if you immediately place your finger or lip on the unobstructed opening of the pipe so that the surrounding air cannot enter, the flesh of these bodily parts is drawn into the pipe or stop-cock, an action which results in the latter's perceptibly pinching and hanging from the flesh. This occurs for no other reason than that the air remaining within the pipe becomes rarefied while the sucking is going on and effectively fills this whole space by itself. When the sucking stops, however, the air ceases to rarefy actively and thus, in occupying less space, draws in that which is nearest to it so as to fill the space that has been left empty. It is for the same reason that the cupping glasses which surgeons are accustomed to use, draw up flesh. For the air, which had previously been heated by fire undoubtedly becomes less dense when the fire is extinguished and as it turns cold and becomes condensed, it contracts into a smaller mass. Thus it leaves an empty space into which flesh is drawn up so as to fill it."

The reply.

Reply: The results of these experiments with pipes and cupping glasses which draw up the flesh with which they are in contact when air has been extracted, have no need of a more detailed reply than that presented in Chapter 32.

The seventh objection arising from the piston which is withdrawn with difficulty from a gun.

7. "If you insert into the barrel of a bronze gun, or a larger piece of artillery like a cannon, a piston that fits the inner surface so precisely that air cannot enter,slipping between the surface and the piston, and then, if you block up the muzzle of the gun, the piston can only be withdrawn with difficulty. Once the piston has been pulled out by the exertion of considerable force, however, it is again drawn back into the pipe."

The reply.

Reply: The same thing happens every day with the piston of an air pump. The reason for this should not be attributed to nature's abhorrence of a vacuum, hut rather to the weight of the cylinder of air which holds back the piston. This air exerts such force that the piston is not only withdrawn with difficulty but even seems virtually impossible to move when the diameter is greater.

CHAPTER 36

THE OPINIONS OF THE REVEREND KIRCHER AND ZUCCHI AT ROME, AS WELL AS FATHER CORNAEUS, PROFESSOR AT THE UNIVERSITY OF WÜRTZBURG, CONCERNING THE MAGDEBURG EXPERIMENTS

The first Magdeburg experiment was shown at Regensburg in 1954.

At the Imperial Diet held in Regensburg in the year 1654, when certain enthusiastic followers of these investigations sought to have me perform several of my new experiments, I was unwilling justifiably to disappoint them. These were the last days of the Diet, however, when it had already begun to be dissolved and the large group of artisans were no longer on hand who, through their efforts and skill, were needed to make and set up the pieces of apparatus again.

The Noble Elector of Mainz brought this apparatus to Würzburg.

It was pleasing, nonetheless, to the most eminent Noble Prince, Johann Philipp, Archbishop of Mainz, to have this apparatus transported to his seat at Würzburg. There the Reverend Kaspar Schott (as he himself records in the Magdeburg Experiment" which he appended to his *Ars Mechanica Hydraulico-Pneumatica*) in the presence of the same most eminent Prince, not once, but on different occasions studied the apparatus, tested it, wrote about it, communicated with scholars at Rome and elsewhere, and sought their opinions. Among this number there were in particular Father Athanasius Kircher, Father Niccolo Zucchi at Rome and Father Melchior Cornaeus at Würzburg and others, as is evident from references in the abovementioned *Mechanica Hydraulico-Pneumatica*.

The diverse opinions of different scholars.

In the first place it is thought that the piston of an air pump finally cannot be moved.

In the first place, all these men came to the same conclusion: "the piston of an air pump, after a number of repeated movements in and out which progressively become more difficult to perform, can no longer be pulled within. the interior of the pump, with the result that one might as well desist from further effort and cease his exertions since the piston cannot in any way be moved further." Now this conclusion is in error, for the piston can in fact be pulled out, although it is done at the end with great difficulty. And furthermore, the larger it is, the more difficult it is. The required effort can be computed as proportional to the piston's diameter, as is explained in Chapter 27.

The reply.

Secondly they believe that this difficulty lies finally in the rarefaction of the air.

Secondly they ask, "What is the cause of this great difficulty?" And they reply, "After the extraction of air, a vacuum is not present. For this would be the complete absence of any body previously present in the vessel, in other words, vacuity, and would be subject to no resisting force. A body, therefore, remains in the vessel which is not further rarefied by the force of the pump's movement, but which has been reduced to such a degree of rarefaction that a further rarefaction is not possible through continuing exhaustion by the use of such force." Or to use the words of Father Cornaeus, "because the air gradually becomes so thinned that it arrives at the farthest bounds of rarefaction and cannot be further extended, expanded, or rarefied." I reply that the Reverend Fathers of the Society of Jesus, in other regards most learned men, wish to ascribe this resistance to the air which has been left inside (of which, they admit, hardly any, or at least, very little, remains). With no offence intended, I hold this to be in error, for the small amount of air which they say is left cannot exert more force than a large amount, had it been left inside.

The reply of the author.

For this reason, the difficulty of pulling out the piston should be attributed not to a very small amount of air remaining in the vessel, but to the external air which by reason of its own weight, holds back the piston with great force.

The third objection that an evacuated vessel can never be completely filled with water.

In the third place, Father Cornaeus recalls in numbers 5 and 6 "that water entering an evacuated glass vessel draws in with it foam and countless bubbles and that for this reason the container is never completely filled with water." I reply that the reasons for this are to be found in the experiments of Chapter 6.

The reply.

The fourth objection speaks out for a vacuum.

Fourthly, Father Schott on page 472, number 7 of his aforementioned *Mechanica Hydraulico-Pneumatica* concedes that a vacuum does exist but adds the following observation, "for when a piston rod equipped with a piston is drawn from the bottom of the tube, A, towards B up to point D, of necessity the space between A and D must be empty of any body, provided that air does not enter from the receiver. Therefore if the stop-cock of the receiver is held closed, the piston rod cannot be drawn by any force at all from A to B, etc." Every day, however, experiments show that the piston rod can be pulled out, although the stop-cock of the receiver is held closed. Therefore that space between A and D is indeed empty of any body.

Fifthly, he asks in the same reference, number 8, "When and how is this vacuum in the receiver made? Is it total in the entire receiver at one time, or does it develop gradually in different parts of the receiver?" I reply that the receiver is gradually emptied of air, not only because of its natural tendency to expand, but also because its natural weight (since the receiver you see is situated in a higher and the pump in a much lower place) that finally makes all air descend into the pump. One should distinguish, moreover, between the rarefaction of air which comes about through heat and the action of fire, and that which occurs from pumping. For the rarefaction which

There is a great difference between a vacuum made by a pump and that created by heat.

is produced by warming (even when you heat a glass vessel until it glows) leaves a fifth part of air in the receptacle, while that produced by pumping, leaves scarcely a fiftieth, or even a hundredth part of air.

In the sixth place, as to the statement made in number 9 about the very deep well dug at Paderborn, namely that "its pipes, although they were of exceptional thickness and were made from the trunks of trees, burst with a terrible noise," I know that this happened for reasons other than those given, because the valves had been too large so that they allowed for the passage of water in the hollowed out tree trunks. I am aware of several other instances of this phenomenon. It is certainly not true, however, that a pipe can be smashed from the drawing in or suction of a pump; it would rather cave in.

The objections opposing a vacuum should be refuted more through experiment than words.

So completely do the various experiments themselves refute the other objections cited here to the existence of a vacuum that there is no need for a more detailed reply. The Würtzburgers, moreover, have not carried out their experiments with appropriate apparatus and for this reason there are many defects and errors in their results.

In conclusion, there are still many other scholars who have written critically of these experiments even before they have seen and studied them. Among this number is August Hauptmann, Doctor of Medicine. In his *Bergbedencken* published in 1658 at Leipzig, he states "that no angel or devil could manage to produce a vacuum. Indeed, if there were such a pump ("vinculum", as he terms it) in nature which could not be broken apart, water would rise from all depths, yes, even from the very center of the earth, and if this "natural pipe" ("filum naturae") (operated presumably in the manner of the air pump in the Magdeburg experiment) were properly set up, heaven itself and whatever else were required to fill it, must bend and sink down, etc., etc . . ." Now we think it is unnecessary to reply to this and other attacks of this kind, for more credibility should be attached to experiments than to ignorance which always is wont to set up preconceived notions against what is evident in nature.

The error of Dr. Hauptmann.

CHAPTER 37

A NEW THERMOMETER, SO-CALLED MAGDEBURG

Thermometers are known to be made in different ways. In some, the rising and falling of water in glass tubes arising from the varying temperature of the air, indicates the increase or decrease of heat and cold in the atmosphere. The different degrees of heat and cold are marked on the glass tubes themselves or on attendant tables. We have already mentioned these thermometers above in Chapter 11. We, however, have devised another method which the Reverend Father Schott describes in his *Technica Curiosa*, Book 11, Chapter 13, page 871 in the following words:

By the Magdeburg thermometer one can ascertain what is the coldest and warmest day in the whole year.

"The ingenious originator of the Magdeburg experiments which have been explained above in Book I, part 1, discovered another somewhat esoteric and far more clever device for indicating the degrees of heat and cold of the surrounding atmosphere for the whole year. At the same time he ascertained the warmest and the coldest day in the course of the year, something that could barely be detected if it was at all, by ordinary thermoscopes or thermometers. The inventor displayed this device on the wall of his home in a place where it could not be reached by the sun's rays. It was designed as follows:

See Plate XVII, fig. 1.

"A is a very large copper globe with a hollow interior, of a size comparable to that of a glass receiver. BC is a copper tube, essentially seven cubits long and a thumb's thickness in width which hangs down from the globe. Alongside of this there is another copper tube, DE, which is joined to it at point C on the bottom. As much wine spirits is poured into this tube as the larger or smaller size of the tube's diameter warrants. Into this is inserted a small tube, KL, $\frac{1}{4}$ of a cubit long, made of a very thin sheet of brass, sealed on all sides, thereby protecting it from the penetration of the liquid. Into this container numerous small lead pellets had been placed before it was closed so that it would be in equilibrium with the liquid and, being suspended, would float on its surface. A waxed string, EM, is attached to this small tube, or weight, and passing out of the tube, ED, it runs over a pulley, F. On its other end there is attached a small figure, G, in the form of an angel or small child made of a thin piece of brass with its hand extended towards the tube in an attitude of pointing. The two aforementioned tubes, BC and DE, are covered by another tube or casing shaped like a prism and made from thin sheets that had been joined together, so that the two aforementioned tubes would not be visible and the thermometer's mechanism might be hidden from onlookers. On the outer surface of this triangular casing the aforementioned small figure moves slowly up and down, indicating with its extended hand the degrees of heat and cold which have been recorded on its surface. See figure 2.

"The globe, A, has a valve, H, on its side which the experimenter calls its navel. Through this, with the help of an evacuating container

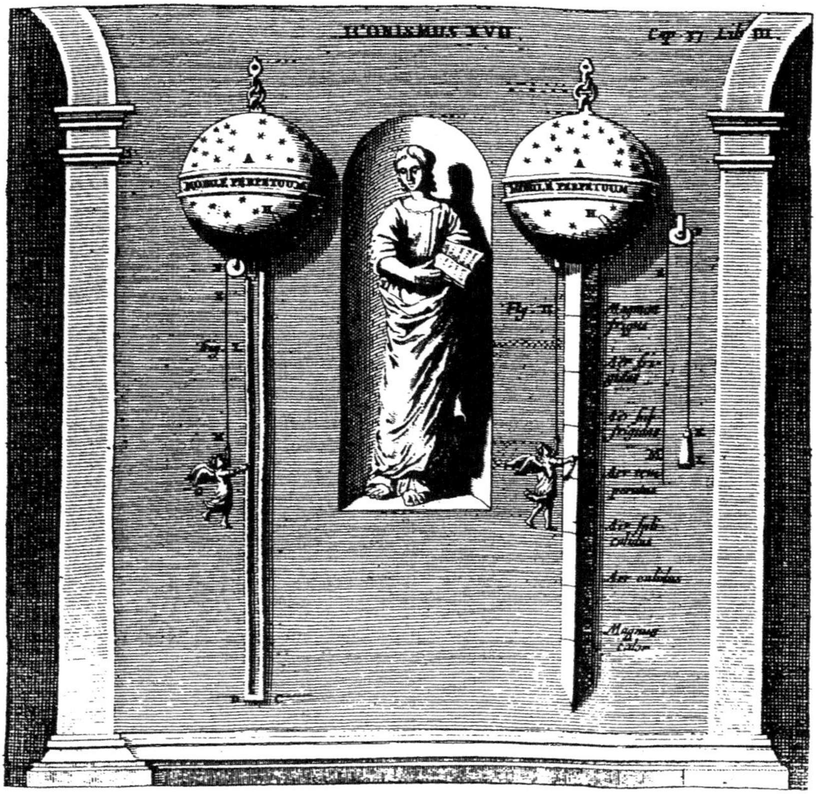

PLATE XVII.

described above in Plate VIII, figure 3, or of another vessel that is frequently used for evacuation (such as we have described in the same place), air must be drawn from the globe, A, until the little figure is suspended at a suitable place on the degree indicator. If one does not know the suitable place, he should extract air when frosty and chilly nights are at hand. For at such a time, the little figure should register about the middle of the scale of degrees recorded on the external casing. Thus, at this time of moderate cold, so much air should be extracted from the globe that the little figure proceeds to this place. Experience and practice will show all this more clearly.

"The designer of the Magdeburg experiments, as I have said, has this globe suspended throughout the year from the wall of his house in the open air in a place where the sun never touches it. It is covered with a blue coating, is studded with gold stars, and bears this inscription, 'PERPETUAL MOTION'." See the aforementioned reference cited in Father Schott's work.

Another Method

I also devised another type of thermometer that has not been known before. When a small figure (such as craftsmen are accustomed to blow from glass with an attached bubble) is inserted in a glass tube two or three cubits long, and when wine spirits are poured into it, then (if this figure is set up as we have suggested in Book IV (ERRATUM, read Book IV instead of Book III), Chapter 4, so that it can hang freely suspended in these spirits), it will indicate warm and cold weather. Indeed, when it is warm, the figure will descend, and when it is cold, it will rise, while in moderate temperatures, it will seek the middle of the tube.

An Additional Method

Hang the glass vessel which is called a receiver (with its stop-cock open) on a balance and you will find that it is lighter in warm weather and heavier in cold. The reason is that it holds more air in cold weather and less in warm.

BOOK IV

MUNDANE VIRTUES AND OTHER ALLIED SUBJECTS

CHAPTER 1

A GENERAL DISCUSSION OF MUNDANE VIRTUES

In our preceding Book II, Chapter 7, we wrote that everything that is created, and earthlike bodies in particular, consists of their own matter, life, and innate virtues.

These virtues are neither substances nor accidentia (non-essential qualities) but are effluvia of mundane bodies which are inherent in them and flow out from them. They should be differentiated, nevertheless, insofar as they originate in one body and flow out from it, or to such an extent as they enter into another and are absorbed by it. They are accidentia of that body into which they flow and are innate virtues of that body from which they emanate. They cannot be lacking in these bodies without the greatest harm and even destruction to their form.

Innate virtues are neither substances nor accidentia.

They are called mundane virtues because they are particularly associated with earthlike bodies, that is to say, the planets, to be sure, and the earth, the sun etc. (as well as parts of the earth). See also Book VI, Chapter 7.

Why they are called virtues of the earth.

Of these, some are corporeal, others incorporeal. A corporeal virtue is a kind of tenuous radiation which flows out from an earth body, not indeed into infinity, but to the bounds of its own sphere which is surrounded by Space. It is called corporeal for the reason that it does not pass through hard and solid bodies, like glass and metal etc.

They are either corporeal or incorporeal.

What a corporeal virtue is.

Air, indeed, is a corporeal virtue of the earth, in other words, it is a material effluvium of all things on earth. We have discussed this in our preceding Book III, Chapters 2 and 8. We perceive this effluvium by its smell, as was mentioned in Chapter 10 of the aforementioned Book. If we did not have an organ of smell (which is located in the anterior part of our heads), we would be unable to perceive or distinguish these corporeal effluvia and thus no smell or any perception of smell would exist.

Through our organ of smell we perceive corporeal virtues.

What an incorporeal virtue is.

As for an incorporeal virtue, this is an extremely rarefied effluvium emanating from a body which extends and diffuses itself in all directions to a finite distance and penetrates all solid and hard bodies (in varying degrees according to the different characteristics of a particular body as well as its own). This expansion we call the sphere of virtue or the sphere of activity.

What the sphere of virtue or the sphere of activity is.

Spheres of virtue are of greater strength and density near the bodies themselves.

Every sphere of virtue, whether it be corporeal or not, is more dense, more compressed, and of greater strength near the body from which it issues than at a greater distance where it becomes progressively more expanded and rarefied until finally it vanishes into nothingness.

There are many different virtues.

Without a doubt there are many different incorporeal virtues which we do not recognize, however, because of the inadequacy of our sensory organs to perceive them. Those that do impinge upon the senses, through which we come to recognize them, have their origin either in the earth or the sun.

The incorporeal virtues coming from the earth.

Thus, there come from the earth (1) the impulsive virtue (virtus impulsiva) (2) the conserving and expelling virtue (virtus conservativa) (3) the directing virtue (virtus directiva) (4) the turning virtue (virtus vertens) (5) the sounding virtue (virtus sonans) and (6) the heat-making virtue (virtus calefaciens), etc.

The perceptible virtue emenating from the sun is light.

Of those incorporeal virtues which come to us from the sun, we clearly perceive the virtue of light, color, etc. (virtus lucens et colorans). It is apparent for various reasons, moreover, that the chilling virtue (virtus gelufaciens) has its origin in the moon.

The vitrues of the planets influence the earth.

There are undoubtedly many other virtues directed toward our earth from the planets whose influence corresponds with their various locations relative to the earth. Astronomers call these virtues influences (influentia) whenever we can identify them with some degree of certainty.

These virtues spread out.

It is the nature of these virtues to spread out. Thus homogeneous virtues attach themselves to adjacent bodies and at the same time mutually affect and influence one another etc. If they come upon a body that is unresponsive to or incapable of receiving such a virtue, however, they go no further and are turned back. The smoother and more polished they are, the more numerous and stronger are the resulting deflections which at length gradually disappear, their virtues having been dissipated. This discussion will be continued at greater length later on.

Virtues have the characteristic of being deflected.

The more matter a body happens to contain, that is, the more solid it is, the more virtue it can take in.

Every body, however, is not adapted to or capable of admitting each and every virtue but only insofar as a virtue is so disposed does it come to occupy an abode of this kind and possess its matter, both with respect to entering and leaving it. There are, however, certain incorporeal virtues that are capable of entering all bodies.

How virtues are excited thorugh rubbing.

These virtues can also be excited through the rubbing, collision, touch or vibration, etc. of those bodies which are capable of producing or absorbing this or some other virtue. Because, indeed, a virtue of this kind lies within the sphere of the earth's virtue, it can be excited through the touch of a responsive body, as the experiments in Chapter 15 will show.

CHAPTER 2

THE INCORPOREAL IMPULSIVE VIRTUE OF THE EARTH

What the impulsive virtue is.

The impulsive virtue is an incorporeal virtue which is present in the earth and consequently can be excited in any similar body that is durable and firm through the motion of "latio" and "circumlatio" (direct and circular translation), through vibration, or "jactatio" (a hurling motion), or through turbulent movement (in other words, when a body is forcefully projected or propelled in flight or made to spin about vigorously or is pulled or driven away, etc.). When this happens, a mundane virtue of this kind is impressed on it by which it is carried from place to place (or here on the earth, through the air). The more vigorously it is propelled, the more force, and consequently the more speed, it attains and the longer its "latio" lasts, other conditions being the same.

The impulsive virtue does not last forever unless its stimulation continues.

An impulsive virtue that has been impressed on a body vanishes when its source has been removed and for this reason, it does not persist, even though there is no air or similar obstruction to prevent its vanishing. It lasts until this virtue that was impressed gradually dies out or that receptive force weakens. For example, the virtue (which we term "impulsive" here) is imposed on a stone when it is thrown upwards by the hand. This motion impels the stone upwards from the earth until the virtue dies out. Then through another virtue, the conserving, as it were, (see the following Chapter 5), the stone is brought down. When this occurs, by reason of the acceleration of the motion of "latio", the stone again and at the same time acquires the impulsive virtue which increases more and more as it approaches the earth. Then when it falls on a hard body similar to itself, and indeed one that is appreciably larger, this virtue is again deflected in the opposite direction, as when a stone rebounds, not once, but many times, after it has fallen. When boys play with marbles, for example, and allow one to fall from a marble table, the marble rebounds, not once, but ten times and more. All this substantiates what we said in the previous chapter: all these virtues have a reflecting characteristic.

Virtues have a reflecting characteristic.

As long as an impulsive virtue lasts, the effect also persists. Thus any body is capable of acquiring greater and greater speed for the duration of the violent motion of "latio" until it is filled with this virtue. (For a distinction between motion and translation, see Chapters 15 and 16 of the following book.) The impulsive virtue, however, cannot increase infinitely for it has its own limits defined in every body according to the size and density of the body.

The impulsive virtue is increased by violent motion.

Now the larger a body is and the denser and more compact, the more it can absorb the impulsive virtue. Because all things finally define their own limits, however, those writers are in error who have written that the acceleration of motion of falling stones and other similar objects increase infinitely and for this reason have extended arithmetical progressions farther, as if a body is capable of acquiring an infinitely greater speed. Now, it is true that it has frequently been proved experimentally that heavy bodies, in their natural descent, are moved to greater and greater speeds with such an increase of velocity as exist between uneven numbers beginning with one, as 1, 3, 5, 7, 9, etc. Thus, if a heavy object falls one stade (about an eighth of a mile) the first minute, in the second it will have covered three and in the third, five stades.

The acceleration of falling stones does not increase infinitely.

One should not assume, however, that this increase continues to infinity so that, according to this projected calculation, a stone could fall 17,916 diameters of the earth in two ordinary days, or, as Father Kircher writes in *Ex Mersenne et Scheinero*, "A stone released into the sky and directed toward the earth would need only six hours to cover this distance." (See Book VII, Chapter 4, number 12.) If this were true, the motion of "latio" would be accomplished in an instant, that is, without any limit and with infinite speed. Since, nevertheless, all things have a limit or end, velocity, too, or the impulsive virtue, must be contained within bounds of its own. It still remains a fact that the greater the amount of a certain substance of the same volume and hardness is contained in a body of a certain strength and hardness and the larger or heavier a body is, the more it will respond to a greater impulsive virtue and velocity. For example, the larger of two bodies composed of the same material and of equal solidity falls more quickly from the same point than the smaller one because the larger body has more substance and is therefore susceptible to a greater impulsive virtue and impetus. Thus a lead ball of two inches in diameter falls much more quickly and strikes the earth sooner than a ball of one inch.

The larger a body is, the more it will respond to a greater impulsive virtue.

Therefore a larger bow can never shoot a smaller arrow than is appropriate, to as great a distance as an arrow of ordinary size. The reason is that a smaller arrow is not capable of so great an impulsive velocity as that imparted to it by a larger bow. In the same way a small stone cannot be projected to so great a distance as one that is in proper proportion to the size of a sling. This is also true of a lead ball attached to a line that is rotated or swung around in a circle. The larger the ball, the more rapidly it moves and the larger the circle it describes. A larger ballista, moreover, can similarly impart a greater impulsive virtue to a larger ball than a smaller one. Thus a smaller ball can never be projected to as great a distance as a larger one. One should remark, however, that the balls must be of similar material, for an iron ball, by reason of its composition, is not similar to a lead ball, and much less like one of gold.

A larger ball can be projected farther than a smaller one.

The impulsive virtue cannot be generated in water to the same degree as in air, both because of the resistance of the water as well as its increased weight with added depth. Thus stones and other heavy objects that have been thrown into the water do not accelerate the motion of "latio" in falling and similarly this holds good for light objects in rising to the surface.

The impulsive virtue cannot be excited in water to the same degree as in air.

When the impulsive virtue is quickly transmitted from a hard body to a fragile object, glass, etc., for example, it fractures and shatters it.

If there were no impulsive virtue, a nail could not be driven into wood.

There is no need to demonstrate the existence of this virtue with many examples. Anyone can recognize it in his own body if he runs somewhat rapidly and finds that he cannot stop at once because the impulsive virtue he has acquired prevents him. In the same way all things that are transported in a ship or cart share in this virtue along with them. Thus, when a stone (or ball) is thrown straight up from either of them, it acquires a two-fold force, one from the ship or car and the other from the hand that throws it. Therefore, they do not ascend and descend perpendicularly but follow the cart or ship in the air and return to the hand of the thrower even though the cart or ship maintains it course.

A small object can produce only a small impulsive virtue upon a larger object.

A small object can transmit only a small impulsive virtue upon a large object. For example, a hammer produces no perceptible effect of this virtue on an anvil. Therefore, an iron horseshoe ("ein Hufeisen" in German) placed on a man's belly who is lying on top of an anvil, can be shattered or broken up into tiny bits by a hammer or sharp chisel without any harm or damage to the man or his belly.

CHAPTER 3

THE NATURE AND CHARACTERISTICS OF THE IMPULSIVE VIRTUE

The nature of the impulsive virtue is such that from one and the same impulse it can function in one direction as well as another.

The impulsive virtue can function in all directions, as the Dutch clearly show who, with one thrust of the foot can glide either in a straight line or a circle on their ice skates. While this virtue has a greater effect on larger rather than smaller objects, provided only that they both are of similar material (as was said in the preceding Chapter), it follows that larger bodies can be projected farther by it and can even describe circles of greater circumference than smaller bodies can.

The impulsive virtue functions in a straight line as well as in a rotating circle.

Furthermore, everyday experience shows that insofar as length and distance are concerned (in a straight line, that is), a smaller ball can never be projected, although all force is exerted, to as great a distance as a larger, if the latter is of appropriate proportions.

Similarly, a smaller ball freely swinging in a circle on a string cannot extend out into a circle with as great a circumference as a larger one.

Experiment with a String Moving in a Straight Line.

See Plate XVIII, fig. 1.

Let a lead ball, a, be attached by a string, ab, to a stick, bc. With the help of the stick a person can make the ball, a, (provided the string is of proper length and the distance appropriate) extend out into a perfect circle. When, however, the length of the string is too long and the distance too great proportionate to the weight of the ball, then the ball does not extend so far and it is either made to revolve with great difficulty or cannot be made to rotate at all.

Experiment with the Turning Motion Itself, Moving in Revolutions.

See Plate XVIII, fig. 2.

Have a section of a concave sphere fashioned from an oak beam of as great a capacity and circumference as possible and fit it with an

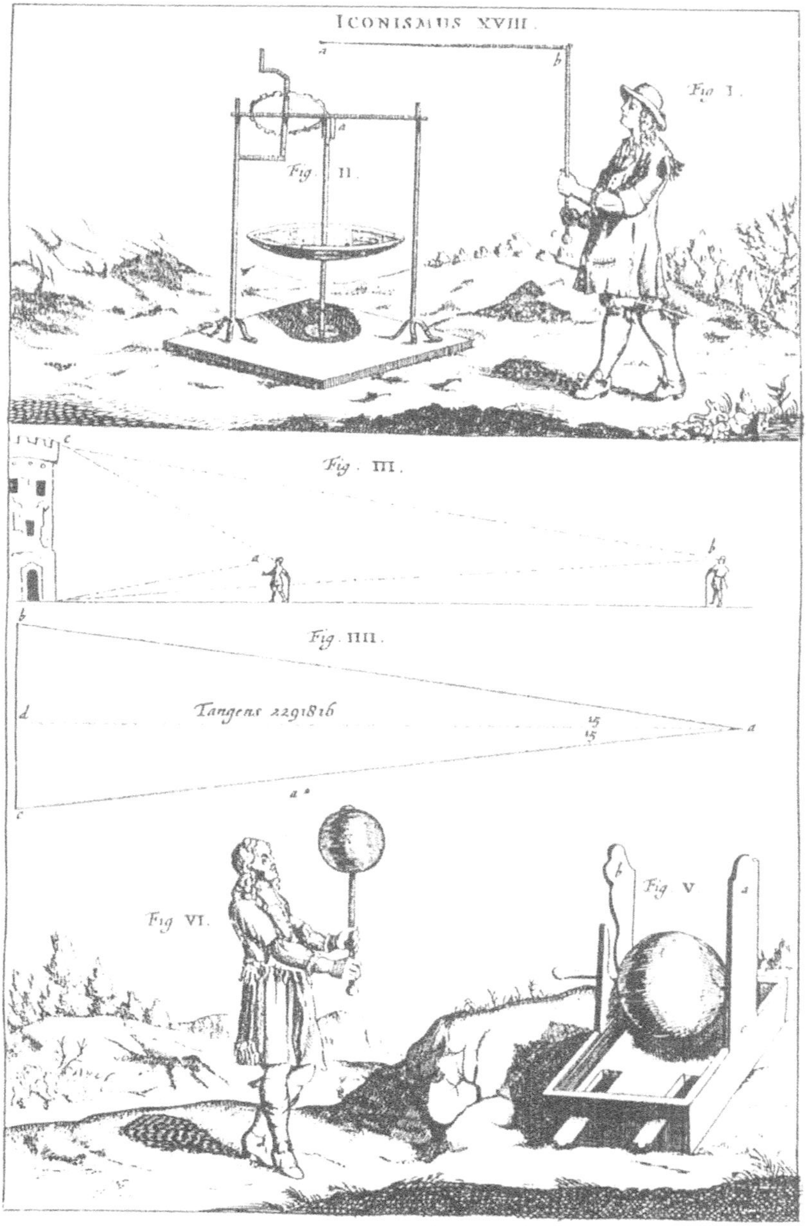
ICONISMUS XVIII.
Fig. I.
Fig. II.
Fig. III.
Fig. IIII.
Tangens 2291816
Fig. V
Fig. VI.

PLATE XVIII.

iron rod, ab, passing through the middle of the concave segment as an axis so that it can spin around quickly like a top or hoop. Then place some marble balls of different sizes, (like those with which boys are accustomed to play) in the container and make it turn. Because of the rapid rotary movement, you will see the larger balls move more in the direction of the rim while the smaller ones for the most part will remain near the center. This is the reason that when different small pellets like peppercorns, hemp seeds, peas, etc. are placed in a shallow container like a saucer (commonly called "eine Mulde") which is in turn rotated, the larger pellets always move from the center and seek the rim, while the smaller tend to revolve nearer the center. As a matter of fact, this order of rotation is visible with the satellites of Jupiter, for the larger a satellite is, the larger its orbit and the more distant it is from the planet. The nearer a satellite is to Jupiter, the smaller it is.

Larger bodies are naturally removed farther from the center than smaller

Each virtue is proportionate to its subject and cannot be extended beyond its limit.

Because every virtue undoubtedly is proportionate to its subject, it follows that the impulsive virtue which is inherent in a body is proportionate to that body and does not extend infinitely but has certain bounds of its own. Thus when this virtue continues to increase in the bodies, it finally (ERRATUM, read "tandem" for "tantum") produces in them the highest rotational speed possible (provided they are free from all obstructions like air, etc.) and projects them to the greatest length or distance it can exert. Then, when they have finally reached the maximum distance, the impulsive virtue remains constant and each body reaches the bounds or limits of its progress and speed. Beyond these it cannot proceed.

The impulsive virtue is the cause of the extended courses and distances of the planets.

See the diagram in Book I, Chapter 19.

As an example, we will assume that the Copernican system is true, that the sun has its place in the center of the world and that it causes the planets to revolve around it within the sphere of its virtue through its rotational or turning virtue. (See the following Chapter 9 as well as Book VI, Chapter 8 in this regard.) It follows that the planets are propelled by means of the impulsive virtue (which they acquire through this turning motion) into the immensity of Space, otherwise empty of all matter, to the regions most distant from the central motive part (that is, as far as is possible within the sphere of the sun's influence wherein a planet can carry on its turning motion in accordance with its own size or mass). The smaller planets, however, cannot be projected or propelled to so great a distance as the larger and consequently they do not turn about in as great an orbit relative to the central turning point because small planets do not embody as great an impulsive virtue as larger ones. The latter, however, take a longer time to complete their orbits because their courses are greater. Contrariwise, the smaller complete their periods more quickly than the larger, For a better understanding of this discussion, see the diagram in Book I, Chapter 19.

Furthermore, because these bodies in such a sphere of activity necessarily are projected through this impulsive virtue when it continues from the central turning point, namely the sun, to a definite distance proportionate to their mass, (we have also seen this occur in the previous experiment when a lead ball, connected to a string is set in motion and the string is gradually lengthened; in the same way these bodies extend out into a greater and greater circumference), it follows that wherever these bodies may be on the sphere, they are always driven to seek a more distant place, the sphere's equator, as it were. And although they slip down or up and verge toward the axis or point of rotation by reason of this innate virtue, they nevertheless always return to the previous more distant point, namely the equator, as will be seen in the experiment in the subsequent chapter.

The impulsive virtue excited in bodies through a sphere of activity projects them to a place at greater distance.

The equator is the more distant point in all spheres and the poles the nearer.

The oscillations of pendula about which certain writers have written are also governed by this impulsive virtue. Galileo, Baliani, Wendelin, Riccioli, etc., have investigated with great care their characteristics and discovered (1) of two pendula equal in all respects except length, the shorter is related to the longer as the square of the swings of the latter is to the square of the swings of the former during the same period of time, and vice versa. (2) of two pendula equal in all things except in weight, that which is heavier continues to be in movement for a longer time and completes more swings than the other within a prescribed time. (3) of two pendula equal in all other respects, that which is only a little heavier but of noticeably greater mass completes fewer swings (I suppose because of air resistance) during the allotted time than the other, etc. In spite of these experiments, however, although they may have been useful for measuring time, inasmuch as clocks with pendula were unknown at that time which could clearly mark first as well as second minutes, those characteristics of oscillation, which are based on the established proportions existing between the weight or mass of a body and the length of its string, are not sufficiently well understood. For if these proportions could be accurately ascertained, then from the periodic motion of any planet, its radius, that is, its distance from the sun, could be determined. And because the proportion of the period of the earth to its radius, that is, the distance from the sun, is known, a more accurate calculation of the distance of any planet from the sun could also be obtained through the Golden Rule.

A characteristic of the impulsive virtue is to recede, but always to return.

The movement of pendula.

The experiment showing that the larger a body is, the larger the impulsive virtue it can embody.

The true proportion which exists between the mass of a body to its distance has not been ascertained.

CHAPTER 4

EXPERIMENT WITH A GLOBE FREELY SUSPENDED IN WATER

The Reverend Father Kaspar Schott recalls in his.*Magica Universalis Naturae*, part 3, Book V, section 19 "that Father Kircher at Rome with indefatigable enthusiasm and persistent concentration had devoted himself to finding a method for this type of suspension until he finally discovered it, as Kircher himself describes in his *Magnes*, Book II, part 1, preliminary explanation 5, method 3, page 157: "Take spring water or some other liquid cloudy with the usual impurities and purify it by boiling for a long time. Then fill a glass sphere half full with it. Add to this some other liquid of dissimilar ingredients derived from a mixture of wine, turpentine, olive oil, or a similar substance that does not mix well with the others and take care that it be of the same color as the liquid lying beneath it. (I myself usually add a tartaric solution to the wine spirits which do not mix.) Then when the container has been filled with such liquids, you will put a small magnetized sphere in it. I maintain that this small sphere, suspended at first on the surface of the liquids, will descend subsequently to the middle of the spherical container and will remain there. For since this sphere is heavier than the surface water and lighter than the lower water, it will descend to the surface of the lower water and will remain there because of its weight relative to the water. Thus the sphere will come to rest in the center of the container.

Father Kircher's method.

'The sphere will rest immobile in the midst of the waters
As the earth rests immobile in the center of the universe.'

"When I carried out this demonstration for the first time, I can hardly describe the amazement which this produced even among those who wished to be regarded as scholars. They could not understand how the small sphere could be suspended in the center of the container with such stability since they could not distinguish between the different liquids in it, identical in color as they were. They kept on asking, "How do you produce this stability? Have you fitted a magnet to the bottom or mouth of the container in such a way that its poles are equidistant horizontally? And when this is done is the small magnetized sphere always drawn by the aforementioned magnet into the middle of the container so that the magnetic poles of each correspond with one another? Is this the reason why it does not shift position no matter how vigorously the sphere may be shaken from the center?)"

The author's method.

So much for Kircher. Father Kircher's industry should indeed be praised, but since we found it necessary to demonstrate the annual "latio" of the earth around the sun by means of such a globe or small sphere which was completely balanced with water and was so alike it in weight that it could now come to rest in the middle, now at the bottom or even the top part of the water, thus even before Kircher's method was made public we sought to prove our theory in the following way.

How to prepare water so that it is not putrid.

First of all, the water used must be demonstrably free of decayed matter. This can be done as follows. Take a jar or some other container that is filled with rain water and expose it to the sun all summer long so that it becomes putrid one, two, or even three times with the resulting generation in it of many varied kinds of organisms. It should then be kept in a storeroom during the winter, protected against frost. The following summer it should again be exposed to the sun and finally poured through paper into another container which pharmacists call a $\frac{1}{4}$ recipient or receiver. In this way it may be preserved without any change and will be free from putrifaction.

How this glass sphere should be made.

Secondly, one should design this glass so that a thin sheet with an iron spike is cemented on its underside. Upon this point it can rotate slowly as on an axis by means of a mechanism made of cog wheels attached to the upper side of the glass.

The globe representing the earth.

In the third place, make a brass or glass sphere (half the size of an egg) in equal balance with the water. (This can be done if hot water is poured into the sphere which is then carefully sealed with wax. After this, small bits of lead should be stuck into the wax so that the sphere is equal in weight to the water. Thus it will have neither too much weight nor buoyancy in the water, but will float freely.) Because this is a difficult procedure to follow, however, the temperature must be closely watched. When you carry out this experiment during warm weather, the sphere readily descends to the bottom of the container while in colder weather, it floats. When it is floating, moreover, and the glass is carried into a warmer place, the sphere slowly descends to the middle and rests there until the temperature of the water causes it again to float or to sink to the bottom.

How this experiment should be begun.

When it is resting on the bottom, should the glass be transferred to a colder place, then the sphere rises. When, therefore, the sphere is at its ascent or descent, or, better still, is at rest, then the container can be slowly rotated and then the sphere finally moves to the greater distance outwards, that is, to the middle circumference, which represents the equator. Although it tends to go either above or below the equator to the tropics, as it were, because of the impetus it has received, it is compelled, nonetheless, to return to the middle circumference because of the reduction in the size of the circle found around the axis or each pole where the impulsive virtue prevents it from being confined.

The course of the planets, its source and manner of functioning.

In the same way, but much more effectively (if the small can be compared with the large), do the planets revolve in the sphere of the sun's virtue (where they all hang freely in pure space without weight) through its continuous turning by which it governs the entire sphere of its virtue. The planets have no possibility of tarrying near its axis, either above or below, for the activity of the sphere is such that they are driven by the impulsive virtue they have received, to a more distant place, namely the equator. Just as the excited impulsive virtue does not remain in one place but always reaches beyond it (as we see with the balls projected over the water from cannons on warships. These do not sink immediately but first made several hops. It is the same with the swings of pendula which are nothing other than projections of this kind of movement originating in the impulsive virtue), similarly do planetary bodies reach beyond, by virtue of this force, and pass over to the place where they belong, namely the equator, until they come to the tropics where the sphere of the sun's activity is more contracted. In this more confined area, however, they cannot remain but are forced to return again to where there is more room. One can easily see that they will seek this time and time again as long as the turning of the sun's sphere of activity lasts.

CHAPTER 5

THE CONSERVING VIRTUE OF THE EARTH

What the conserving virtue is.

The conserving virtue of the earth is an incorporeal virtue which causes all earthly bodies to come together in harmonious unity. It therefore cannot properly he called "attraction" but rather the "appetite for", the "conjunction", the "union", or "preservation" of itself; it occurs when a body is joined to another by the beams of an outflowing virtue peculiar to themselves which produce a single unified body from two contiguous bodies. This virtue is imposed on all things by nature so that they may preserve their identities in themselves either by shunning the opposite or by uniting.

By means of this virtue the earth, together with everything related to it, maintains itself in tact in the ether and does not leave anything behind nor is dissipated or dispersed by the erosion of the rays of the sun and the attendant impulse and forceful motion of translation (or locomotion) and other causes.

The conserving virtue also has its own sphere of influence up to a certain distance,as we have said of virtues in general in Chapter 1. Within this sphere, the earth attracts or even rejects or in a certain way preserves whatever is advantageous or useful to itself. Because of this virtue in our sphere of air, it appears to us that all bodies are heavy, a characteristic, nevertheless, which stems from nothing more than this virtue. Indeed, nothing in the world is either heavy or light in itself, but whatever is held to the earth by this conserving virtue is difficult for a man to separate, remove, or set apart. For that which we usually call heaviness or weight and which we assume is present in a body itself, has, in actual fact, its origin in a property of the earth, for all earth bodies would remain at rest or be freely suspended if the earth were not there. A scholar, therefore, gives false instruction concerning weight and lightness when he attributes weight to bodies themselves but completely disregards that virtue which pulls them down. As a matter of fact he ascribes weight to the sphere of the earth and in this assumption lies the source of many errors. For even though there is a virtue in the earth which makes it appear to us that bodies separated or removed from it are of weight, it does not follow that the earth itself has weight. Take for example a strong man who takes a staff from the hands of another. The fact that he draws it away does not make it heavy, for as attraction cannot give weight to the human body, neither can the virtue of attracting all earth bodies make the earth itself heavy.

The conserving virtue also has limits or bounds.

There is no weight in bodies

The earth itself has absolutely no weight.

You will see a visual demonstration of this virtue which is a natural characteristic of the earth and is an integral part of nature in the sulphur globe experiment discussed above in Chapter 15. Furthermore, no one has ever demonstrated that there is any natural weight in earthly bodies. If there were, they would show a marked appetite or inclination to seek out the lower place, a disposition we do not perceive in our experience and one which we find to be quite otherwise. For, when bodies, etc. fall, they seek out the lower place against their will and very often, contrary to nature, suffer death.

It is therefore clearly apparent that this virtue of attracting and uniting is a natural phenomenon, but that heaviness or weight in bodies is obviously contrary to nature. As for the fact that the sulphur globe tends to attract light bodies, it is easily understandable that this occurs through a resistance to the earth's virtue which draws all bodies to itself. As for the bodies this globe attracts, they must draw away, as it were, from the earth's virtue because the globe must exert more force than the earth and for this reason the bodies must be lighter.

How authors differ in their computations of this imaginary weight.

Although the sphere of the earth is imagined to be very heavy, nonetheless, this opinion should be corrected since the total weight of the earth is not equal to that of a grain of barley, indeed not even the lightest feather. The calculations were gratuitous, therefore, which were made by Simon Stevin in his *Statices*, Book III, where he sets the weight of the earth at 2,000,000,400,000,000,000,000,000 pounds and by Forerius in *Viridarius Philosophicus* who sets the weight at 899,564,914,285,314,283,714,285 pounds. Scipione Chiaramonti likewise made a calculation in his *De Universo* of 2,711,787,896,857,350,340,185,118 pounds and Mersenne in his *Veritas Scientiae* and the *Comentaria in Genesis* of 65,923,634,426,652,872,385,072,000 pounds.

One should not presume that a stone can fall to the center of the earth.

One should not presume, however, that this virtue extends down into the depths of the earth, for it, in fact, exists only in the upper part. Thus, it would be impossible for a stone that was dropped (should there be an opening in the earth through to the antipodes) to fall to the center of the earth. Inasmuch as nature does not admit that which is not compatible with it but rather expels it (as is recorded in Book V, Chapter 3, number 3 where mighty rocks are hurled up to the earth's surface from the depths), it does not tolerate, therefore, a stone which has its natural origin on the exterior surface of the earth to be found in the interior.

The limit of this virtue.

Finally, since this virtue, like the others, has its own bounds and a limit to its sphere of activity, it does not extend out to the rest of the planets, much less to the fixed stars. Thus those scholars have made a useless calculation and have engaged in purposeless labor who have tried to measure the time required for a stone to fall, if it could, from the so-called starry sphere to the earth. We have discussed this point previously in Chapter 2.

Facts relative to the antipodes.

From these facts, then, one can easily understand the nature of the antipodes, for their seat is no more below us than ours is below them, nor should we be afraid that they will fall down into space any more than they should fear our being projected into the concave heavens. For the earth turns its face on all sides upon the sky equally and holds all bodies in such a way that none may be lost.

CHAPTER 6

THE EXPULSIVE VIRTUE OF THE EARTH

Aristarchus, who for many reasons was of the opinion "that the earth is endowed with a spirit capable of sensation", examines the opinions of those who thought that among the many other capabilities in which it abounds, the earth particularly delights in attraction and expulsion and exercises this virtue to the limits of its influence etc.

It seems reasonable to assume that since the earth has the power to attract those bodies which are congenial to it, it should also be able to expel those things which are harmful or are not acceptable.

The experiment involving the expulsive virtue.

This expulsive tendency of the earth will be substantiated in the sulphur globe experiment which follows in Chapter 15. For when this globe is stroked or rubbed several times with the hand, it not only attracts all light bodies to it, but occasionally repels them at its pleasure only to attract them again. When it does not so desire, however, it does not attract these bodies. Concerning these reactions the third part of the aforementioned Chapter should be consulted.

When the sulphur globe has attracted a small feather, it not only repels it occasionally, but at other times retains it in the sphere of its virtue, not allowing it to come in contact with its surface until the feather has touched some other body, (perhaps so that it might derive some sort of stimulus from the latter). In this way a small feather, cast into such a sphere of virtue, can be carried throughout a room and transported to any object or the top of any object you may wish. After this, however, the sulphur sphere again may attract the feather and draw it to itself.

The influence of the planets.

From these and other unexpected results derived from the sulphur globe experiment, one can easily see how the influences of the planets vary according to the differences in their forms and their proximity to one another because on occasion one exerts opposing influences on another, while the other resists these with its own impelling virtue. On the other hand, since the earth and moon are always associated to the extent that the moon follows the earth in its annual movement around the sun (one can see that Jupiter's satellites function similarly), one must conclude that they influence one another greatly through effluent virtues of this kind. Among these, without doubt, is the chilling virtue imparted by the moon (which in winter is periodically received and rejected according to the nature and state of the earth). This virtue cannot have a greater effect than at the time when the rays of the sun are weaker and the earth is deprived of all heat. (See also the end of Chapter 23 of Book V.)

The chilling virtue can have a particularly great effect in winter.

CHAPTER 7

THE DIRECTING VIRTUE OF THE EARTH

What is the directing virtue?

The directing or directive virtue of the earth is likewise an incorporeal force (because it passes through all bodies, even glass, as we see when a magnet, outside of a glass container in which a magnetic needle has been carefully sealed, nonetheless directs the movement of the needle). This virtue is inherent in a stone commonly called the lodestone, but it is also present in the whole sphere of the earth, directing it by the so-called poles, with the result that they provide its natural boundaries and the earth in turn is oriented toward these points. By way of example: if the earth were tilted from its own proper position, the directive virtue would make it return immediately to that location it has or should have with respect to the whole world system.

To what end the directive virtue has been ordered.

Because the earth hangs freely in the air and swings about the sun every year and at the same time makes one rotation itself every day, it has need of some virtue through which it can maintain itself at the points of its axis, namely the arctic and antarctic poles (like a wooden globe turned on a lathe), to prevent its deviating from this position and from oscillating in its daily motion through the friction of the sun's rays, thereby changing the times of year as well as the days, etc. For these reasons God endowed the earth with this directive virtue which, although it is diffused through all bodies everywhere in the whole earth, is inherent and abides in none but the lodestone, constituted, as it is, of iron which can retain this virtue.

An iron body alone retains the magnetic virtue.

The earth is not a large magnet.

The earth, however, is not a large magnet, as some contend, but it does have the virtue to orient and direct itself towards definite regions of the world.

By reason of the magnetic virtue, therefore, the position of the earth's poles are set or rendered fixed, and this orientation is not accidental, but is as it must be; for nothing has in itself the propensity for self-propulsion unless it leads to something useful to itself.

To be sure if the earth were not subject to oscillation because of the friction of the sun's rays in its annual rotation as well as its own daily motion, this virtue would clearly be superfluous and without purpose. Since, however, nature does nothing in vain, the daily and annual movement of the earth should certainly be presumed to be derived from this virtue that is so very apparent to our eyes.

The earth is subject to fluctuation or oscillation.

Furthermore, experience shows that this virtue is not inclined toward uses other than that of direction. For this reason some scholars, being unaware of the rarity of magnetic powers, when they were unable to carry out new and reliable experiments, falsely filled the world with baseless conjectures and fantasies and went so far as to postulate the existence of a magnetic ointment and other medicinal products derived from a magnet. On this subject, see Kircher *De Magnete* and others.

There is no basis to the postulation of a magnetic ointment.

This virtue, diffused in magnetic veins in the earth, as it were, not only has its sphere of influence outside the earth in the surrounding air, but also can be excited by one iron body's being struck forcefully upon another body made of iron. Take an iron wire, for example, a finger in length, and strike its ends with a hammer upon an anvil oriented north and south in such.a way that one end is pointing north and the other south. Then hang this wire freely in the air and you will see it orient itself like a magnetic needle. In the same way those iron instruments which artisans use to perforate iron also take on this quality through repeated blows of a hammer and attract iron filings, commonly called "barley grains" in great number.

Magntic virtue exists outside of the earth, and can be excited through rubbing.

As a matter of fact, all the iron bars we are accustomed to see fitted on windows, take on this virtue after being exposed fifteen and more years to the air, indeed, not only those which were in horizontal rows facing south but also those which which were set up vertically and remained perpendicular to the ground. (Apparently the direction the windows faced was of no significance.) Indeed, the lower ends of the bars, which are directed towards the ground, attract the south pole of a compass needle and shun the opposite point and contrariwise the upper ends which are turned toward the sky, attract the northern pole of a compass needle's point and repel the south. Because we dwell in the northern hemisphere, the lower ends of the bars have a magnetic attraction for the north.

Iron bars that have been set up ten and more years take on a magnetic virtue.

The lower end is oriented towards the north and the upper end to the south.

From these and many other experiments, it is clear that this earth of ours is endowed with a directive virtue. Although certain scholars seem to feel that this virtue holds the earth immobile and prevents its wobbling from north to south (while others believe that this kind of movement is prevented by the weight of the earth) and from being moved by external forces like strong winds, storms, etc. swirling about its center, they do admit that it has a certain mobility which is not contrary to nature. This is also the case with the moon whose oscillation we perceive with our own eyes. For a further discussion of this subject see Chapter 20 of the subsequent Book. Here it will be shown that the earth cannot be moved by external forces of the kind mentioned above because they constitute one body together with the earth and even at that, are only a very small part of it. It is set in motion, therefore, by another moving influence, like the rays of the sun, perhaps, as well as its own.intrinsic turning virtue.

Unless the earth were capable of movement, the magnetic virtue would be useless.

The reason for the deviation of the magnetic needle.

Furthermore, it should be borne in mind that the magnetic virtue is more powerful on a high point of the earth than in lower regions. This is evident from the fact that a magnetic needle turns aside in most places either into an easterly or westerly direction, whereas it is only infrequently that it adheres rigorously to a north-south orientation. The reason is that this virtue is properly a constituent of a magnetic stone, which, when it is in a somewhat close proximity, makes the compass needle deviate from its usual position, as we see in all compasses which can be directed at one's pleasure into whatever direction one wishes whenever a magnet is exposed near them. For the present, let us pass over an observation made by various scholars that magnetic variation is not always stable and constant, for a compass does not always register the same number of degrees from the same place.

The magnetic stone has two facets, one facing south, the other north.

Finally, as far as a magnetic stone is concerned, it should be remarked that it adjusts itself to the directing movement of the earth and has, as it were, two facets to its nature, one oriented to the south and the other to the north, or to put it another way, there is one single and simple virtue diffused throughout the whole substance of the stone which has a twofold property enabling it to maintain its poles or axis parallel to that of the sun and to hold a unique position or place in the world. One observes a similar example in the case of man in whom his parts are not only organized in an orderly manner but are also oriented in the same way toward the earth, with his head, as it were, lifted to the sky, his feet planted on the ground, and his whole body standing in an erect position. Should the Most Gracious Reader wish to ascertain the poles of any magnet, he should take a needle or magnetic indicator from a solar clock and place it on a pin so that it can move freely. Then, in turn, he should place this device on a magnetic stone. Thereupon the north end of this needle will point to the southern part of the stone while the other end of the needle, the southern, that is, will indicate the northern part. At this point the stone can be honed into an oval shape about its poles and will he ready to make new magnetic needles, when, as it were, the southern end of some as yet unmagnetised needle is rubbed on the northern part of the magnet and the operation is reversed.

How the poles of any magnetic stone can be ascertained.

CHAPTER 8

THE DIFFERENCE BETWEEN THE CONSERVING AND DIRECTING VIRTUE OF THE EARTH

The drawing together of magnets at their compatible poles is not properly, but improperly, termed attraction. For a magnet does not actually draw another magnet or a piece of iron to itself; rather the magnetic virtue enters or comes into the iron and for this reason directs the one towards the other.

There is a great difference, therefore, between the conserving and directing virtues. The former draws to itself or attracts all bodies, not only at their poles, but throughout, and these bodies that have been attracted are not altered, but are held by the conserving virtue, while those at a distance are carried in a straight line toward the center of the earth. The latter, on the other hand, only gives direction to the magnetised body polewise, that is, about the poles, while it affects the other parts only obliquely and transversely and at times even repels them completely. The conserving virtue, therefore, is the source of the earth's capacity to collect and the directing virtue to order and direct.

The writers who have written about magnetic attraction always associate it with electric attraction when, as a matter of fact, there is a great difference between the two. Gilbert, for one, in his work *De Magnete* holds that electric attraction arises from the emission of moisture and that since moisture attracts moisture, this attraction has its origin here.

The reason they ascribe to be the sources of electrical attraction do not agree with experimental evidence.

Cabei, however, in his *Philosophia Magnetica*, Book II, Chapter 21, criticizes Gilbert. Although he grants that magnetic attraction arises through the transmission of an effluvium, he does not agree, however, that it is through moisture, but rather through a pure emenation which stirs up air and forces it to return in a circle as in the case of amber which carries small bodies along with it. He finally concludes with these words: "I say, therefore, that a very tenuous effluvium flows out from amber or any other body of similar powers of electrical attraction, which thins and disperses the air and can even stimulate it violently. Then the agitated and tenuous air returns to the electric body which is its source and takes along with it straws and whatever other small bodies it meet with."

Kircher in his book, *De Arte Magnetica*, Part 3, Chapter 3, is essentially of this same opinion.

The author's opinion of electrical attraction.

As for us, who in the preceding chapter assume that the attraction of the so-called sulphur globe is the same as electrical force and observe that it has its source and origin in the conserving virtue, are not able to concede, however, that this attraction is activated through air because experiments visually prove that the sulphur globe (when it has been rubbed beforehand) can exercise its virtue through a linen thread an ell and more in length and attract a body to its end. This experiment will be further discussed in Chapter 15, article 3, number 4.

CHAPTER 9 (ERRATUM, read 9 for 13)

THE TURNING VIRTUE

What is the turning virtue?

The turning or rotating virtue, whereby a body turns about its own axis or center, is logically dependent upon the directing virtue. For indeed, for what purpose would the directing virtue be in an earthly body unless it were on account of its rotation? That the turning may take place in the way it should, a body must be maintained in its own proper position. If, however, there were no rotation, then there would he no need of a directing virtue, for a body which is in itself immobile has no need of alignment.

In everyday life we see the turning virtue evident in the tops with which boys play when they make them rotate by whipping them in such a way as to conserve this virtue for some time. The more solid such a body is, moreover, or the more it can encompass the impulsive virtue, the greater effect it has on the body. The fact is that this virtue not only encompasses the whole earth but even attracts all things surrounding it, those that are on earth as well as those that are in the air, indeed, even the air itself (insofar as it is able to partake to a certain extent of this impulsive virtue), and the moon as well. Although, once more, it cannot be demonstrated visually that this is the case for any one such body individually, still, that this is so, will be demonstrated in many ways in the following Books V, VI, VII.

This virtue should be considered from a twofold point of view; in the first place to the extent to which it is naturally present in an animate body and then to the extent to which it accidentally influences another body in its sphere of influence.

By reason of this virtue, the moon is also made to turn.

Just as the spheres of influence of all virtues gradually become fainter and weaker at a greater distance, so also do the more distant bodies in this turning sphere revolve more slowly, like the moon, for example, which is carried around in $29\frac{1}{2}$ days by the daily twenty-four hour rotation of the earth.

Animate bodies are so created as to excite this turning virtue in themselves (which is as a vital flame contained deep within them. Thus man can move and transport himself freely in the air without the support or setting down of a foot. Nor is great energy required, for the impulsive virtue immediately comes in to aid, with the result that a single impulse makes the rotation continue that much more easily and quickly. This can be seen in the tricks of performers who can turn about on their heels like a top ten and more times, being propelled, however, by a single impulse.

The turning virtue in the earth has its origin in an inherent spirit.

Round or spherical bodies are the most suitable of all to exercise this virtue and perform such a rotation. Of this type are the sun and planets although their revolutions vary greatly from one another. For a discussion of this subject see Book V, Chapter 15.

Round bodies are most suitable to exercise the turning virtue.

By means of this virtue not only is the rotation of these bodies, together with their spheres of influence, activated about their own individual axes but even the rotation or transportation (ERRATUM, read "sed" for "seu") of other bodies existing in their spheres of influence. Otherwise, the last-mentioned would not participate in the movement of "latio", and thus there would be no movement of translation or revolution of the heavenly bodies if the sun did not lead around the planets by means of this virtue and the planets do not do the same with their accompanying satellites. See Book V, Chapters 15 and 16, *De Motu*, on this subject.

The turning virtue activates a whole sphere of activity including all things existing in it.

The sun is the cause of the planets' course.

Because this virtue is natural and constant, we do not perceive it nor the extreme speed with which the earth turns about its axis and carries us in a mere second of time, a moment, so to speak, 384 ordinary paces with it.(See Book V, Chapter 15.) Nor should this be a source of wonderment for who of us can perceive the impulsive virtue in his own body or that which we call the conserving virtue which manifests weight or attraction? Or who is there shut within a ship who perceives its progress (for even though it may be propelled swiftly, the motion is quiet)? Or who can perceive with his senses the pressure of the air surrounding us, the existence of which, nonetheless can clearly be demonstrated?

The rotation of the earth cannot be perceived.

CHAPTER 10

THE SOUND-PRODUCING AND ECHO-PRODUCING VIRTUE

Sound, din, noise, clanging, hissing, voice, etc., are also aspects of an incorporeal virtue which can be stimulated through the rubbing or collision of bodies. This occurs when they mutually intercept or impinge upon another, as, for example, when a hammer strikes a bell, a staff a table, or when a body of any kind strikes another – then a sound is produced which is diffused throughout a defined sphere. When we are positioned within this sphere, our ears or any other receptive body perceives this sound.

Sound arises from the rubbing together of bodies.

Sound is varied.

The capacity for producing sound or the sound-producing virtue varies in bodies according to the diversity of their material, location, and temperature. For just as various colors result from the differences in the light-giving virtue of diverse bodies, so a variety of sounds arises from the varied sound-producing virtues of different bodies.

Although a faint sound is not heard, it nonetheless exists.

Although our ears many times do not perceive the sound of bodies that have been moved relative to one another, nonetheless, there is a sound that has been produced. Indeed we are often aware that dogs hear sounds which we do not hear, and at night we discover that because of the quiet, a noise can be heard which is not perceptible during the day. Just as our eyes do not see extremely small objects, so also do our ears not perceive a very faint sound, when, in fact, it exists. Indeed air, being corporeal, gives out a hissing noise or crashing sound when it strikes itself or anything else corporeal, especially when it has been separated by an empty space and is again reunited. At this juncture, it produces a strong and terrible sound. See Book III, Chapters 25 and 26 on this subject.

Sound is not actually carried by the air.

It penetrates all things.

It can be transmitted through a beam.

Furthermore sound, din, crashing, noise, etc., are not actually carried by the air as scholars generally hold, for as incorporeal virtues penetrate everything (under otherwise similar conditions), so does sound also pass through all corporeal bodies which, indeed, air cannot penetrate. The experiment of ours in Book III, Chapter 15 demonstrates this fact. Here, even if the receiver is carefully closed up and sealed, the ringing of the bell can be clearly heard through the glass. There is also another experiment which is frequently performed in which sound and even voice, is immediately transmitted through very long beams. In this experiment, if a person puts his ear near one end of the beam and another person strikes the other end very lightly with his finger, scrapes it with his nail, whispers very softly, or makes any other noise, this sound is carried to the ear of the listener in a moment and is not intercepted by the conversation of people standing in the vicinity.

The thinner and more porous a body is, the more easily and readily it can be penetrated by sound. Thus sound can pass through air (because it is a very thin body and closely approximates a vacuum, as can be seen in Book III, Chapter 1 of *De Aeris Natura*) much more easily than water. It can also penetrate stone more readily than metal, etc., provided only that both materials be of the same thickness.

It is certainly true that when the weather is fair, (that is, when the air is not in motion), a sound can be heard more clearly than during a storm. If, however, sound is carried by means of the air, how can it be transmitted to our ears when the air is not in motion at all? From this, then, one is forced to conclude that sound is not carried through the agency of air. When the wind blows, however, sound is indeed carried by the wind. The reason for this is that sound also strikes air as it does any other body and then the air, being struck, is carried by the wind, together with the sound, just as the sound of a bell can be carried from one place to another.

A sound strikes other bodies in the vicinity.

That a sound, however, can also strike other bodies in the vicinity is to he seen from the fact that when one string of a musical instrument is touched and produces sound, another produces sound of the same tone even when it is not touched. Mersenne describes the following phenomenon in his *Harmonia*. "In Paris, in the Church of the Franciscans, when the organ is played, the floor of the church shakes with such vigor around the region of its diameter, on the side of the room which the organ occupies, that one might have grave fears that the earth were going to split open. This sensation becomes hardly perceptible if you either approach the organ or retreat from it at a distance." As a matter of fact, Father Athanasius Kircher recalled that in Rome a stone of enormous mass always trembled at the sound of a certain organ pipe. For the same reason a very loud sound can make a person deaf. Furthermore, just as the sound-producing virtue penetrates all bodies, though in different degrees, so also does it penetrate water, as we have described in the experiment with a stone which produced sound when it was lowered to the bottom of a mountain torrent. Divers, moreover, lowered to great depths in the water hear loud noises, and fish can be conditioned by the sound of a bell to gather together. It must be concluded from this that sound is not transmitted through the agency of air. But as the impulsive virtue is almost completely obstructed in water, similarly is sound, produced in water, confined and allowed little scope to expand.

Sound penetrates water.

Sound has a limit to its sphere of influence.

Now just as all these virtues are not propagated to infinity, so each in accordance with its nature and condition has a limit to its sphere of influence, at which point it vanishes. That is, the farther it is removed, the more it is weakened until it completely disappears.

Sound is reflected.

Sound is not diffused in a moment, like light, but grows in successive stages and spreads out spherically. Similar to light, however, it is reflected like the light-giving virtue and the impulsive virtue as well. We commonly call this reflection "echo", although this terminology is not precisely accurate because it actually is resonance, or the reflection or repercussion of sound.

For although sound, falling on an object or objects, is reflected according to the composition of the different bodies and their various locations and thus is perceived as an echo, still it would be more properly termed a resonance and repercussion of sound rather than an echo.

What an echo is.

An echo, however, is the sound-producing virtue present in a body which is capable of receiving sound with all its qualities and of returning it again intact.

The difference between resonance and an echo.

On the other hand, resonance or the ability to resonate is the reflection of sound by a body unsuited to receive the sound-producing virtue. Sound, therefore, is appropriately reflected (not, however, with all its qualities as we find in an echo), but in different ways, in accordance with the character of the place and the bodies by which it is reflected.

Therefore, just as we see one reflection of light to be that made by an opaque body unsuited, as it were, to receive the light-giving virtue and another reflection of light by a translucent body capable of receiving light, so also there is one reflection of sound which is made by bodies unsuited to receive the sound-producing virtue in themselves and another which is made by a receptive body otherwise called an echo.

There are, for example, the lodestone and iron which are bodies receptive to the directive virtue, while all diaphanous bodies are bodies receptive to all the qualities of the light-giving virtue. Thus there are also, without a doubt, certain bodies in nature which are capable of receiving the sound-producing virtue in themselves and of sending it back with all its perceptible characteristics. Furthermore, even though all reflections of virtues are usually found to be stronger in flat or smooth bodies and weaker in those of a rough surface, nevertheless, we frequently discover that an echo resounds with great clarity in woods and other rough and rocky places, even those covered with snow, where otherwise one would expect no, or at least very little, reverberation of sound. For example, in a woods near Helmstadt (where Julius University is) there are several valleys from which the noise of a shot being fired reverberates with multiple echoes that can be distinctly heard and counted. See also Book V, Chapter 8 at the end.

An echo can reproduce all the

Elsewhere there are places which reproduce or send back voices so perfectly and completely that a person might believe that he was hearing not only a record of, but even an improvement on his own voice. Indeed, he would hear words enunciated more perfectly than he himself had originally spoken them, for echoes return not only ultimate and penultimate syllables, but even whole utterances and speeches, the sound of musical instruments, and a complete range of voice sounds and songs in the same order that they are delivered. Lucretius in Book IV of *De Rerum Natura* describes it in this way:

characteristics of a sound.

"When you once call out, six or seven different places
seem to return the sound
In such a way do the hills make the words rebound on other
hills and repeat the words that are returned."

This phenomenon clearly ought not be ascribed to the locality but to some innate concealed characteristic of a receptive body. Of what composition is this material – is it a stone-like quality or something else? No one knows as yet, but it will undoubtedly come to light at some future time, just as formerly there was no one who believed that the light of the sun or fire could be absorbed by a stone (namely the "lapis bononiensis") and could be preserved in it for some time and even be carried from place to place.

It is not known in what material an echo has its seat.

Who then can seriously deny that the "os durum" or the stony bone in the ears of animals is not of such a material as to be suited to receive in itself the sound-producing virtue and to cause an echo in the ears? It does not follow from this, however, that this stone-like bone is the material which alone can absorb the sound-producing virtue. For just as we cannot affirm that the clear humor in the eyes of animals is the only matter which is suitable for absorbing light or brightness (to be sure we see that crystal, glass, and other hard materials are also very well adapted to this), so should we not ascribe the virtue of producing sound or echo to this stone-like bone alone. For certain other stones or materials can also be receptive, but as yet we have not identified them.

It is also evident that there is, in addition to sound itself, another individual characteristic of the extraordinary quality of sound and dissonance which stirs the hearts of men and animals through the harmonic tones of the human voice and musical instruments.

Sound moves animals.

We will not ourselves discuss this subject but will leave it to others. It is enough for us to say that sound is an earthly and incorporeal virtue which, together with other such virtues, penetrates hard bodies and within its own sphere of influence exercises its effect on a suitable body.

CHAPTER 11

THE HEAT-PRODUCING VIRTUE

What the heat-producing virtue is.

The heat-producing virtue is incorporeal and like the other earthly virtues, can be produced in receptive bodies, with the exception of water, by means of rubbing. After a period of time it enters into bodies (because it is not something introduced) and penetrates them, unless they are too thick.

Just as the aforementioned virtues are present in earthlike bodies, so one must assume that this virtue is also present in the other planets as well as the earth.

The more frequently and forcefully friction is produced and the more the heat-producing virtue is increased, to this extent does fire finally follow. We see this occur in pieces of wood, stones, and iron, etc. that have been rubbed very hard.

Fire is produced from the virtue of subterranean fire.

It must be concluded from the foregoing that our earthly fire has a similar origin because it arises from an internal virtue that is excited through vigorous rubbing. For since fire undoubtedly exists beneath the earth, its virtue also must be dispersed and excited outside of the earth through the production of a more violent kind of friction. Just as the virtue of the magnet beneath the earth is dispersed outside of the earth and can be excited by rubbing in the same way our fire is also produced.

One must suppose that this subterranean fire has borrowed something from the sun, but differs from it as a secondary from a primary source, or as light from brightness or as an illuminating body from one that is illuminated.

Through the friction of the sun's virtue, bodies are made hot.

Thus when the earth and the other planets are touched or rubbed externally by the virtue of the sun (see the following Chapter), at the point of contact they are made hot, (and indeed, the more directly they receive this virtue, the hotter they become). For the radiation or penetration of the sun's virtue is the same as friction and consequently is a producer of heat, especially when it is collected and focused by means of a convex glass. At this time the very intense heat produces fire in a receptive body like sulphur or grease unless an opposing element like water prevents it, other things being equal.

In an empty Space where there is no body, no friction or any other effect of the sun exists.

The thicker or more solid a body is, the more susceptible it is to this virtue and the more rarefied, like air, the less receptive it is. Thus in the ether or in empty Space because there exists no corporeal body, no friction can be produced by the virtue of the sun. Consequently, no heat results and certainly not fire. Therefore a Fiery

Sphere does not exist and is nothing more than a figment of the imagination.

In the sphere of a fiery body's virtue, the flame is closest to it which clings and keeps it afire when the moisture (composed of grease and air, as it were) has dispersed and disappeared. When the heat-making virtue or the heat of fire enters a body, the latter stretches and expands and at the same time becomes enlarged to the point where it dissolves into minute droplets through liquefaction. In this way those tears or drops of glass, recently produced in glass factories (I have heard of them rather often but have never seen them) are made, when red-hot globules of melted glass are dropped into chilled water. Externally they become smooth and solid, as accords with the nature of glass, while internally, they are less dense and porous. The reason for this is that with the application of heat, all the glass particles are separated from one another and after that their bodies expand, becoming porous. Then, when they are plunged into cold water, the glass particles immediately solidify and cannot come together as they did before or consolidate quickly enough because of the external hardness of the glass globules (as they would otherwise after an interval of time). Thus the heat, being an incorporeal virtue, passes out and leaves those small spaces previously filled by it empty and all the innermost parts of the body porous. When a droplet of this kind, moreover, is broken. external air with its weight and concomitant noise rushes into these small empty spaces (as the experiments of the foregoing Book demonstrate) and reduces to dust all parts of this body that had previously been made porous by heat.

What a flame is.

Gläsern Springhornlein (glass droplets).

CHAPTER 12

THE LIGHT-PRODUCING AND COLORING VIRTUE

Among the many other virtues of the sun (which is clearly a fiery, life-giving body whose flames are very close to the borders of its sphere of virtue), another virtue flows out (indeed, without interruption),which is called light or the virtue of light, commonly termed the "solar rays".

What is light or the light-producing virtue?

When this force rubs against or touches a body, it not only warms it (as we have described in the foregoing Chapter), but at the same

What brightness is.

time illuminates it. This illumination is called brightness or the primary reflection of light and differs from it as an illuminating source differs from that which is illuminated.

Primary and Secondary.

The reflection of brightness, moreover, can be called secondary brightness, then tertiary, quaternary, etc. Indeed light can be reflected in various ways. For example, the virtue of the sun's light, in other words, light when it touches the moon, illuminates it and makes it bright. When the moon's light, in turn, is reflected on the earth, this can be called secondary brightness, etc.

The light-producing virtue of the sun, or light, reacts according to the nature of the objects it touches and upon whose surfaces it is reflected (when these bodies are opaque). As for transparent bodies, it enters and even penetrates them.

What is a transparent body?

We call a body diaphanous or transparent or clear which is receptive to the light-producing virtue or one in which or through which an entrance or passage is furnished to the light-giving virtue or brightness. Included in this group are air, water, glass, crystal, and other transparent stones, as well as the transparent humor in the eyes of animals.

Likewise an illuminated body?

We call that an opaque body which is impervious to the virtue of light or one in which this virtue or bright cannot dwell or penetrate. When such a body is touched by light or even brightness, it is called a luminous body.

The reflection of light can occur many times.

The reflection of light from one opaque body upon another opaque body or bodies can occur, as we have said, not only once or twice, but three, four, and even ten times, etc., according to the composition and location of the bodies, with the angles of reflection being the same. The light, however, becomes progressively weaker until it finally disappears into nothingness. Thus if someone were positioned at the bottom of a well sufficiently deep that all reflections of light ceased, he would experience the shadows of night there and consequently would see the stars of the heavens even at mid-day.

The virtue of light does not penetrate opaque as it does transparent bodies.

The virtue of light or of brightness can indeed exist superficially in all bodies but it cannot be present intrinsically except in a receptive body, that is, a transparent one. For that which is not transparent cannot take light or brightness and its reflections into itself. Thus if animals did not have eyes, transparent organs through which light or brightness can be taken in or perceived, they could not recognize it nor would they be aware that such a virtue exists in the nature of things.

If no body is present, there can be no friction causing brightness or light. Thus if a person were standing alone in the atmosphere and there was no other body lighted by the sun either underneath or before him, he would see nothing other than shadow unless he turned around and looked at the sun.

There are shadows in the ether.

The light-producing virtue radiating from the sun passes without activity through the ether or empty Space until it touches the encircling planets as well as the earth. Thus in permeable air, unobstructed by the presence of bodies where no friction of light or brightness exists, there must be shadow.

Unless the light-producing virtue, radiating from the sun is intercepted by an intervening body, it continues to spread out spherically in surrounding empty Space to the extent of its sphere of influence. Because of this, it does not dissipate in the course of time but is instantly reflected from body to body.

The light-producing virtue does not dissipate in the course of time, but remains.

A strong light intercepts a weak light; thus a lighted candle is as nothing compared to the light of the sun.

The light-producing virtue can be retained for some time in the stone, "lapis bononiensis". When this has been exposed to the sun or even fire for a period of time, it absorbs the light-producing virtue and retains it for as long a time as a living coal does, once it has been removed from a light source and concealed in some dark place.

The light-producing virtue can somehow be absorbed by the stone "lapis bononiensis".

The colors of light and reflections of brightness accord with the nature and location of a body.

What may colors be?

Some are pure colors which originate in different reflections of light and brightness in themselves and are evident in the rainbow and in triangular glasses. Others are material colors that are mixed with matter, for light takes on the characteristics of the body to which it adheres.

What is the source of the cerulean colour?

The cerulean color in the upper part of the air has its source in black and white, for at the point where air is devoid of the most rarefied aqueous humors and is absolutely pure, there the white ends and the black begins, (for pure air freely transmits light and brightness without reflection or refraction and therefore appears black, that is, devoid of all color). At the mid-point between black and white, the cerulean color comes into existence, just as a drop of milk and a drop of ink set next to one another mingle and produce the cerulean color at their point of contact. Thus it is that a cerulean shadow can be clearly projected on a white paper at dawn when a finger or some other object is so inserted between the paper placed beneath a lighted candle that it casts its shadow on the paper. This shadow will then appear not as black but pure cerulean.

The light-producing virtue in the case of the sun and fire is an innate virtue; as far as illuminated bodies are concerned, it is a non-essential quality. See the beginning of this Book and Chapter 7 of Book II.

CHAPTER 13

THE NATURE AND CHARACTERISTICS OF VISION

The source of vision.

Inasmuch as the crystalline humor in the eyes of animals is a suitab medium for absorbing light and brightness, thus through the com munication of brightness (or the reflection from objects) in the crys talline humor, vision is produced.

All visual perception takes place through an angle.

See Plate XVIII, fig. 3.

All visual perception takes place through an angle, a visual angl as it were, and the farther we recede from an object, the smaller th visual angle becomes, relative to this same object, and contrariwis the the nearer we come to an object, the larger the angle become For example: when a man looks at a tower from point "a", whic is nearer the tower, he sees it from a larger angle than if he shoul look upon it from a more distant point, namely "b". For angle "a is much larger than angle "b".

An object appears smaller when the distance is greater.

Therefore, the farther away the eye is from a figure of visible siz the smaller the angle that is made. Furthermore, this become smaller and smaller until finally the whole angle disappears into straight line with the result that no angle appears in the eye and n object.

The size of an object varies with varing distances.

From this it likewise follows that one and the same object (th tower, for example) now appears larger and now smaller as often a its distance from the viewer changes. Thus objects nearer at han appear larger because they are seen from a larger angle. Furthe more, this tower becomes smaller or appears in a smaller angl because it is at a greater distance.

A star which appears to be of the same size as the moon or sun is 229 of its radii distant.

Thus a star in the sky, which is 229 of its own radii distant fro us, must necessarily appear under an angle of 30 minutes (regardle of those stars which are to be discussed in the following Chapter If, for example, we see a star in the sky whose radius amounts t one mile, which subtends an angle of 30 minutes, then it would b 229 of its radii distant from us. And vice versa, every body which perceived by us under an angle of 30 minutes is 229 of its own rad distant. The sun and moon are perceived by us under an angle o 30 minutes (that is, the diameter of the sun and moon is found b an angle measuring gauge to be 30 minutes). Therefore they are 22 times their own radii distant.

The basis for this computation.

The basis for this computation follows:

See Plate XVIII, fig. 4.

Postulate an equilateral triangle ABC whose angle BAC is 30 m nutes. It follows, therefore, that half the angle is 15 minutes an (because the angle ADB is a right angle) the complementary angl

Book I, theorem 32. The tangent of this angle is 2291816. Therefore as the 10,000,000 parts of the radius BD are to the tangent DA with 229.1816 parts, so is the line BD to the distance DA, 229.1816.

From these calculations it follows: if somewhere in the ether we were distant 229 radii of the earth from here (which is the equivalent of 196,940 German miles), then we would see the earth under an angle of 30 minutes and it would appear to be of the same size as the moon and sun.

If one were 229 radii of the earth distant from here in the ether, then the earth would appear of the same size as the moon.

Should you want to know the distance at which the earth would appear to be of the same size as Venus (which we see from an angle essentially of 2 minutes), first take half of this angle, which is 1 minute and then find the tangent of its complement up to 90°. Divide the number you have found, namely 34,377,466,738 by 10,000,000 parts of the radius. The quotient will be 3,437, which will give the distance being sought. Consequently we should be distant so many diameters of the earth for it to appear of such a size as Venus (other considerations should be borne in mind, however, which must be discussed in the following Chapter).

If someone were 3437 radii of the earth distant from here, then he would see the earth of such a size as Venus.

On the contrary from any given distance, we can find the angle under which the earth may be seen. For example: I would like to know in what size angle the earth would appear from a distance of 3437 of its own radii. We would find that as the given distance of 3437 of the earth's radii is to 10,000,000 parts of the radius, so is the radius of the earth,1, to the tangent of the angle, which, when doubled, the earth subtends.

When the distance of a star is known and its radius given, then its apparent size can be found.

EXAMPLE

The Distance Given – 3437 radii of the earth	Parts of the radius 10, 000, 000 10,000,000 divided by 3437 equals 290.9	Radius of the Earth 1 The tangent of the angle is found to be under 1′. The earth subtends this angle doubled (or 2′) in the distance given

Finally, one ought to know (ERRATUM, read "sciendum" for "secundum") that the farther any body recedes, the smaller (or of lesser magnitude) it will appear. In earthlike bodies it happens that the farther away we are, the greater the curve of the body's surface our eye perceives until it finally arrives at a complete picture of a

From a greater distance we arrive at a larger picture of a star.

hemisphere. Thus the more the visual angle is lessened, the greater the portion of the hemisphere's curve is encompassed and the smaller the surface of the globe becomes, the more of its body that is visually perceptible.

CHAPTER 14

THE DIFFERENCE IN APPEARANCE BETWEEN STARS AT GREATER AND LESSER DISTANCES

The difference between illuminating and lighted bodies.

There is a great difference between an illuminating and lighted body.

According to optic principles, illuminating and lighted bodies do not decrease like opaque bodies.

The following points should be noted here: (1) Just as there is a difference between light and brightness, so do illuminating and lighted bodies differ from one another. Illuminating bodies like the sun and fixed stars have light in themselves. Lighted bodies, however, have no light in themselves but receive it from another source, like the planets or their satellites. Consequently one can easily understand why the wanderers have a less bright light encompassing them than the fixed stars. There is indeed a great difference between the illuminating sphere of the sun or a fixed star which is diffused through its own virtue in the ether and the bright sphere of a planet which is merely a reflection. (2) It is characteristic of a lighted and illuminating body that its diameter does not decrease as its distance increases to the same degree as that of an opaque body. For a torch is visible when an opaque body similar in size and at a comparable distance is not. Thus a light and bright body like this, removed at an immeasurably greater distance is preserved in its own size. In other words, light and brightness make far distant bodies seem proportionately much greater than they ought to appear according to optic and geometric principles. The reason for this is that the illuminating and bright sphere is also contracted, with the result that that which previously was otherwise invisible, becomes visible at a distance. (For the more constricted the virtue of a sphere of light or brightness becomes, the more vivid and visible it is; and whatever is more vivid and visible, this is easier to perceive and is more apparent to the eyes.) Consequently although the stars which are visible to us despite their great distances away, ought to seem smaller, they are not so reduced in size as geometric or optic principles demand. Their bodies are joined to the illuminating and bright sphere encompassing them and thus they necessarily appear far greater than they actually are.

As illuminating and bright spheres are invisible when close at hand, so they become visible at a distance when they are joined with the bodies they encompass.

It will not be difficult, therefore, to understand how it happens that when Mars is in opposition to the sun and thereby is six times nearer the earth than at its nearest point in conjunction, according to our computation in Book VI, Chapter 13, the planet seems, nevertheless, hardly two times greater in this position than in the other when, in fact, it should appear to be five times larger. The same occurs with new stars which, when the earth comes nearer, do not increase in size to as marked a degree as one might imagine they would. (3) Because the nearest sphere of stars giving illumination or light is made brighter by constriction, we also see in the case of a planet that its sphere appears similar to a star because of the increasing distance and contraction of its body as well as its attendant lighted sphere. (Thus there can be no doubt that the sun sheds a brighter light in our eyes than if we lived on one of the planets nearer it, Venus and Mercury, for example.) Indeed, even clouds that are high in the heavens, because of constriction, appear to be of the same color as the moon when we see it during the day, but in the evening, when they are lighted by the sun, they seem to be on fire, which would not be the case if they were nearer. Thus one can also easily understand that a mass of air (when it lies quite far from here in the ether where it can take light from the sun at night) appears to us as a comet. When it is removed at even a greater distance, however, one can see that it appears as a new star because of the constricted light received from the sun. (4) It is further demonstrated in the aforementioned Book V, Chapter 10 that because of our earth's sphere of air, all stars appear larger to us than they actually are. They seem larger when they lie low on the horizon and somewhat less large when they are higher in the heavens. (5) We have discovered that we do not observe stars simply and clearly, but rather with radiating rays, some of which are short, others long, with the result that their bodies seem to be many times larger than they are in actuality. We see a similar phenomenon with candles and torches that are raised and extended on high, which appear larger at night when seen from a distance surrounded by many rays of this kind radiating from them. Near at hand, however, they show a defined and small flame. The reason for this is that the shorter rays of light (and brightness, too) are reflected in the smooth membranes covering the pupils of our eyes. This reflection cannot be perceived close at hand because of the bright sphere; it is, however, visible at a distance when there is more shadow there. Thus it seems to us that there are rays of this kind in candles as well as stars, provided only they are found to be above our pupils.

Why Mars does not always seem to be of the size it should.

How is it that planets (which to be sure are opaque bodies) have the appearance of stars?

A mass of air in the ether may appear to us as a comet.

Why do stars on the horizon appear greater?

The author's opinion as to why stars are perceived with radiating rays.

As for the longer rays, however, they have their source in the eyelids and the more the eyes are squinted, the longer the rays that are produced. Otherwise when the eyes are open, these rays are absolutely imperceptible. Thus when we look through a telescope at

Sirius, for example, a very beautiful star, larger than all other fixed stars and one which discloses many rays of this kind to the naked eye, these projecting rays are removed from it by means of the telescope and its disc appears many times smaller. It should seem many times larger, however, because a telescope enlarges objects greatly. Galileo, therefore, in his third *Dialogus Systematis Cosmici* declares that the apparent diameter of a fixed star of the first magnitude which is commonly held to be of two minutes and even of three minutes according to Tycho's calculation, is not greater than five seconds which is a twenty-fourth or thirty-sixth part of that which others had assumed. Thus, with stars removed at a considerable distance, not only those that are extremely far off but also those that are nearer, the radiation of the rays is essentially imperceptible. For either because of the great distance and the feebleness of their light, or (if, to be sure, they are not far enough distant proportional to their size) because of their bright spheres, the radiation of these stars' rays has no effect on our eyes. We have already described this in the case of candles that have been placed near at hand and see this in the sun and moon which have no radiating rays because their stronger light outshines the weaker. (6) It is the case, moreover, that Venus which should appear much larger when it approaches the sun in its evening conjunction than in its morning conjunction, does not appear even two times larger at that time. The reason for this is that except for the aforementioned radiation, there is also its transformation into the shape of a sickle causing its light to become meager and weak because of its horns, making its radiating properties less strong and vigorous than when it shows us its whole bright hemisphere. Our telescopes give proof of this fact. (7) We ought not assume, however, that the apparent magnitude or smallness of this or that fixed star should be determined by the same measurements we use to determine that the planets become now larger and again smaller. In view of the incomparable greater distance of the fixed stars, the diameter of the great orbit (which the earth carries out each year about the sun) is not of such size that the diameters of these bodies can make any difference, one way or the other, as to the visibility of the fixed stars when our earth draws nearer them. This increase in proximity indeed, should be counted as absolutely nothing. (8) Here one should take Jupiter into particular account because its diameter may appear to us greater than it actually is compared to the other planets. The reason for this is that Jupiter is surrounded by a larger atmosphere. (9) In conclusion, fixed stars ultimately cease to be visible at increasingly greater distances. Even if we could perceive them (despite their being incalculable distances

Stars seen through the telescope have no rays.

Stars seem greater than they ought.

The different aspects of Venus.

The difference in appearance between stars and planets.

The diameter of the great orbit is as nothing with respect to fixed stars.

The aspect of Jupiter.

Stars finally disappear from sight at a great ditance.

away), not from the size of their bodies but rather their luminous spheres, still at some point they finally would have an outermost limit to our eyes and would become completely invisible.

CHAPTER 15

THE EXPERIMENT WHEREIN THESE AFOREMENTIONED IMPORTANT VIRTUES CAN BE EXCITED THROUGH RUBBING ON A SULPHUR GLOBE

The construction of this globe.

If one is so minded, he should take a glass sphere, a so-called phial, the size of a baby's head and pour in sulphur that has been powdered in a mortar. Then, by heating it, he should cause the powder to melt. After cooling it he should break the glass sphere, extract the ball which remains and store it away in a dry place of low humidity.

Section 1

The impulsive virtue.

As for the impulsive virtue, anyone can produce this through a stimulus imposed upon any body, and the heavier the body's material is, the better can this virtue be impressed upon it. Thus this globe, being almost of the weight of metal, is very clearly and to a very great degree, susceptible to this virtue. Thus if a virtue of this kind is transmitted to it by the hand or arm, it will be projected farther than a piece of wood or other similar lighter bodies can project it. I shall refrain from further discussion of this virtue because it is evident and quite obvious to all.

Section 2

See Plate XVIII, fig. 5.

How the device which is to demonstrate the conserving virtue is to be constructed.

In order to demonstrate the conserving virtue present in this globe, one should set it up with a rod through its core on two supports, ab, on a stand labelled, abcd. This should be a palm's width in height from the base and all kinds of shreds or bits of leaves, gold, silver, paper, hop plants and other tiny particles should be strewn beneath it. Then one should touch the sphere with a dry hand and rub or stroke it two or three times etc. At this point it will attract the aforementioned fragments to it. If the globe is rotated on its axis, it will carry these bits along with it. Now we can visually perceive how the sphere of our earth holds and maintains all animals and other bodies on its surface and carries them about with it in its daily twenty-four hour motion.

It also attracts air and smoke and only in a certain way, water.

When this globe is placed in contact with drops of water they become swollen and enlarged. It likewise attracts air and smoke etc.

From these examples, it can be seen that a virtue of this kind exists in the earth for the sake of the conservation of its bodies. It can also be produced through rubbing an especially receptive body like this globe and functions with more force here than in the earth itself, (for whatever this sulphur globe attracts, it pulls it away, so to speak, from the earth and takes it to itself). Among the ancients there was also a stone known to have particular attracting and expelling virtues. We shall discuss this in Book V, Chapter 4.

Section 3

A demonstration of the expulsive virtue.

See Plate XVIII, fig. 6.

One can clearly demonstrate the presence of the expulsive virtue in this globe when it is removed from the aforementioned stand and being held in the hand, is rubbed or stroked in the manner already described. Then it not only attracts, but also again repels from itself small bodies of the kind mentioned above (depending upon the prevailing weather). Once it has touched these bodies, it does not attract them again until they have subsequently touched some other body. This virtue can be seen particularly clearly in its effect upon very soft and light feathers,a, (because they fall to earth more slowly than other bits and shreds). Thus when the feathers are propelled upwards and hang in the sphere of virtue of this globe, they can float for quite a long time and be carried around the whole room with the globe wherever one wishes.

A small feather is driven and attracted here and there.

It can be carried through an entire rom and may come to rest on any point.

It shuns fire.

A small feather does not change the side it presents.

These further points should be noted in this regard: (1) A rather soft feather of this kind not only spreads itself out on the globe but also on the air itself and seems almost alive. It readily attracts everything that is near it or, if it cannot do this, attaches itself instead to those bodies in its vicinity. For this reason it has an affinity for the tops of a variety of objects (if it is carried about a room in the manner already described) and can even come to rest on a person's nose. But if someone should set a lighted candle on the table and should direct this feather toward it (although it is somewhat more than a palm away), it suddenly draws back to the globe and seeks it almost as a refuge. Otherwise this globe does not attract the feather as readily as it attaches itself to other things. (2) The side of a feather which first comes into contact with this sphere and subsequently is driven from it, remains unchanged in the globe's sphere of virtue so that if a person raises the globe above the feather, the latter turns over in the air and presents its original side to it. (Perhaps it may be for the same reason that the moon always presents the same face to the earth and, at a distance in the sphere of the earth's virtue, is also repelled by it and held there. We shall discuss this later in Book V, Chapter 20.) (3) If the ends of a feather are spread out on a globe and if you extend a finger or some other object to it, the feather is attracted to it but subsequently withdraws to the globe. This result pertains, however often the experiment is repeated. If, then, you hold out a linen thread to the feather, all its little ends become forthwith attached to the sphere and for a long time these ends, thus applied, lie as if dead until they again rise and extend themselves. In the same way this feather shuns fire so that if the flame of a candle is moved near when it is spread out, it withdraws to the globe. (4) If you lower a linen thread hanging above the globe down until it almost reaches it, and then try to touch it with a finger or some other object, the thread draws back and does not tolerate the approach of the finger. (5) If you attach a linen thread of more than a cubit in length to the point of a sharpened piece of wood, which is fastened to a table or block, and let the thread fall in such a way that it can lie near another body beneath it, about a thumb's width away, then whenever the globe has been rubbed and is moved near the pointed end of the wood, the end of the string becomes attached to the body lying near it. (6) If this globe is suspended on its axis in the stand we have already described, in such a way that it can be turned about in it and then if it is stimulated by hand in the usual way and a down feather is placed beneath it, the globe attracts and repels this feather to and from itself many times in the space beneath. This movement can continue for several hours. (7) It is very frequently the case that the tiny tips of a feather (when one is lying on a globe that has been rubbed and is spread out on it, seem naturally to cling to that on which they are spread. Suddenly, however, they may constrict and give evidence of an antipathy which they themselves or the globe itself projects at that time. Furthermore in the same way as these feather tips that are spread out do not allow themselves to be touched, should you try to touch them with your finger, they may instead turn away and attach themselves to the globe.

A conjecture concerning the unchanging side of the moon.

A feather is attracted to bodies that are nearer at hand.

It shuns a linen thread.

The globe does not permit the union of bodies outside of itself.

The virtue operates in a string up to a cubit in length.

The globe attached to the stand we have already described repels a feathre lying in this space beneath it and appears to draw out someting from the wod.

The tiny tips of a feather show both antipathy and sympathy.

Section 4

As for the directing virtue, since it is inherent in no other bodies than a magnet and iron, it cannot be produced in this globe unless a magnet is inserted in it.

The directing virtue is not inherent in this globe.

Section 5

Nor can the turning virtue be demonstrated with this globe because the feather is drawn below the globe toward the earth.

The turning virtue cannot satisfactorily be demonstrated by the rotation of this globe through its center because the feather (as it slips down perpendicularly from the sides of the globe) is continually being drawn to the earth by its conserving virtue and thus is prevented from circulating around it. This virtue, nonetheless, can be produced, insofar as the motion of turning in the feather itself is concerned, by revolving the globe about the feather, as has been described in the preceding Section 3, number 2.

Section 6

The virtue of sound can be brought forth in this globe.

The virtue of sound is also inherent in this globe, for when it is held or carried about in a warm hand and then is moved to the ear, noises and sounds are heard in it.

Section 7

The heat-making virtue can be produced in this glove as in all other bodies (with the exception of water).

Just as the heat-making virtue can be produced in a body through harsh rubbing so that fire is finally brought forth and generated from the very intense heat that has been made, so can one assume that this heat-making virtue is also able to be produced in this globe through hard, continuous rubbing.

Section 8

How the light-producing virtue can be excited in this globe.

As for the light-producing virtue, this is produced in a similar way in this globe. For if you take it with you into a dark room and rub it with a dry hand, particularly at night, it shines in the same way as crushed sugar.

We shall dispense with a discussion of the other virtues which are manifest in this globe since they demand a somewhat deeper investigation. For nature (as Kircher says in his *Ars Magnetica* often exhibits wondrous miracles in the most lowly of bodies, which, nonetheless, are little understood except by those who, with great perception and an innate capacity for detail, teach that experiment is the best of all teachers. As regards this globe, M. de Monconys in his *Journal de Voyages par Allemagne*, published at Lyons, recalled that he had seen properties of the globe demonstrated among the other new experiments carried out by the author in his home at Magdeburg, but had lost his notes.

CHAPTER 16

OTHER CORPOREAL AND INCORPOREAL VIRTUES

There is no doubt that there are in addition a variety of other different virtues of this kind, unknown though they may be to us, corporeal as well as incorporeal, which other animals use and enjoy and which are the sources of the preservation and destruction of their lives. As for the corporeal virtues, even though they be odors and effluvia of air, nevertheless, they are not sensed and perceived with equal sensitivity by all animals but by each in accordance with its own nature and characteristics; for indeed, a man has one sense of smell, an eagle another, a dog another, and deer yet another, etc. See Book III, Chapters 10 and 30 also on this subject.

It is undoubtedly the case that there are many more kinds of incorporeal virtues which are unknown and uninterpretable to us because we do not have the organs with which each of these virtues can properly be perceived and identified. Thus that virtue is not known which irritates the skin through rubbing or scratching, thereby causing an itching sensation. We do not actually know what sort of a virtue it is, although all animals frequently experience it.

The virtue producing itching is unknown.

Furthermore, among the incorporeal virtues, there is one which produces fermentation and decay in bodies as well as heat (as we see in quicklime, grain, and other bodies impregnated with water). Likewise there are virtues which cause particular diseases and afflictions and those which bring about digestion in the bellies of animals, etc. We are more ignorant than knowledgeable about such virtues as these. Moreover, just as we are aware and perceive that many virtues emanate and flow from our earth's sphere, diffusing themselves far and wide in surrounding space, so also must we assume that the rest of the earthlike bodies or planets individually release different virtues consonant with their own natures and characteristics that mutually touch or affect one another according to the nearness or farness of their positions. Furthermore, these virtues vary with the regions in which the different earth bodies or planets exist, as when, for example, they look toward one another from opposition or conjunction or quadrature or sextile. Astronomers are wont to call these virtues, influences, which are either good or evil depending

The virtue of fermentation and decay.

The influence of the stars.

The influences of the moon and sun on our earth are indispensible beyond all others.

upon the different times as well as the interactions of these earth bodies upon one another which affect each one in a great number of ways.

First of all, the influence of the companion to our earth, namely the moon, is so indispensable that if it were taken away, the earth together with all its creatures, could not exist in the perfection of its being. At this point we shall say nothing at all about the sun, the king or ruler, as it were, of all these bodies.

Furthermore, although different effects and influences from the other planetary bodies are exerted upon our earth as well as upon these bodies themselves and one another, still they are not of so individual a nature that astronomers can indicate with any certainty the weather, fertility, sterility, rainfall, winds, prevalence of diseases and war in one or another state, region, or population nor predict certain effects which may result.

Astrologers are not able to prognosticate future events with any certainty.

"For who, (as Father Anton Rheita writes in his *Oculus Enoch et Eliae*, Book IV, Chapter 2, page 243), "has come to understand thus far the hidden effects of such things as eclipses upon our earth? Why then should not these poor little deceiving astrologers blush to hazard guesses at future events from the stars when they know so little of many of them and thus make blind and false prognostications? Since they always make ambiguous statements in their prophecies, there is little wonder that simple people have faith in their words even when events take place which in actual fact only occur by chance and often are contrary to the intent of the prophecies. Prudently and cautiously, then, shun the nonsense of these ignorant fabricators of nonsense."

Francesco Farnese in his *Philosophia Naturalis*, Book II, Chapter 13, number 756 says "The prophetess called Astrology is false as are the dreams which forecast the future from the configuration of the stars as from letters. Still it is not a wonder, although these dreams have lied many times, that some particle of truth emerges once in a while from their portents."

Mersenne agrees with this statement in his notes on *Aristarchus* page 61 where he says "Because vapors and exhalations are discharged from the earth spontaneously or by some expulsive virtue, in vain do the miserable little weather prophets torment themselves to learn the causes of rains, thunder, winds, and other natural phenomena of this kind from the planets or other heavenly bodies. They ought not look far beyond themselves, for this should be sought from the earth which they inhabit."

According to Riccioli in Book IV of his *De Luna*, Questions, near the end, "The differences in the moon's appearance and its oscillating motion,not understood for so many centuries, as well as the movements of the satellites of Jupiter and Saturn, unknown before this present century, or changes in the sun spots, are potent elements in questioning the veracity of the astrological prophets. It is indeed very likely that the different celestial influences on the earth here below derive from the diversity of the aforementioned factors."

BOOK V

THE SPHERE OF LAND AND SEA AND ITS COMPANION, THE MOON

CHAPTER 1

THE SPHERE OF LAND AND SEA, OR THE EARTH, AND ITS SIZE

We are not going to describe the earth here in terms of its superficial characteristics, that is, its various kingdoms, regions, seas, rivers, mountains, islands, cities, inhabitants, etc. as it is usually presented in geography books.

The sphere of land and sea is to be treated here as a celestial body.

Our earth is a round sphere, certainly not in a strict geometric sense, but still essentially spherical inasmuch as its mountains and ocean depths are of no appreciable magnitude when compared with the whole. Thus during a lunar eclipse, the shadow of the earth falling on the moon appears to be perfectly round. As for the seas, they are all levelled out horizontally on the sphere's surface and lie at an equal distance from its center, as we shall show in Chapter 5.

It is round.

The seas are levelled out horizontally.

Antiquity has taught that this globe is 5,400 German miles in circumference, the equivalent of 21,600 Italian miles, and more recent computations are consistent with these figures. Its diameter, therefore, contains $1718\frac{4}{22}$ German miles (inasmuch as the proportion of the circumference to the diameter is always as 22 is to 7), or to simplify these figures, 1720 miles, and its radius, therefore, is 860 miles. If one does not understand this method of measurement, he should consult the geography books and more particularly the *Geographia* of Dr. Bernhard Varenius where he will find what he seeks on page 42. He should disregard the error, however, in which occurs Varenius' calculation of the earth's volume, since the circumference of this sphere is 5,400 German miles, which, when multiplied by the diameter, will give a result of 9,288,000 German miles, or the number of square miles on the earth's surface. This is the extent,

How great the size of the earth's circumference and diameter are. Its radius contains 860 German miles.

The surface of the earth

then, of all kingdoms, regions, provinces, mountains, rivers, streams, seas, islands, lands, inhabited as well uninhabited, known, as well as unknown to us, which are in the whole earth.

as well as the capacity of the whole earth.

Then, when this total surface of the earth is in turn multiplied by a sixth part of the diameter, the result represents the internal capacity of the whole earth, or 2,661,940,800 cubic miles. This number is clearly of such magnitude that it is incomprehetlsible to human imagination and understanding and is in no way comparable with the aforementioned surface.

How we should regard the internal capacxity of the earth.

Now if we were to divide this internal capacity by the number of square miles given for the area of the kingdoms, regions, seas, etc., on the earth's surface and then should allow the height of one German mile for each cubic mile (instead of the true depth of its sky), the internal capacity of the surface would be 286 times greater than the external surface with all it's kingdoms, regions, provinces, seas, islands, etc., or, if we should wish to attribute a higher sky to some region, say of two miles (though indeed, the clouds do not extend up that far), still the internal capacity of the earth would be 143 times greater than the external surface; that is to say, under our earth there could be 143 more regions, mountains, seas, rivers, etc., than there are above the earth (with the height of the sky, or the over-estimated height of the sky, be;ng two German miles).

Who of all mortals has knowledge of this enormous space under the earth's surface, much less of the composition of things hidden therein? For indeed, whatever our senses do not perceive, with these things we do not concern ourselves. Thus many seholars, inasmuch as they cannot conceive what may be in the vast recesses beneath the earth's surface, judge that the earth's sphere is nothing more than a single and compact mass of earth. Thus they may persuade themselves that, except for certain holes, pits, ravines, and caves near the surface, there is nothing else beneath, other than stones or rocks, crammed together into a solid mass. No one, however has penetrated even half of a quarter mile, reckoning from sea level. For gold mines, caves and other subterranean holes are always found near mountains and other heights and do not extend far down beneath the surface The greatest depths to which men penetrate or excavate are the salt mines which are dug out on lowland plains. Their depths, however, do not extend down as far as an eighth of a mile.

The earth should not be regarded as a simple compact mass.

The earth has never been penetrated to the depth of an $\frac{1}{8}$ mile.

The sphere of land and.sea, therefore, is a body of such size that it remains incomprehensible to the human intellect.

CHAPTER 2

THE SIZE OF OUR SPHERE OF LAND AND SEA RELATIVE TO THE PLANETARY SYSTEM

The earth should be regarded as a point relative to our planetary system.

What seems an enormous field to an ant is tiny to the eyes of a giant. Similarly the earth also seems to us to be a body of incredible vastness, but compared with the rest of our planetary system bounded by Saturn's orbit, it is no larger than a point. Indeed, we would say it was nothing, were we to compare it with the orbit made by the fixed star nearest to us.

Father Christopher Clavius, the renouned mathematician, who shared in the common opinion that the fixed stars exist in a sphere about the earth and that they all, moreover, are equidistant from it, asserted in his treatise, *Sphaera Johannis de Sacrobosco* that the sphere of our earth relative to the aforementioned starry sphere is as 1 is to 11,652,340,095,703$\frac{1}{8}$. As a proof of this hypothesis, on page 147 of this work, he quotes the opinion generally held by astronomers, namely that the radius of the starry sphere is equivalent to 22,612$\frac{1}{2}$ diameters of thc earth and because spheres contain in themselves a capacity of three of their own diameters (according to Book 12, proposition 18 of Euclid), he compares this given value for the volume of the earth's sphere with that of the heavens embraced in the concavity of the firmament. Since 1 is as nothing, to 11,562,340,095.703$\frac{1}{8}$, so the earth is as nothing to said starry sphere. For this reason Cleomedes in Book I of his *Cyclica Theoria* says, "The earth, seen from the sun, is of the size of a point, but from the fixed stars is not visible at all, etc." One can find actual proof of this observation in the preceding Book IV, Chapters 13 and 14.)

Clavius' comparison.

Cleomedes' comparison.

From the foregoing it is quite clear that the earth is indeed imperceptible and not even of the proportions of a point relative to the postulated Starry Sphere.

In actuality, however, since there exists no sphere of fixed stars and they are not equidistant from the earth, and since the fixed star nearest us is much farther away than Clavius supposes, the earth certainly should not be compared in any way with a hypothetical sphere of stars of this kind. It should rather be compared with our planetary system in the following way.

For example, Saturn's orbit is the most distant sphere in our planetary system and contains, according to Riccioli, 73,000 earth radii in its radius (as is to be seen in Book I at the end of Chapter 25.). This figure cubed (according to Book I, Chapter 20) is 389,017,000,000,000. The relationship of the earth, not to the afore-

Riccioli's comparison of the earth with our planetary system.

mentioned Starry Sphere, but to this planetary system of our is as 1 is to 389,017,000,000,000 (a numher which is 35 times greater than that posited by Clavius for the Starry Sphere). Our earth, therefore, can scarcely be regarded as even a point relative to Saturn's orbit in our planetary system. What, then, would it be with respect to the so-called Starry Sphere? Indeed, nothing at all! One must conclude, then, that our planetary system (according to the aforementioned Chapter 20) has the same proportion to the so-called Starry Sphere that our earth has to its own system.

As the earth is to our planetary system, so this system is to the sphere of some fixed star.

"This mere point in our planetary system, as Pliny writes in Book XI of his *De Natura*, "is the substance of our glory; this is its seat. Here we win honors, here we exercise power, here we amass riches, here we engage in wars, even civil wars, and make the earth emptier by mutual slaughter. To pass over other manifestations of humanity's madness, I will cite the example of the man who drives back his neighbors and secretly digs into his neighbor's soil from his own boundaries so as to extend his fields as far as possible and thereby acquire some of his neighbor's land. In how large a part of the earth does man find satisfaction? And when he finally has increased his portion to the measure of his avarice, what will he finally have when he is dead?"

Pliny says that for the sake of lowly gains on earth we forfeit the eternal.

Seneca agrees.

Seneca agrees with these sentiments in his preface to the *Naturales Quaestiones* when he says, "This is that point which divides so many peoples by fire and sword. Oh! how ridiculous are the boundaries of mortals: this is the point wherein you sail, wherein you fight, wherein you set up kingdoms, etc."

If only harsh and heartless men would bear this in mind who act unmercifully toward others and, being discontented with the lot in life assigned to them by God, seek in every way to transfer the goods of others to themselves, whether by force or under the appearance of right. Through such actions as these in this narrow and restricted dwelling place, they lose the future richness of the abundant joys and delights to be found on God's right hand for eternity

CHAPTER 3

THE EARTH IS COMPOSED OF AN INFINITE VARIETY OF THINGS BOTH EXTERNALLY AND INTERNALLY

Since God and Nature create nothing that is without purpose or use or is imperfect in itself and since every complete and independent body is composed of matter, life, and certain innate characteristics, as was said in Book II, Chapter 7, it mnust also be concluded that this sphere of land and sea, being a complete and perfect body, not only embodies this perfection externally, but to even a greater degree internally and should be described and examined with particular interest. Thus the most important element of the earth's body lies within itself, not outside and therefore, should be looked into with more attention than its surface, and similarly all bodies must be assessed principally as to their interior qualities. For indeed, when one observes animals externally, who does not associate all their external qualities with the body inside?

Our sphere of land and sea should be examined not only externally but to an even greater extent internally.

In the same way, when we regard this sphere of land and sea, we see nothing other than its external surface splendidly equipped, without a view into its depths (for the caverns, mines, caves, passages, canals, etc., which we call subterranean are not acually subterranean, but superficial because they do not extend down one tenth of the way into the earth's covering, much less to the interior of its whole body). Since, therefore, all circumstances relative to its position in the world have been considered as well as its size, motion, and superficial appearance, we must now concern ourselves with this, its other aspect.

As individual bodies are protected by a shell or skin covering them, so must the earth body be similarly enveloped.

Thus, we can examine the surface or covering of the earth from a threefold point of view:

How the covering of the earth can be examined.

(1) Externally, extending up to the higher parts of the earth where air, clouds, winds, as well as storms, snow and other meteorological phenomena are found (which undoubtedly have their source in the sea, being turned into precipitation by the force of the solar rays' vaporizing effects).

(2) Superficially, where we see lands, mountains, forests, stones, grasses, fruits, animals, etc., as well as rivers and springs perpetually flowing and debouching into the seas. These sources, however, do not increase the seas' expanse because the waters are always absorbed through various passages in the earth in as great a quantity as they pour forth. Similarly the blood in the veins and arteries of animals, originating in the middle part of the body, rises to the head and again passes down intold the feet, etc., through its continuous circulation.

Seas and waters pass through the earth as blood flows through the bodies of animals.

(3) Intrinsically (although in a geometrical sense no surface can be called intrinsic because surface has no depth, still since man's penetration of the earth's surface is negligible in proportion to the thickness of the sphere, nonetheless, it is the place where its covering begins. For this reason, whatever takes place on this surface, whether extrinsically, superficially, or intrinsically, we say that this occurs on

the surface of the earth), Here we find openings, passages, caves, canals, chasms, crypts, mines, volcanoes and other subterranean fires from which warm springs issue when subterranean waters enter them. When water and fire come into conflict within the confines of the earth body, they give rise to smokes, vapors, blasts, etc., which are then expelled through fissures and openings of the earth's surface, Indeed, when they reach colder parts of the earth's surface, they are again changed into water and thus they again become springs and streams.

Fire in the earth.

One also hears and perceives in the aforementioned clefts and passages the mighty roar of falling water, the fearful crashing of thunder, the flashing brightness of lightning, and thunderbolts and similar meteorological manifestations, with the result that almost every one of these phenomena which is evident in the external air is also produced by nature beneath the earth. One must suppose, therefore, that violent storms, winds, thunder, lightning, and other atmospheric disorders are generated in the caverns of the earth, especially beneath the mountains, and that in view of their diverse and fearsome effects they must be animate essences which project themselves on high, unite with particles of air, and thereupon produce effects consonant with the nature and quality of each, after which they proceed to die out again. There are many examples, especially evident in windstorms and flashing bolts of lightning which demonstrate that wondrous manifestations of this kind could not occur unless they were caused by some living essence or else by some animate spirit. In Psalms 148 it speaks of "the spirit of storms fulfilling His word." Sailors call this "Orkan" which seemingly destroys, breaks up, and in other violent ways attacks all things. Furthermore, in the recesses of the earth there are also to be seen various forms of evil demons, pigmy men and other similar figures. If these are in the superficial regions, what, then, will one find in the deeper caverns of the earth?

Under the earth different noises are heard.

In the caverns of the earth animate essences are generated which thereafter produce storms.

There are various creatures under the earth, corporeal, wraith-like, and also of mixed characteristics.

In the sea there are whirlpools and raging eddies which the ancients called "Euripi" and the sailors "Maelstrom." The most famous and greatest of these is near Norway and is described as being thirteen miles in area. In its middle is a rock called "Mouske." For a period of six hours, this whirlpool sucks in all things approaching it or in its vicinity, water, whales, freighted ships, and such like, and for the same number of hours, it belches and vomits forth all that is unpleasing to it with great and violent thrashing of water and uproar. As to whether it belches forth whales, that is a question in doubt.

In the sea there are whirlpools and eddies.

We discover, moreover, that mountains come into existence and that enormous rocks (which we find deposited one upon another in an amazing fashion in many places, arise from the earth; likewise that metals, stones, gems, and countless other different materials are formed in this covering of the earth. It is also a remarkable fact that very often in random excavations for new mines, wells, and ditches (where it is certain that no one had lived before) men discover mussel and snail shells in their diggings under the earth. Actually hills, composed of mussels and other marine creatures of this kind that have been reduced to stony material, are also found in sites far removed from the sea (which are nothing other than swellings from lower-lying stratifications). Thus Ovid writes in Book I of his *Metamorphoses*:

Mountains grow and enormous rocks emanate from the earth.

"I saw seas where once was solid land
  And lands made from the sea
And at a distance from the shore sea shells could be seen
  And an old anchor lying on a mountain top, etc."

Furthermore in 1663 in Quedlinburg which is situated on the Seveckenberg, the site of a limestone quarry, there was found a skeleton of a unicorn in a cave. It was resting on its hind quarters, as animals are wont to do, with its head reared back and a horn the thickness of a man's limb projecting from its forehead. The latter was almost five cubits in length. At first the skeleton of this animal was improperly handled, owing to the ignorance of those who found it, but subsequently it was able to be dug out and brought to light piece by piece, until finally the head with the horn, some ribs, the spinal column, and a few bones were gathered and given over to the keeping of the Most Reverend Abbess Superior who lived there. From what has already been said, it must be concluded that with a long passage of time, the earth increases in size and its growth is similar to that of the vegetation on its surface. Thus the ruins of the city of Babylon which without a doubt stood exposed at one time now lies hidden for the most part under the earth. And who is unaware that the Rome of old is to be found under the Rome of today, for the Pantheon which one now enters by descent, formerly was approached by many steps?

A skeleton of a unicorn was found in a rock.

It must be assumed that the earth like other living bodies increases in size.

These, then are the facts we wished to discuss briefly about the external surface of this sphere of land and sea (into whose depths, however, no one has ever penetrated as much as a half of a German mile).

But what, indeed shall we say about the internal depths 100, 200, 400, 800, and more miles down? Who knows what unsuspected creatures are shut up and lie hidden away down deep in these re-

The internal area of the earth is completely unknown.

cesses, of forms and internal functions not to be imagined in our thought? Since, to be sure, God and Nature have so wondrously designed our microcosm, both externally and internally, with all its living creatures and has omitted nothing nor added anything superfluous, it must indeed follow that this internal space, incomprehensible in extent, was designed for the ether of this creature, as it were our microcosm, and not as a rocky or muddy mass. It would certainly be ridiculous to decorate the exterior part of a king's palace and to design the space inside for no specific use. The same can be said of the space beneath the earth that has remained closed to the human intellect.

CHAPTER 4

THE EARTH SPIRIT

There are those who hold that the spirit of the earth is animate and perceptive.

The foregoing and other observations about the sphere of earth describe so many functions and effects that they recommend themselves to not a few philosophers as aspects of an animate and perceptive spirit.

Thus Chilon, in his discussion on Mount Olympus, held that the earth was similar to man in that it is endowed with an animate and intelligent spirit, but one which accords, however, with its own nature.

The Stoics thought that the earth is a living animal.

The Stoics, moreover, believed "that the Earth is a living thing, endowed with mind and spirit and possessing certain openings in the ocean floor, like nostrils, through which it breathes in and exhales water, the source, so they hold, of the tides."

Reasons.

Even though the earth does not take steps (as the animals on its surface are wont to do), there is a mind within, nonetheless, which moves its mass. Surely anything in nature that reaches such heights of development must have a spirit. We, therefore, worms of the earth as we are, should not hold the noble celestial bodies in low esteem since the omnipotence of God is particularly manifest in them.

Spirit.

The activating force of an organic body is called its spirit. Not only is this impulse diffused throughout the whole mass of the earth, but even all birth, growth, and further development, etc., in actuality derive from this.

If you deem organs necessary for this spirit, look to the poles which are maintained stable and immobile by means of the magnetic virtue for no other purpose than to enable our sphere of earth and water to turn about its axis or center in a fixed position without any vacillation and to look upon the sun (by which it is refreshed and about which it revolved in a large annual circle). In this way it can receive light by turning itself so as to preserve its own safety as well as maintain the life of all things born and living upon its surface.

That this movement of the earth is not derived from an imaginary First Mobile existing very high in the heavens is very apparent from the well confirmed theories set forth in the text of Book VI regarding the true system of the world.

The movement of the earth is not derived from an imaginary Mobile.

The earth's motion, therefore, most certainly has its source in the earth itself! This movement is its spirit which resides in the innermost parts of the earth and which undoubtedly takes its fire or fiery flame from that of the sun.

The life of the earth is taken from the sun's flame.

It is thought to be fed in part by water (for we see that the smith can activate fire by sprinkling water on it) and, as a prime activating agent, this is believed to give form to and animate all the rest. Thus we find the continuous energy of water and other different effluvia emanating from the earth and exuding vapors in a constant flow.

This hypothesis agrees with Hippocrates' statement that "an animate body has the ability to breathe everywhere and is disposed to emit or admit whatever is of a spirit-like nature."

Hippocrates' opinion as to what an animate body is.

Similarly Aristarchus, the Samian, whose work *Systema Mundi* has once again been revived in France, writes in this treatise under the subsection, "De Cometis": "We said initially that for many reasons we believe that the earth is endowed with a perceptive spirit and is a highly organized body. We subsequently added the belief expressed by those philosophers who think that numbered among its many other qualities, this spirit embodies the ability to repel and to attract: thus it can either attract or repel that which it innately feels is serviceable or harmful to it, as various animals and plants are wont to do as well as certain stones which exercise these powers at an appreciable distance away from one another, etc." In this regard the sulphur globe experiment described in Book IV, Chapter 15 should be reviewed.

Aristarchus' opinion about the earth's spirit.

There are stones which have the ability to attract and repel

The earth turns, therefore, by virtue of its innate life and revolves about the sun. As it turns, its position is maintained by the magnetic virtue which prevents its wavering or wandering in the wide expanse of the heavens. Were it not to revolve in its daily course, the sun with its constant light would be directed towards one particular side of the earth and would scorch it if it remained there for a long period of time. The whole race of animals on that side would perish while in the other regions, everything, veiled in deep shadow and eternal night, would be raw and stiff with the persistent cold. Since the earth

The daily movement of the earth is extremely necessary.

could not endure either one of these wretched and miserable states, it revolves in an orbit by virtue of its astral spirit and life:

"Its inner spirit nourishes it and its all pervading
  intellect
Causes it to move."

If the earth grows, then its motion becomes slower. This accounts for the precession of the equinoxes.

Furthermore, if the earth is an animate body, then it will also possess the potential to move and grow. If it grows, however, then its daily motion and annual revolution will not always be uniform, but will be slower with the passage of time, as we described in Book IV, Chapter 3 and the end of the subsequent Chapter 16 of this Book. This, without a doubt, is the origin of the precession of the equinoxes which otherwise is ascribed to a particular motion of the fixed stars and is called by astronomers retrograde motion.

CHAPTER 5

THE SEA AND ITS TIDES

The seas would be dispersed over the earth's surface if there were no barriers.

The seas, together with the land, form a sphere, one, indeed, that is round in a geometric sense, for if all seas were united and intervening barriers were removed, they would conform to the spherical shape of the earth. This phenomenon can be easily established. For example, let us imagine that water is poured from a vessel into the sea. This water is immediately dispersed equally over the sea's surface and as many times as the water is poured out this situation prevails. Thus the courses of all rivers follow the earth's curve, provided only that their source is slightly raised, even though it be but one cubit high. Similarly a spring does not cease to flow until it reaches a river which follows the earth's curve and throughout is equally distant from the center of the earth. The Roman aqueducts also show this to us for even if there is a slight drop at their source, they can carry water along, following the earth's curve, for a number of miles.

As for the tides or the surges of the ocean, its ebbing and flooding, there is scarcely a phenomenon of nature which so heavily taxes the ingenuity of the natural philosophers and eludes their efforts at explanation.

The flood and ebb of the sea is relatively small compared with

An ocean tide is a definite, yet insignificant, motion of that whole body of water which appears to us as a great surge ebbing and flooding or a mighty swelling and deflating. There are always two floods of this kind in opposite regions of the earth at the same time. The interval between the flood and ebb of such a tide is 12 hours and $24\frac{3}{8}$ minutes. Two of these periods are completed within 24 hours, $48\frac{3}{4}$ minutes, or essentially 25 hours. Thus high tide occurs almost a whole hour (50 minutes) later each day at the same longitude.

the size of the earth body.

The time required for flood and ebb.

The sea floods in, therefore, for six hours and ebbs for the same period although varied contours of the shores, declivities, situations of different localities and seas, obstructing islands, changing water and wind currents, varying distances of regions from the lunar path, etc., quite often make for variations in different places. The greatest flood tide in any given place ordinarily should occur when the moon is in the meridian of its position (either the highest or lowest point) unless one or another of the aforementioned conditions prevents this. Thus when the moon is vertical to these regions of the sea, there one finds the greatest ebb and flow, but in those areas at a greater distance from this point, the ebb and flow is generally less marked. The experience of sailors bears witness to the fact that the flooding is greater during full and new moons than during the quarters.

The full and new moon caused the largest tides.

Those regions in the torrid zones lying between the tropics experience a greater tidal increase and decrease since the moon has a greater effect there and the motion of the ocean on the equator is stronger than at the poles.

At the time of the equinoxes, a greater flooding of the sea is perceptible than at other times. This alternating motion of the ocean has been variously explained, but we feel that a catalogue of all the various theories is superfluous.

For the most part, writers attribute this motion to the movements and influences of the moon and sun. Although the moon has a great effect or power on the sea insofar as flooding is concerned, still the tides are not completely dependent upon the moon, for if they were, the sea would only flood in that region subtending the moon, which is directly below the moon's highest point or meridian. Flood tides also occur, however, on the other or opposite side at the same time and therefore they cannot be dependent only upon the moon. Thus, indeed, it is quite obvious that if the earth moves in its orbit, it is impossible for the ocean, being fluid, as it is, and great in extent, not to take on some vibration from its motion. In the same way as it is impossible to move or carry around a vessel full of water, without the water's moving or shifting within, so is it also impossible for the sphere of earth (which is, as it were, a vessel or container of the ocean) to be carried around in its orbit without some movement of the sea being evident. Although this movement is insignificant

The reason for the tides.

It is the motion of the earth.

and essentially imperceptible in the sea, still, because of the ocean's vastness, it seems great to us on the shores.

Since, therefore, the earth, in its orbital course always pulls away from the ocean slightly in the east and leaves a little of it behind in the west, by reason of its fluidity, even though the nature of water tends towards levelling out, some motion necessarily results. This can only be brought about by these two simultaneous floodings and ebbings on the opposite parts of the earth, (because the whole ocean is divided in two by America or the New World). Thus when we see flooding about midday, this occurs at the same time in the antipodes in the middle of the night and when the ebbing occurs in our parts six hours before midday, this also happened in the antipodes six hours after noon, except when differences in regional contours, of which we spoke before, prevent this.

The moon is not the cause of the tides but is an element in the flooding.

As for the fact that there is a greater flood during tne period of the full and new moon, this arises from a greater force being exerted by the lunar body upon the whole earth then and consequently upon the ocean as well. Because of the ocean's vastness, the moon can hetter exert its influence upon it than upon small seas and those which are farther removed from the moon's zenith. Water has a natural tendency to flood, as we have seen in Book IV, Chapter 4 with the globe suspended in water, and Chapter 15, section 2. Inasmuch as the sea floods more abundantly when the influence from the moon is greater, it is consequently susceptible to even greater flooding at the time when this flow should otherwise be normal.

How to extract water from the bottom of the sea.

At this point it comes to my mind that some scholars claim that the water at the bottom of the sea contains no salt. This can easily be verified by the use of a glass (or better, a copper) vessel, a so-called $\frac{1}{4}$ recipient, which we described in Book III, Chapter 5. When this vessel has been completely evacuated of all air and is lowered by a line with a weight attached to the bottom of the sea and its stop-cock is opened by means of an auxillary line, the vessel necessarily becomes filled with the water which is at the the bottom of the sea. (This can easily be done with the help of a small iron armature attached to the auxiliary line, which operates the key of the stop-cock in such a way that when it comes in contact with a certain obstruction there, the stop-cock opens itself.)

CHAPTER 6

THE AIR SURROUNDING THE EARTH AND THE ELEMENTARY FIRE THAT HAS BEEN POSTULATED

Inasmuch as it is clear from Chapters 19, 20, 21, 30, etc., of Book III that the air surrounding the earth has weight equivalent to a column of water 18 to $19\frac{1}{2}$ Magdeburg cubits high and presses down upon the earth with this weight, while at the top of towers and mountains this pressure rapidly changes and becomes lighter, it follows that in the space extending above and beyond the air to the planets and fixed stars, nothing else corporeal or material can exist. Consequently no Elementary Fire will be found there. If there were such a fire, indeed, its weight would not be small, but rather so great as to be intolerable to man and the other animals because of its incredible height. Furthermore, its pressure would remain constant and would not become lighter with the slight variation in height found at the tops of towers and mountains; neither would it change at the onset of winds and rains, as experiments prove.

There is no Elementary Fire.

That a postulated Elementary Fire or celestial matter extending up to the fixed stars would be of such intolerable weight, is to be concluded from the Peripatetics' statement that the so-called Sphere of Fire as well as the planetary spheres and that enormous space up to the fixed stars is material or solid. Since these scholars hold that all solid bodies have a certain weight, therefore the weight of so great a mass pressing upon this earth of ours, considering the height involved, would be so heavy as to be insupportable to animals and cannot be compared with the weight of water a mere 19 or 20 cubits high. Furthermore there could be no such variations in the weight of air as we mentioned in Book III, Chapter 31, for how and by what means could so many solid spheres extending to the fixed stars quickly change with the onset of rains and winds and appear now heavier and again lighter? In the third place what is the height of a tower or a mountain as compared with the height of the planets and the fixed stars? Nonetheless on the top of any mountain or tower, according to Chapter 30 of the aforementioned Book, a lessening of weight is immediately perceptible. Reason dictates, therefore, that there is no other weight than that of this sphere of air surrounding the earth and that all other spheres, that of the presumed Elementary Fire as well as those of the planets and fixed stars, together with all heavenly matter, are figments of the imagination and are incompatible with reason.

If there were an Elementary Fire it would be of insupportable weight.

Neither could the weight of air change because of rains

nor on the top of a tower or mountain.

The spheres of the planets are not material.

As a matter of fact, if such a fiery sphere did exist, it would require nourishment and would undoubtedly burn up all the air beneath it. It would also shine at night, either by itself or by the absorption of solar rays, for whatever is made of matter attracts to itself the rays of the sun. We see proof of this from the air which creates twilight for us from the sun's rays. Why then should we postulate the existence of a fiery sphere when we see no indication of it?

It they were material they would absorb the rays of the sun.

What the sphere of air is.

There is, therefore, nothing more than a sphere of air around and above the earth. According to Chapters 8 and 10 of Book III, air is an exhalation or a kind of corporeal force flowing out from land or water into that space (otherwise empty) which is nearest to the earth and with which the earth is encircled as by a mantle. It cloaks it with a soft covering and protects it from the shafts of Phoebus by mitigating the all too violent attack of his rays. Or to put it another way, the sphere of air is nothing other than a fabric compounded of many tiny corporeal elements arising from the sphere of earth and sea (not, ideed, extending into infinity but dissolving only at the point of nothingness) which, possessing definite weight, adheres to bodies in the same way as the fuzz surrounding a quince. It is rightly considered, therefore, to be a part of the earth and consequently constitutes a single body together with the earth, follows it wherever it goes, and is carried along in its annual forward motion around the sun.

Air follows the earth.

The earth in its course perhaps leavs a slight residue of air behind it.

Nevertheless, it is not unreasonable to assume that just as a rose which one carries about leaves an odor behind, so also does the earth, in its annual course, leave a little of its odor, or air, behind in empty space. In quantity, however, this residue should be regarded as negligible compared with the whole sphere of air. The same assumption must be made about the other planets, because (as we wrote at the end of Book I, Chapter 19), we discover, upon looking through very powerful telescopes, that these, too, are surrounded by air.

The earth together with the air is an independent and free body.

If, then, the earth together with the air, is a body individual and distinct from the sky or ether, it follows that the earth exists independently in that intermediate and empty space between the stars and is supported by no matter from any other sphere. Otherwise if the air were surrounded by other solid and contiguous spheres, then the earth, together with the air and all higher spheres, indeed with the entire so-called universe, would constitute one body. In one and the same body, however, there can be no weight, much less a fluctuation of weight as we discover in the case of air. The earth, therefore, together with the air, constitutes one body, but independent of the whole universe, for it is individual and is separated from the rest by intermediate Space.

CHAPTER 7

THE HEIGHT OF THE AIR AROUND THE EARTH

The true height of the air cannot be measured with accuracy because, as we showed in Book III, Chapters 30, 31, and 33, air is greatly compressed in these lower regions near the earth and higher up expands more and more in otherwise empty space until it becomes imperceptible and finally dissipates into nothingness.

It is impossible to measure the height of air with accuracy.

It is a common characteristic of the different virtues that they are stronger around the body from which they emanate and subsequently hecome weaker and weaker. This is apparent, for example, in bodies which are odorous and which send out the virtue of their smell to a certain defined limit. Similarly, the air of our earth does not extend out infinitely but spreads out little by little and by degrees until it completely vanishes.

It is characteristic of all virtues of effluvia to be stronger around the body from which they emenate.

At what height, then, extenuated air can still be felt (that is, at what distance from the earth we can still perceive the essence of air), this can be determined from the beginnings and ends of dusks and the angular depth of the sun below the horizon. This height (since according to daily observations the angular depth of the sun at dusk does not exceed 24 degrees but usually extends at the most up to $21\frac{1}{2}$ degrees and many times at 18 and sometimes even 16 degrees) is calculated according to Riccioli's method, Volume I, Book VIII of his *Almagestum Novum*, Section 1, Chapter 14, questions 4, 5, and 6, to be 24 German miles. In Book X, section 6, however, he concludes that the atmosphere of the air (the elevation at which it is commonlv believed the stars' rays are refracted) is much lower and does not exceed a height of 4 miles, but he reduces this estimate at times to 3 or even fewer miles, especially in cold regions or during the winter. (This calculation does not pass unchallenged, however, as Varenius in his *Geographia* computes this number to be one mile and Kepler one half mile.) The perceptible height of the air, therefore, is not great (although our experiments seem to indicate that, in view of the expansive nature of air, very rarefied air extends not only 24 miles, but even 1,000 or 2,000 miles where it becomes increasingly rare and would be imperceptible to us here on earth), for if it were greater, the limits of dusk would also necessarily have to be adjusted to a much greater spread of sunlight. Daily observation, however, does not admit this.

Air is still perceptible at the height of 24 degrees above the horizon.

The atmosphere of the air cannot be perceived beyond 4 miles in the refraction of the stars.

It must be concluded that very rarefied air extends up to 2,000 miles from earth.

Indeed, if perceptible air were much heigher than 24 miles, it would always be dusk at night (for if air extended perceptibly up to distances of 100, 200, or 300 miles), it would retain the light or refraction proceeding from the sun by reason of the air's expansion.

If this were the case, the nights would continually glow as dusks because of the illumination of the upper layer of air being penetrated by the sun's rays on all sides. In actuality the nights do not project light and consequently no air is perceptible to us beyond the 24 German miles we have already cited. Much less have we reason to believe that there is anything else corporeal above the air's limit because this, too, would necessarily absorb some light from the sun. Inasmuch as this is not the case, Space there is pure and empty of anything corporeal.

Perceptible air is not always of the same height because at times it absorbs a large amount of water and at others its temperature varies with heat and cold.

It is clear from our experiments in Book III, Chapter 33 and elsewhere that air, or more particularly perceptible air, is not always of the same height, since at times it becomes heavier (and lower when it is saturated with a large amount of water) and at others it fluctuates in height because of its temperature (for in warm weather, air is more expanded and consequently is higher than when it is cold, as we see with the air in thermometers which expands with heat and contracts with cold).

The odor of the earth can undoubtedly be perceived at a great distance from here.

Furthermore, just as we detect the smell of bodies at a distance that have a particular odor, for indeed, we perceive large cities at a distance of 4 German miles and in sailing, we become aware of land at a distance of 6 miles, so there is no doubt that if a person were to set forth from the moon or any one of the planets through the air and to direct his course toward this earth of ours, he would become aware of its odor, or its surrounding air, at a distance of 500 to 600 miles, while a dog or an eagle would perceive it at even a greater distance away.

CHAPTER 8

AN OBSERVATION MADE IN THE CARPATHIAN MOUNTAINS IN HUNGARY BY DAVID FRÖLICH WHICH SEEMS TO MAKE A SIGNIFICANT CONTRIBUTION TO AN EXISTING OPINION ABOUT THE HEIGHT OF PERCEPTIBLE AIR AND STRATIFICATION

"Of all the mountains in Hungary," he says "the Carpathian are of first importance. The whole range of Sarmation mountains which divides the Hungarians from the Ruthenians, Poles, Moravians, Silesians, and that part of Austria which is on this side of the Danube is commonly called by this name. Their more rugged and higher ridges which reach up into the clouds are visible in the county of Zips near my beloved hometown, Caesareopolis, commonly known as Kesmarckt, which gives its name to the Kesmarcker Gebirge, also known as the Schnee-Gebirge (because they are almost always covered with snow). The Slavs call them the Tatrian or Tarezal Mountains, being as they are bald and eroded.

"These rocks in their ruggedness and precipitousness are far higher than the Italian Alps, the Helvetian and Tyrolean, and are essentially impassable. They are explored only on very rare occasions by lovers of nature.

The Carpathian Mountains tower above the Alps.

"Thus in June of the year 1615 when I was a youth, I sought to explore the heights of these mountains with two school companions. Climbing with great difficulty, I reached the summit of the first peak and believed that I had scaled the highest of them. Another rocky height, however, much higher than the first, presented itself as I struggled across vast and insecure rocks (if one of these were dislodged by a climber or were hurled down or rolled into the valley, it would take hundreds of larger stones along with it, with an accompanying noise so great that a person might have fears of the whole mountain's falling apart and overwhelming him) and again another greater height appeared. Thus with great danger to life and limb, I continued to climb many of the smaller peaks throughout the valleys, the ones already scaled always appearing lower than those lying ahead, until I might finally reach the highest of them all. When I looked down from the sides of the mountains upon the low-lying valleys filled with trees, I could see nothing but dark night and blackness, like the depths of the air (although the sky is generally termed bright and clear), and it seemed to me that if I should fall from the mountain, I would land, not upon the earth, but into the middle of the sky. Because of the height of the slope, all visible objects were small and dim. When I sought a higher mountain, I found myself in the midst of very dense clouds. Once I had pushed my way through these and after an interval of several hours when I was resting not far from the summit, I looked out and saw in those places where I seemed to have been surrounded before by clouds, that white compact clouds moved. For several miles above these and beyond the bounds of Zips, I was granted a good view. I perceived, however, some clouds even higher and others lower with not a few being at an equal distance from the earth. From this I learned three things: (1) I had passed through the lower limits of the middle region of air. (2) The distance of the clouds from the earth is not the same, but is in some places higher and in others lower according to the

When rocks are dislodged and are hurled into the valley, a fearsome noise is heard.

Forests and trees appear black.

The author has a view over the clouds.

The clouds are not equally distant from the earth.

condition of the vapor. (3) The distance away of the clouds nearest the earth is much less than certain natural scientists have described, for certainly they are not 72 German miles distant; rather they are only a half German mile away. When I had reached the highest point of the mountain, I found the air there so tranquil and soft that I could not even feel a hair move when, as a matter of fact farther down the mountain, I was aware of a very strong wind. From this I gathered that the summit of this Carpathian mountain was one German mile distant from its base and extended up to the highest region of air where there are no winds.

On the top of a mountain there is no wind.

The height of this mountain is judged to be one German mile.

"I fired a shotgun on the summit and it produced no louder noise than if I had broken a stick of wood. After an interval of time, however, the sound increased and filled the lower parts of the mountains, the valleys, and forests.

When a shotgun is fired, there is a small noise.

"As I descended through the snows which had accumulated over the years within the valleys, I again shot off the shotgun and a greater and more fearsome sound arose than that produced by a very heavy gun ("tormentum"). It made me fearful that the whole mountain might fall down upon me with the concussion. This noise lasted for 7 or 8 minutes until it had penetrated the most remote ravines on whose sides the air reverberated many times. Inasmuch as the were no hollows of this kind on the mountain's summit, the reverberation of the first shot I fired was almost imperceptible until it descended nearer to the caves and valleys and struck them with more force than on the summit. It is also the case that snow and hail fall very often in the high mountains even in the middle of the summer at a time when it is raining on the land below, as I myself experienced. The snows of different years can be distinguished by their color and the hardness of the crust."

Subsequently it becomes louder.

Echo.

CHAPTER 9

THE STRATIFICATION OF THE ATMOSPHERE

We admit that the atmosphere can be divided into regions, but under certain understood conditions. Although air maintains a certain weight about the earth, the lower part is always being pressed upon by the upper, therefore making it heavier, and the higher the upper part is, the lighter and more diffused, still it can be divided into three strata, the first or lowest, as it were, then the second or middle, and the third or highest.

Air can be divided into regions.

The first of these regions can in turn be divided into three different layers. The first is the limit up to which the stars' rays are perceptible because of the presence of the vapors of the earth (like smoke) and water, as well as other exhalations and fumes. This is usually called the earth's atmosphere and its height does not exceed four German miles. The second stratum of this first region is the extent to which air can absorb the sun's rays. Its height is roughly 24 German miles, as was seen in the foregoing Chapter. The third stratum of this lowest region is that in which there are extremely rarefied water vapors, but which nonetheless absorb some whiteness or brightness from the sun's rays. This stratum can be perceived only in its very dark color, which we discussed in Book IV, Chapter 12, and must be considered the bounds of the lowest region and the beginning of the second.

The lowest region has three grades.

The cerulean color.

The second region is that where the cerulean color ends and where pure air, free of all water vapor begins. It extends up for several hundred miles and in becoming more and more rarefied and lighter, it exerts less and less pressure downwards through its increased diffusion.

The second region of air.

The third or highest region of air is very clear. It consists not only of very light and pure air or ether but extends, as one presumes, for 1,000 to 2,000 miles. It gradually becomes so completely dlffused that absolutely no air or odor or corporeal effluvium from the earth remains and pure space, empty of all matter, begins.

The third region of air.

The experiments in the preceding Book III prove and visually demonstrate that air is constituted in this way and can be compressed. Furthermore they show that it can be more and more expanded until it disappears into nothingness (when it encounters greater space and is not subjected to pressure). This upper region of air, then, is completely imperceptible to us mortals, although one can easily surmise its existence, unless, under favorable conditions, we actually perceive the tails of comets there (which are nothing other than refractions of solar rays in this ether). More will be said about this in the Appendix of this Book.

The tails of comets.

Although air is divided into different regions of this kind, not only insofar as its weight and density, etc. are concerned, (as can be seen with clouds which are found at different altitudes, the heavier being lower and the lighter higher) but also in temperature, still no one variety of air penetrates solid bodies more or less easily than another. This is true because all air, even the most rarefied, is a corporeal effluvium, and one corporeal element cannnt penetrate another. As for temperature, air does not become diversified from its presence in the different regions we have mentioned but from the qualities of the earth it covers, for the air which is nearer to the earth varies

Why clouds are formed at different altitudes.

More rarefied air does not penetrate hard bodies or glass.

What accounts for the diversity of air?

with the temperature of its surface. Experience shows that the temperature of air is not always the same at the same time in all lands nor is it of the same temperature no matter what the time of day on the top of the same mountain. When Zarabella reached the summit of Mount Venus near Padua in the summer, he did not find greater cold there than in the region below, a condition which, in popular opinion ought to be the case. Indeed, although greater cold is usually felt on very high mountains than in the lowlands, still, in our experience we have not found the same temperature to prevail on all mountains at the same height. Indeed, we have seen that all very high mountains are not covered with snow, while more often some of the lower ones are snow-covered throughout the whole summer. This makes one think that this cold emanates from the earth and that the mountains themselves and the earth admit frost (which may be an incorporeal virtue which comes from the moon) according to the differences of time and place, and in greater or lesser degree according to its nature or properties. This, then, is the source of the air's variety, especially in the middle of this lowest region where heat and cold come together from elevated ground, together with frost. One has no reason to doubt, however, that air in the highest part of this lowest region is usually colder since the reflections of the sun's rays reach there with less strength.

Frost in the mountains does not properly have its source in a cold region of air but in the effluvia of a mountain.

Frost perhaps originates in the moon.

CHAPTER 10

THE REFRACTION OF AIR AND THE CONSEQUENT DIFFERENCE IN THE APPEARANCE OF STARS, BOTH AS TO THEIR LOCATION AND SIZE

Air refracts the rays of stars and makes their sizes fluctuate.

Because the air of the first region always contains some water, and the higher or denser this stratum is, the more water it holds and the lower, the less, and because air is itself corporeal, as we have said in our preceding Chapter 7, it follows that air not only receives the light of the stars but also refracts their rays in various ways, dependent upon the difference in their height, thus causing the apparent size of the stars to fluctuate.

For when air absorbs more moisture, then it expands more and occupies a larger space. The less water it contains, however, the more it is contracted and the less space it occupies. Thus, contracting, it renders the refractions smaller. For this reason, whenever it has rained, the stars appear brighter and less large and have smaller refractions, and consequently this is much the best time for observing stars with accuracy.

After a rain, stars appear brighter.

Furthermore, variations in temperature, either heat or cold, or the presence of frost make a difference in the air and in this way change the aspects of the stars.

Stars also appear higher.

Common experience shows, moveover, that because of the refraction of air, stars (1) appear to us higher above the horizon than they actually are. As a matter of fact, this only appears to be the case when in actuality they lie hidden below it. Thus many of their characteristics are visibly altered (namely, their elevations, risings, settings, longitudes, latitudes, declinations, etc,) and even the equinoxes themselves and the solstices are rendered uncertain and unequal. (2) Their diameters appear larger on the horizon and the higher the stars are, the more contracted they seem in size. Furthermore, for the same reason one sees at times several suns and moons appear and not infrequently the sun and the moon (when they rise and set near the horizon) are elliptical or oval in form, with the result their diameter horizontally is 30′ while vertically the diameter is 36′.

The reason for the varied dimensions of the stars.

All these phenomena have their source in the variations in air and its greater or lesser depth or height. Thus it is manifestly clear that we see the stars not through their direct but rather refracted rays, and that in astral observations, errors can easily creep in unless the quality of the air is well taken into account.

The source of astronomical errors.

As for the first observation it is certain that because of the earth's sphere of air the stars always appear somewhat elevated above their true positions.

Tycho Brahe compiled two Tables of Refractions, one for the fixed stars and planets and another for the sun (which he claims is also applicahle to the moon to some extent). He gives these refractions for the sun up to 45 degrees in altitude and 20 degrees for the fixed stars. For indeed, if one were to assume the stars to be higher, he thought it would be superfluous to take in consideration their refractions because the rays of the stars would penetrate the vapors almost vertically and their refractions would not be observable. I find, for example, using a quadrant, that the height of the sun is 5 degrees above the horizon. This height corresponds with 13 minutes 30 seconds on the Table, a number which must be subtracted from the 6 degrees to arrive at the true height. The same process is applicable to fixed stars.

Tycho Brahe compiled Tables of Refractions.

But although the refractions of the fixed stars are somewhat less than that of the sun, both in their positions near the horizon as well as at other altitudes (because the stars are observed particularly at night in the winter while the sun can be seen only during the day and in the summer), nevertheless according to the Laws of Optics, the refraction of the fixed stars ought to be of essentially the same size as that of the sun, and not only at 20 or 45 degrees, but at any height, for indeed, the sphere of air is everywhere about the earth and consequently refraction is too. In the year 1600 when Tycho was in Bohemia and was carrying out his experiments he found this to be the case and discovered that for Spica Virginis at a height of 30 degrees the refraction was only one minute.

For there is refraction at all heights.

For this reason I say that neither by careful examination nor reliable observation, can the refraction of stars be extended to any definite degree as a limit of height, for there is some refraction at every height which makes observations differ on account of the changing height itself of air.

Because of the variation in the air, the Refraction Tables are also rendered uncertain.

Secondly, it should be understood as regards the variation in size of a larger or smaller astral sphere, that generally speaking, (for we are not discussing here the approach of Venus, Mars, Jupiter, or Saturn toward the earth), the nearer a star is to the horizon, the farther away it is and the thicker the layer of air through which we would see it and the greater refraction we would have. For this reason the stars near the horizon appear larger than the rest and smaller when they appear directly overhead, not because they are nearer to us than at their zenith but because they are seen through a greater thickness of air which makes objects appear larger to us than they actually are, for much air produces greater refraction and enlarges objects more than a little air.

The difference in size of a star.

And since the lower and nearer air is to the earth, the more it is compressed, and the more it is compressed, the denser it is, it follows that stars which are seen horizontally through a greater mass of air, appear to be much larger. For example, we said in Chapter 7 that Riccioli deduced that the atmosphere of air, where the stars' rays are refracted (that is, the point to which astronomers extend refractions) has a height of 4 miles. When, therefore, we see a star horizontally, the observation is made not only through these four miles of the so-called atmosphere but also through another 4 miles which is commonly represented as the visible horizon from the earth's surface. Thus the depth of the air through which a star is observed on the horizon will be 8 miles. Indeed, as for these last-mentioned 4 miles (namely the distance at which the horizon commonly is believed to be visible to our eyes on earth), they encompass air that is the most compressed and moisture-laden of all and for this reason the stars there appear much larger.

The reason for such different appearances.

The deeper and more compressed air is, the larger the stars appear through refraction because of its aqueous vapor. In the same way as coins seem larger when submerged in water because of the curved surface of the water, there is a similar result with air.

The sun and moon appear larger when they are in the southern signs of the zodiac.

It follows from the foregoing that the sun and moon vary in size according to their height at mid-day in the course of their daily travels. When they are in the northern signs of the zodiac, they appear smaller, but the lower they are at mid-day, that is, when they are in the southern signs of the zodiac, they seem larger to us. This difference in size, then, is not caused by their greater distance from the earth but rather by the refraction of the air.

Furthermore, because the height of the air fluctuates as its weight and temperature varies, (as we observed at the end of Chapter 7), the size of the sun and moon also vary, which is the source of their different appearance and sizes. This subject was addressed in Book I, Chapter 24.

CHAPTER 11

THE ECCENTRICITY OF THE SUN AND MOON

The sun does not admit of apogee or perigee.

Because the eccentricities, or the so-called apogees and perigees of the sun and also the moon,under appropriate circumstances have their source in the refraction of air, as well as the variations of its depth, something must be said about them here.

That eccentricities of the planets should also be postulated is demonstrated at certain times by their diameters, which are sometimes larger and again smaller. This is sufficient proof that they are not carried precisely around the center of the earth but deviate from it.

Nature does not admit of irregularity in itself.

Since we have all been convinced that we should beiieve those bodies which belong to and appear in the heavens are of the greatest regularity in movement, thus astronomers in times past similarly supported this opinion vigorously and always tried to assign regularity to whatever motions seemed irregular. The nearer a person came to defending the regularity of motions in setting forth his hypotheses, the more credible his theories were held to be. No one, therefore, will find it remarkable if he sees us also leaning toward regularity in nature.

The planets are not eccentric in respect to themselves, but with respect to the earth.

As for the planets or the five wanderers, namely Saturn, Jupiter, Mars, Venus, and Mercury, it cannot be disputed that they appear to us at certain times now larger and again smaller. This arises because they have a different center from that of the earth, but the

The earth is not carried around the sun irregularily or eccentrically, nor is the moon around the earth.

sun is the rotational point for all of them, as can be seen in Plate IV, in Book I, Chapter 19.

Now, we see the sun and, under appropriate circumstances, the moon, appear large at one time and small at another. This is not caused by these bodies moving toward or away from the earth, but it is the air, which either because of its varied composition, both in the upper reaches or the lower, or its greater or lesser thickness, that accounts for these differences in appearance, as we said in the preceding Chapter. We have an example of the air's ability to change rapidly in Kepler's *Optica Astronomica* where he asserts that on one and the same day, March 2, 1588 he found that the difference observed in the moon's diameter at the highest point had been at one time 31′, then $32\frac{1}{3}$, and again $30\frac{1}{4}$, and on the preceding day to be 33′. Furthermore, when the moon was at middle distance on February 22, 1591, he observed it twice at 31′, six times at 32′, seven times at 33′, and six times at 34′. On one and the same day, then, the moon's size appeared to vary with such frequency! Furthermore, Gemma Frisius states in his *Radius Astronomicus* that on December 15, 1542 a little after the first quarter of the moon, its diameter was observed to be 30′, at which time, according to Ptolemy, it should rather be 50 minutes and more (because this astronomer computes a distance of $33\frac{1}{2}$ earth's radii to the moon in perigee during its first quarter from which it follows that the apparent diameter then would amount to 56′, which is almost two times greater than the apogee of the moon in conjunction and opposition). Many others, however, have observed that the moon in every quarter has a much smaller diameter. See Riccioli, Book IV, Chapter 14, number 2.

Examples of the change of air from different heights.

A greater thickness of air makes stars seem larger.

Furthermore, a greater or lesser thickness of air brings about changes in the appearance of the sun and moon. For when they are in the southern signs of the zodiac, and thus are lower, we see them through a greater thickness of air and consequently they appear larger, as we said in the preceding Chapter. Thus, in winter when the sun is in Capricorn, it appears larger because of a thicker layer of air between it and our vision. When it is in Cancer, however, and more directly overhead, it is observed through less air and consequently appears smaller. This also accounts for the different appearances in the moon and for the fact that astronomers never agree in their observations of the diameters of the sun and moon nor of their apogees and perigees, as one can see ln Book I, Chapters 22 and 24.

The reasons used by astronomers to

They seek, however, to prove such an apogee of the sun and moon:

(1) through eclipses, which Riccioli himself in his *Almagestum*

note 2 of Book III, Chapter 10 and also the note of Book IV, Chapter 16, admits is completely unreliable. One must likewise consult the following Chapter 24 of this Book which treats of eclipses;

prove the eccentricities of the sun and moon.

(2) through the parallax of the moon found to be now smaller, now greater at the same distance from its zenith. But even if this were so, and we do not wish flatly to deny that the sphere of earth at times does exert a modicum of influence on the moon through a particular disposition on its part to be attracted to it or slightly repelled from it (just as we see with the sulphur globe that propels a feather farther away from itself at different times. See Book IV, Chapter 15, part 3 in this regard), still it does not follow that the moon admits of any center other than that of the earth and that the time of its apogee and perigee can be forecast with complete certainty. As for the parallaxes seen by astronomers in different places and at different times it is clear that such differences as they have observed had their origin in the difference of refraction in a star, when, as it were, the air was higher or lower, or the moon was observed from a more northerly or southerly point of vantage. This is also the reason that Tycho and Copernicus, who were northerners, observed a smaller parallax than Ptolemy and thought that the Ptolemaic observation should be corrected;

Even though the moon is occasionally nearer the earth, it nevertheless does not have an eccentric course.

(3) these same astronomers say that it is clear from observations of the equinoxes that the sun lingers in its northern semicircle (that is, from the beginning of Aries through Cancer to Libra) 186 or 187 days, but tarries in its southern semicircle (from the beginning of Libra through Capricorn to Aries) 178 or 179 days. Therefore they attribute this discrepancy to the apogee and perigee of the sun, holding that the sun in apogee lingers in that quadrant or semicircle in which it makes longer delays while in perigee it does the opposite. From the time of Hipparchus, however, until that of Albattani it delayed more days or hours in the first quarter of the ecliptic than in the second and in the second more than in the fourth and in the fourth more than in the third, with the result that in the ascendant semicircle, that is, (from the beginning of Capricorn through Aries to the beginning of Cancer) it delayed for a longer time than in the descendant semicircle (which extends from the beginning of Cancer through Libra to the beginning of Capricorn). Therefore the sun's course has changed these days so that the delay which it showed from Hipparchus' time to Albattani's in its movement through Capricorn to Aries, it now has from Aries through Cancer to Libra.

Does the sun linger in the northern constellations for a longer time than in the southern?

In the Astronomical Tables the sun's motion is found to be uniform.

Furthermore, the sun's movement according to the Astronomical Tables is consistent and uniform. Thus its motion, which seems irregular to us (who observe it from the earth) in actuality, is not,

Its apparent

irregularity of motion, therefore, as observed by astronimcal instruments, must be ascribed to the air.

but only appears to be so. We conclude, therefore, that this apparent variation in motion must be ascribed to the air and if this is the case, there is no apogee or perigee of the sun. For when the sun is observed in the vernal equinox, it will still be in the sign of Pisces; on the other hand, in the autumnal equinox when it will be visible at the beginning of Libra, it will be at the distant end of Libra, depending upon the amount of air, or refraction, which is present. Astronomers have not yet explored this problem sufficiently. Chiaramonti in his *De Universo*, Book III, Chapter 12, and Fracastoro, and others, moreover, agree on this point.

CHAPTER 12

THE MOTION OF CIRCULATION OF THE AIR

The air revolves with the earth.

Inasmuch as the sphere of air envelops the earth with a weight equivalent to a column of water 18 to $19\frac{1}{2}$ cubits high and in this way forms a unity with the earth, it is a necessary corollary that if the earth spins, as Copernicus believes (See Book IV, Chapter 9), the whole system of air at the same time accompanies the earth undisturbed.

Just as we are carried around with the earth by the force of gravity (or in other words, the conserving virtue), so does the air weighing down and clinging to the earth with its mass likewise move, thereby being transported along in the earth's course.

Air carries along with it all things projected into it.

For this reason air carries along all things projected into it and other things encompassed by it in a swift, but nonetheless constant and uninterrupted course. This is done in such a way that these objects do not jump, impinge upon, or even touch one another. We shall take as an example a stout and water-tight ship that sails its peaceful course on tranquil waters. When a container of water like a watering can is suspended within the ship and drops of water are allowed to drip down, all fall to the ship's bottom and strike the same place. The ship's course in no way alters their fall even though it sails at top speed. The reason for this is that the air, shut up in the ship. is borne along with it and carries the drops of water and whatever else may be there with it. It does not matter with what sorts of actions these bodies are associated. In the same way targets can also be hung up to which weapons can be directed, no matter where a ship is being sailed. For a ship with its interior air and everything else shut up in it constitutes a single body and that which

That which constitutes a whole in conjunction with a body, follows lthis body.

is carried along with it is included in this body and as a part of it, follows its whole.

Thus it necessarily follows that all things which are above the earth and in the air, rotate with the earth and its system of air, whether they are falling drops of rain, clouds, winds, meteors, stones, weapons, or even balls discharged from large guns, etc., because all these, together with the earth, constitute one body.

The objections, therefore, of certain philosophers are of little weight in which they dispute the existence of the earth's motion, basing their theories on the movements (or rather the course) of a stone and all other such objects falling to the ground. They do not recognize these motions are carried out in one and the same body of air by which they are transported and are compelled by necessity to follow its course throughout. It is quite impossible for these bodies to have independent motions (as these philosophers wish to understand them).

Although, as we have said, air is uniformly carried around with the earth, still in some places, particularly below the equator and the tropics where the earth's revolutions are somewhat more rapid because of the greater circumference there than elsewhere, a certain slight slackening is perceptible which indicates that the air does not completely obey the pull of the earth's revolution. The preceding Chapters 5 and 15 should be looked into regarding this phenomenon and its source, for it presents no light argument to bolster the Copernican system.

The turning of air below the tropics is somewhat slackened.

Further it must be assumed that inasmuch as we have established that there is a certain slowing down of air at the equator, similarly must air following the earth's revolution in the higher regions, be more sluggish. Thus in the very high regions where the air is extremely rare and light (and at the same time is unable to impart an impulsive virtue) scarcely any revolution can be perceived.

CHAPTER 13

THE SPHERE OF EARTH AND WATER DOES NOT LIE AT REST IN THE AIR

Since we see that the earth is surrounded by a covering of air which is in no way supported on columns or suspended by ropes, we think that it hangs at rest in the air, or as they say in German, floats.

Reasons why the earth does not lie at rest in the air.

Now it has been more than adequately demonstrated that air is only an effluvium of all things, in particular, the seas, waters, and land which rests upon the earth's sphere with a certain weight. It follows that the earth as a whole cannot lie at rest in the air as it is a part belonging to it.

Furthermore, were we willing to concede that the earth rests in the air, it would follow necessarily that it is of the same weight as the air, for whatever lies at rest in any surrounding matter, this is of the same weight as the matter itself (as, for example, we see that air rests in air, water in water, and all other matter in those substances similar to it, because they are of the same weight. We have quite a clear example of this with the small sphere suspended in the middle of the water, described in Book V, Chapter 4). According to the theory of the Peripatetics, however, the earth lies at rest in the air which surrounds it; therefore it is of the same weight as the air. This theory, however, is not valid because they say that the earth is a heavy body but the air is light. It follows, then, that this belief that the earth lies at rest in the air has no validity.

A further proof that the earth cannot rest in the air lies in the demonstrated weight of air around the earth; this is a necessary consequence either of the conserving virtue of the earth (which was discussed in Book IV, Chapter 5) or in the so-called weight of lower bodies, namely the air itself and Elementary Fire, etc.

If this weight is the result of the conserving virtue, which is a characteristic of the earth, it is impossible, as we have already said, that anything can be at rest in the air which this virtue holds or attracts to itself. Consequently the earth cannot lie at rest in the air, but together with the air which it holds, hangs freely in all-encompassing Space.

Objections and their resolutions.

The air is one body together with the earth.

But if it stems from the innate weight peculiar to air, as the Peripatetics say, through which all heavy things tend downwards, it follows that air, pressing down upon the earth from all sides with its weight, is, together with this sphere, one single compact body. A compact body, however, and indeed one which is freely suspended as the earth with its air is, hangs in its entirety, for indeed one cannot be separated from the other. If one were suspended or lay separated from the other (as the earth, for example, from its air), it would necessarily follow thnt the air, in turn, would rest in some third body (some so-called celestial matter). That air does not and cannot rest in some third body is clear from the fact that it lies upon the earth with weight and therefore is independent of a third body (for that weight indicates, as it were, a separation from an external third body and an inclination toward another, namely the earth), and is one with the earth. If, then, air is bound to the earth, it follows that the earth cannot lie at rest in the air because these two elements constitute one body. As to whether the air together with the earth is a heavy body outside the center of the world or rather a light body in the center of the world that is freely floating, and likewise whether, in any case, it is in the center of the world at all, see the following Chapter.

CHAPTER 14

THE SPHERE OF EARTH AND WATER IS NOT IN THE CENTER OF THE WORLD

Longstanding opinion concerning the weight of the earth and its position in the center of the world has put down such deep roots in the minds of men that they do not dare contradict this without trepidation. Thus so as to support this conviction they held that the earth was in the process of losing its weight and of being removed from the center of the world at the time when Pythagoras had first began to say that it was mobile. This theory not only attacked the beliefs of the common man but also seems to have outgrown the theories of almost all scholars who regard as holy the words of Aristotle in all matters.

Thus they have not only attributed weight to the earth and all other things which seem heavy to us, but also a peculiar innate disposition in objects to fall to the earth as to the supposed center or lowest point of the world. Up to the present time they have consistently accepted as true Aristotle's statement found in *De Caelo*, Book I, Chapters 8 and 9 that "None of those things which exist in the world can be outside of the world because each and every one, of necessity, seeks its own proper place. In the same way, no body of earth, water, or air can exist outside of the world since there are other places in the world which are theirs by nature. Therefore, they cannot have being if they exist contrary to nature, for each thing itself must seek its own individual place."

This opinion of Aristotle is essentially true.

All things seek their own individual place.

In accordance with the Aristotelian thesis, some men have sought to assign to the earth a certain point, as it were, in the world from which it could not be moved (a statement, however, which is not made in the text). Furthermore, they devised various strange paradoxes about the gravity of the earth, likewise an assessment of its weight and a calculation of the time required for a stone, released from one of the fixed stars or from the outermost bounds of the world to fall down to the earth. They further asserted that if there were an opening through the whole earth to the Antipodes and if a stone were dropped through it, the stone could not fall farther than the middle of the opening and that after a series of vibrations, it would finally come to rest in the earth's center, etc.

Objections to the mobility of the earth.

In these and similar discussions, they reach wondrous conclusions about the motion of projectiles and use these to disprove the motion of the earth. Among these Father Anton Maria de Rheita in his book, *Oculus Enoch et Eliae*, Chapter 3, presents some remarkable and abstruse proofs with which he believes he has imposed silence upon the whole Copernican system. (1) He imagines (if motion is appropriate for the earth) that a stone, falling from the top of a tower would not follow a straight downward path, but rather a spiral, curved, or circular one." (2) He tries to prove that "the motion of a stone, then, cannot be accelerated but will always be the same and uniform," a statement, however, which is contrary to empirical knowledge. (3) He thinks that "a stone, released through an opening In the earth to its center would describe a still more remarkable movement because of the earth's rotation, etc. and that, therefore, it is impossible that the earth move."

Resolution of these objections.

To be sure, there is no need to refute these and similar conjectures since it is more than clear from the preceding Chapter 12 that air carries all bodies of this kind with it and accordingly that nature does not use a circular forward thrust ("latio"), but rather a perpendicular. Furthermore spiral, curved and circular motion, etc. (or rather "latio" which is discussed in Chapter 16) is not contrary to nature as we find with snowflakes and drops of rain and often hailstones, which rarely fall perpendicularly.

What is more, as for the Aristotelian hypothesis that the earth is indeed a heavy body and that all objects of weight, wherever they may be, are drawn towards the center of the world in accordance with their nature – this theory has never been proved or demonstrated – but rather is contrary to considered reason and the actual disposition and order of earth bodies relative to one another.

Should one assume that this Aristotelian hypothesis is true, however, then this tendency to be drawn to the center of the world has its origin either in a natural innate quality of each body or in the forward motion of surrounding matter, or in a unique ability or virtue of each body's being attracted to the center of the world, a focal point which they say is the earth.

Arguments against gravitation toward the earth.

(1) If this tendency has its origin in the innate nature of a body to occupy its own place, for example, because a stone (which may be on a star or high above the outermost bounds of the world) has the natural tendency to descend to the earth, one may ask how it is possible that this nature, intellect, quality, or virtue to seek the

world's center would be present in a stone when it is so many hundreds of thousands of miles away (where the earth cannot be counted even of the proportions of a point)? One may ask further with what help or by what means or aid a stone can set out on its long journey and attain enough strength to descend to the center of the world. Then a further question arises: if a person were on a star, Sirius, for example, (where, because of its being so many hundreds of thousands of miles away and because our earth is completely invisible since this person's position is turned away from us, he does not know where to look for this earth of ours, whether up or down, or to the right or to the left). Furthermore, if he were to drop a stone from his hand, where would it come to rest? Would it stop somewhere in the ether? Would it fall from this star directly toward the center of the world (which, indeed, cannot be seen and whose position is not known)? Or would this stone go around the star's surface that is turned away until it came to the other side and then finally would pursue its course to the invisible center?

(1) *If this tendency has its origin in an innate quality.*

(2) If this tendency arises from the forward thrust of surrounding matter (for indeed we see water, bubbles, inflated bladders, wood, and all rather light objects rise upwards because celestial material pushes all these bodies to their own individual places), it follows that celestial material is heavier than the aforementioned stone and that all weight in these lower regions is light, that is to say, all things which seem heavy to us are extremely light and by means of some natural virtue, namely the weight of celestial matter, are pushed to a lower place which we call the center of the world. If, then, a stone is shoved to the earth by surrounding matter in the same way, then the earth would have to be the lightest body of all. The veracity of this conclusion, however, they dispute.

(2) *If it arises from the forward motion of surrounding matter.*

(3) If this tendency arises from a certain singular characteristic or quality of the world's center which, as it were, attracts all things, then necessarily this center must be something of substance. You say, this is the earth. Then the question arises (1) why do we wish to attribute such a characteristic or virtue to the earth inasmuch as it is the center of the world? Should not rather this quality be attributed to the body itself whether it be in the center of the world or outside? For indeed, it is contrary to reason that the center of the world should have such a characteristic because in itself it is not a thing of substance. (2) One further asks how it is possible that the earth can have an innate capability to attract all things to itself from great and incomprehensible distances (when the earth, as compared with the vast and infinite Space of this world, is to be counted as a grain of sand). Another question arises as to how the earth can

(3) *If it has its origin in a quality of the world's center.*

extend this capacity to attract so widely that it fills the whole Space of the world and reaches its outermost bounds and perceives all bodies herein and directs them toward itself?

In view of all these arguments which are contrary to reason as well as the great number of attendant inconsistencies which have been demonstrated throughout, one must conclude that our sphere of earth and water is not the center of the world and accordingly that it does not attract to itself bodies which are compatible with it from the farthest boundaries of the world but only those a certain distance away, for the sake of its own preservation.

According to Tycho's hypothesis individual planets can be regarded as centers of the world.

In the same way, moreover, as Tycho's hypothesis places the earth in the center of the world, so also can Saturn, Jupiter, Mars, Venus, and Mercury be placed there and the same phenomena will prevail. For all rational animals (if such exist on these planets) can assert that the planets they inhabit are the centers of the world, and just as we see the sun in one year run through signs of the zodiac and Saturn circle the sun in thirty years, so could a man on Saturn see the sun run through the zodiac once in thirty years while the earth circles the sun twenty-nine times. Furthermore, all other phenomena described in Tycho's system would also prevail, as can be seen above in Book I, Chapter 9 as well as in our subsequent Book VI, Chapter 16.

CHAPTER 15

THE ROTATION OF THE EARTH

What motion is.

Motion is the subject to be discussed here: "latio" or forward motion will be the subject of the following Chapter.

Motion is the activity of a body which results from an inner force in the body itself which causes motion directly. That is to say, it functions without help from anything else, as for example, a living body which moves itself and its parts, eyes, ears, feet, etc. independently.

In the same way earth bodies like the sun and planets turn themselves about in a circle, a motion which is called revolution.

The earth's motion of revolution, whereby in twentyfour hours it turns completely about its central axis and moves from west to east while maintaining its orientation unchanged toward the pole, has been described by Plato, Nicetas, Ecphantus, Heraclides and other writers of antiquity and later on by Copernicus and followers of his like Tost, Argoli, Longberg, Muliers, Gilbert, etc. As a result of this motion we go to meet the stars and see them pass into the west, just as sailors who are sailing toward a certain region see the shores and banks move and recede into the opposite direction, although in actual fact they are standing still.

The motion of revolution.

The motion of revolution is the lightest and most natural of all motions and has no need of an external stimulus. Thus anything that hangs freely can make itself revolve by means of a certain virtue of its own spirit. This revolution is perpetuated by means of an impulsive virtue (See Book IV, Chapter 2) whereby it is carried out and continued with uniformity and consistency. See Book IV Chapter 9 where this motion is discussed.

The source of the stars' twinkling.

The revolving motion of the planets is significantly different from that of the stars.

All stars which are animate possess this motion, as is visually evident to us in the sun and the twinkling of the fixed stars and similarly in planets, particularly Venus and Mercury since they are nearer to us than the others. There is a great difference, however, between the revolving motion of the sun and fixed stars and that of the planets, for the former are fiery bodies of very gaseous natures whose motion is extremely animated and rapid. The latter, on the contrary, are solid bodies which embody an impulsive virtue through which their motion of revolution for the most part can be achieved.

The circulation of the air slows down somewhat in the tropics.

Moreover, the fact that the earth rotates is confirmed by the journeys man makes by sea. For although air may be held in by the earth's conserving virtue and revolves along with it, still bodies of little solidity, as air is, are less able to receive this impulsive virtue than others that are more solid, and for this reason air is somehow left behind the earth and does not follow it along as quickly as it moves. This phenomenon cannot be as easily observed in our latitude, however, as much as it is under the equator or in the tropics because the rotation of the earth is not as rapid here.

According to the observation of sailors, air slackens somewhat below the tropics.

In the tropics, however, sailors have observed that the wind always blows from east to west with so constant a force that they can make a voyage like that of Columbus from Spain to the West Indies in essentially a month's time. For the return trip, however, four months are required. On the other hand Portugese sailors complete the journey to eastern India via the Cape of Good Hope to Calcutta much less rapidly than the return trip, although winds of equal strength blow. It has been further observed by skippers that the voyage from Palestine to Spain is completed more quickly than that from Spain to Palestine. Indeed, one can see with one's own eyes that the Atlantic Ocean washes in on the shores of America, while contrariwise it is evident that the Pacific Ocean on the other side of the continent moves away from its shores. This is particularly visible from Cape Corrientes between Panama and Lima. The air (since it is not connected to the earth but is only contiguous to it and its matter is more rarefied, making it less able to receive the impulsive virtue and consequently unable to follow the rotation of the earth so quickly) and likewise the sea (on which air presses with its weight) leave a little of their substance lagging behind in the west when the earth rotates towards the east. Nevertheless the sea does not move itself from one part of the earth to another; rather, it is the earth which leaves behind it parts of the water, one after the other.

Even though the Peripatetics seek to ascribe this lag in the air to some Tenth Heaven or a First so-called Mobile, nevertheless, because it has been demonstrated that above the air there is nothing connected or contiguous but only Space without any matter whatsoever, or, in other words a vacuum, one must easily conclude that this imagined First can have no influence upon air that is so far separated and rests so heavily upon the earth and cannot carry the air around with it in any way.

The magnetic virtue also indicates the turning of the earth.

Another very clear indication of the earth's spinning motion is the stability of the earth's poles which are maintained immobile through the magnetic virtue, enabling the earth to turn about its axis. For what other purpose would this virtue serve? We have discussed this question in Book IV, Chapter 7.

How far the earth turns in one hour and in one second.

Any point on the earth's equator in one day passes through 5,400 miles, which is the earth's circumference. At this rate, in one hour it travels 225 miles, in one minute $3\frac{3}{4}$ miles and one second $\frac{1}{16}$ of a mile. The nearer a point is to the poles, the slower the motion on the earth's surface is, with the result that any point in Germany, or more particularly, one located in Magdeburg with a latitude of 52°, travels 3,324 miles in one day, $138\frac{1}{2}$ miles in an hour and $2\frac{3}{10}$ miles a minute or 23,000 regular paces (10,000 of which constitute a mile) and in one second, 384 paces.

CHAPTER 16

"LATIO" OR THE FORWARD MOTION OF THE EARTH

What forward motion is.

Forward motion is the transportation of a body from one place to another. It is also called the motion of place, or motion to or from one place to another, and occurs in different ways.

A body does this by itself, but with the attendant help of something else. For example, when an animal runs or walks, the movement of his feet, indeed, is made by himself but his forward movement occurs through the help of the ground upon which his feet are set one after another, thereby enabling his body to move forward from place to place. If there were no ground to help, the foot motion could not result in forward motion. In the same way birds, when they spread their wings on the air, provide forward motion for themselves by this motion. If there were no air to help, however, they still could move their wings but they could not move forward. The same can be said of fish in the water etc.

Forward motion is brought about in different ways.

This motion can also be produced by means of a certain implement or instrument, like a cart or other similar vehicles, etc. For example, when men are carried in a ship, they are transported, not by their own motion but by a ship and at the same time participate in forward motion. In the same way the animals on the earth do not embody motion in themselves (even though they are carried around with it) but participate in the earth's turning.

Forward motion is also brought about by an impulsive virtue (concerning this, see Book IV, Chapter 2 and following). For example, when a stone is thrown down, it has no motion in itself but is flung with the help of the hand. At the same time a virtue is impressed upon it whereby it is carried down, and when this occurs, it does not embody in itself forward motion, but only participates in it. This is the case because the stone does nothing by itself (See Book IV, Chapter 5 in this regard) but is attracted by the conserving virtue of the earth. For the same reason, when the sun carries the earth and the other planets around with its virtue, these bodies thereby do not themselves embody this motion but are at rest. The sun on the other hand which carries them around, does have a separate and indeed, a revolving motion in itself. See a discussion of this subject in the preceding Chapter.

At this juncture, we believe it is necessary to discuss forward motion and to set forth the difference between it and ordinary motion.

The difference between motion and forward motion.

As for the forward motion of the earth it is known that the sun is seen from day to day at various points of the ecliptic and progresses in such a way that this phenomenon is completed in the space of one year. Ptolemy asserts that the sun is the cause of this movement, while Copernicus attributes it to the earth, with the result that we believe that the sun runs through the zodiac as we ourselves are being carried on the earth and look upon the sun from opposite sides of the earth's orbit.

Let us accept this assumption, although it is true that appearances in both cases seem the same in the heavens. It is confirmed on very reliable evidence, however, that it is not the sun, but the earth that is moved through the ecliptic. Now this revolution of the earth takes place in empty and very clear space in a simple and natural way, easily and without strain. This can be seen in Book VI, Chapter 8 which deals with the sun and its turning virtue, by means of which it moves the whole sphere of its virtue (in which the earth and planets are included) and makes it revolve. Chapters 3 and 4 of Book IV should likewise be reread which discuss the nature and the characteristics of the impulsive virtue. For just as the small balls described there are carried around by this impulsive virtue as well as the globe floating freely in water, so even more freely and naturally are the planets carried around in the very clear space of the heavens.

How the forward motion of the earth is produced.

As the sun turns the whole sphere of its virtue, the earth and all the planets are also forcefully driven at the same time, not only to the farthest possible distance from the sun (insofar as the impulsive virtue, stimulated in any one of the planets, can extend itself in accordance with the mass of each, into the largest possible circle, otherwise called its equator, at any one time) but also with the greatest speed possible in its periodic revolution.

The source of revolution of the planets.

When each one of these earthlike bodies, filled as it is with the impulsive virtue, is carried out in so great a circle, it cannot help but go beyond the bounds of the equator and slip a little in this sphere of the sun's virtue towards one or the other of its axes, (that is, the Tropic of Cancer or the Tropic of Capricorn). (This slippage naturally occurs obliquely in the planet's following the zodiac, for the motion cannot properly be carried out in a straight line from one Tropic to the other.) But when the planet arrives at the smaller diameter of the sun's sphere, the Tropic of Cancer, for example, then this inherent impulsive virtue (which seeks after greater distances), directs the planet back, and, in the same way as we have said, carries it out obliquely to a circle of larger diameter again, namely the equator. Then, when it has returned, this same virtue (because it always goes beyond, as we see in the oscillations of the pendula in Book IV, Chapter 3 which are nothing other than excesses of the impulsive virtue) directs it to the other somewhat smaller circle, namely the Tropic of Capricorn. It does not permit it to come to rest there, however, for the reason already given but again directs it by degrees, following the zodiac to a circle of greater diameter, the equator as it were, and this procedure continues as long as the turning of the sun lasts. In this regard see the end of Chapter 4 of the preceding Book.

The source of the forward motion.

In the same way the earth carries out its annual forward motion obliquely within these Tropics following the zodiac. This path is called its ecliptic. In the course of this forward motion, the earth (since, to be sure, this motion occurs in conjunction with its own individual daily motion) produces its varying seasons, summer and winter, for example, day and night, etc. without the introduction of any third motion. For it is enough that in summer and winter, and with the increase or decrease in the length of days, that the rays of the sun should now touch the equator of the earth and now either this or the other Tropic perpendicularly.

If it is true, moreover, that the earth's sphere is growing larger very slowly (as we have already touched upon in Chapters 2 and 3 of the Book), then the difficulties inherent in the different motions will easily be resolved which the Ptolemaics attribute to the sun and other planets as well as to the so-called sphere of the fixed stars through this forward motion of the earth.

The source of the so-called retrograde motion of the stars.

Since the impulsive virtue has a greater effect on larger bodies, projects them farther, and draws them into a larger circumference than smaller, it follows that with the earth's increase in size, its course around the sun also grows. If, however, its orbit is greater, so is its annual period lengthened, with the result that the earth in a tropical year (which consists of 365 days, 5 hours, 48 minutes and 45 seconds) does not reach a fixed star in the ecliptic with which at the beginning of the year it had been in conjunction, but remains about 50″ behind. If then the earth does not reach this point of the ecliptic, it must necessarily be the case that each year the points of the equinox and solstice move forward 50″. For this reason the Ptolemaics attributed individual motion to a so-called Eighth Starry Sphere which they termed retrograde motion. (This motion, nevertheless, is present on the earth.) We discussed this question cursorily at the end of Chapter 4, Book I. See also Chapter 3 of Book IV, "We will take, as an example" – as well as the end of Chapter 4 of this Book.

That which astronomers call the Second Mobile does not exist.

Finally, because the great orbit or circle of the earth is completed within a year by the earth's revolution, a distance which otherwise the sun is assumed to cover each day according to the Ptolemaic hypothesis, with the result that on any given day it would travel 14,293,714 German miles, as reported in Book VI, Chapter 9, it follows, according to the same Chapter, that on any given day, the earth will travel 39,160 miles, 1,631 miles in an hour, 27 miles in a minute and almost a half mile in a second or moment. If, however, the sun followed that course, it would cover almost 10,000 miles in one minute and in a second 166 miles which, indeed, seems unreasonable.

The extent of the earth's revolution

and contrari-wise.

If the sun covered this distanced what would it be?

CHAPTER 17

THE MOON

Orpheus, Thales, Anaxagoras, Democritus, Heraclides Ocellus, etc., and Plutarch in his *Placitum Philosophorum*, Book II, Chapter 25 believed that the moon is another earth which has many mountains on its surface. Thus Agostino Nifo says in Book II, section 49 "De Caelo", "Perhaps it is not unreasonable to say that the regions of the moon are diverse, as are the regions of the earth. Some sections have valleys, others mountains – configurations which produce differences on the face of the moon. This is not at all unreasonable, for geometrically speaking, the moon is an imperfect sphere."

See Plate XIX.

Writers who agree that the moon is similar to the earth.

Blancani and Longberg in Book I of his *Theorica* page 315 agree with this, as does Kepler in the *Astronomia Optica* where he writes, "The body of the moon is dense like that of the earth." He likewise agrees with Poseidonius' statement recorded in Macrobius that "the material of the moon is similar to that of the earth."

Kircher, furthermore, in his *Itinerium Ecstaticum* states "Not only do the Holy Fathers of the Church agree that the spheres of the stars are composed of four elements, but they clearly demonstrate that the results of the experiments they observed are valid, the veracity of which they tried to establish with such avid enthusiasm. Let us begin with the moon which no one of the modern astronomers will deny is compased of rugged earth matter, peaked with mountains, etc."

Mountains are seen on the moon.

That there are mountains on the moon is apparent in the first place from the fact that when the moon is observed with a good telescope, there appear on that part of the moon's face which is not yet torched by the light of the sun and which lies completely in shadow, certain points of light which shine outside of the boundary of brightness until they join with the rest of the illuminated body. Indeed, the shadowy intervals which separate those illuminated peaks from the rest of the lighted body gradually lose their darkness and admit the light of the sun.

Clearly this phenomenon could occur in no other way than if the peaks or heights of mountains thrust themselves upwards and receive light before the other parts, just as we see the light of the sun on the peaks of our mountains which always receive its morning light before the lower regions of our earth. A second argument for the presence of mountains on the moon is the fact that a telescope reveals rough and uneven boundaries on the illuminated part of a crescent moon. Using only our eyes we see many bright points in

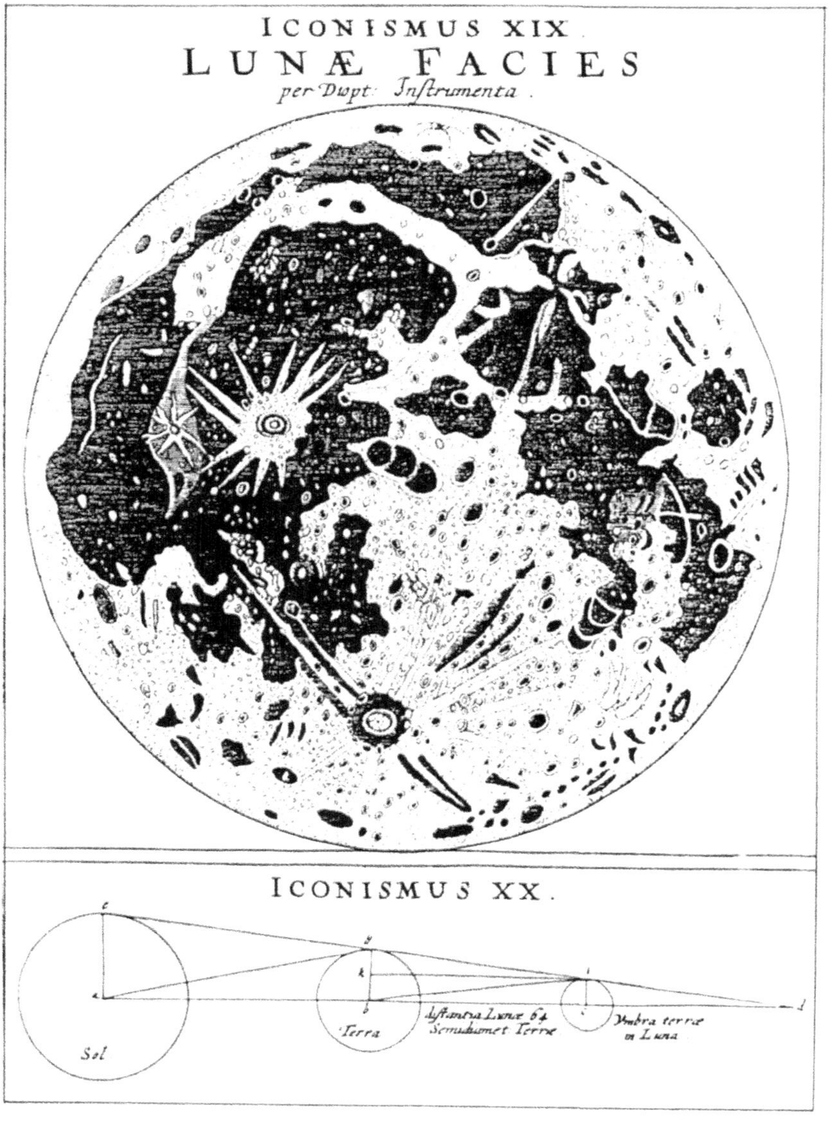

PLATES XIX AND XX.

the dark part of the moon which are completely separated from the lighted section. We also see the shadows between them diminish as the brightness grows until they completely vanish and are evident nowhere on the lighted sphere.

Thus Father Riccioli in his *Almagestum*, page 206, number 12 says, "The shadows of the moon's mountains, projected now on this region and now on that give evidence of an uneveness similar to the surface of the earth." He confirms this statement with many arguments on page 207, question 2.

In the same way, then, as our earthly sphere is essentially round in shape but nevertheless uneven and rough by reason of valleys, upward thrusting hills, mountains, forests, etc., so we discover, with the help of telescopes as well as the eyes of our intellect and body that the moon has caverns, mountains, seas, etc.

The moon is a solid and not a diaphanous body.

This solid and opaque body, therefore, is not in any sense diaphanous and reflects on us the light received from the sun in the same way as our sphere of earth and water undoubtedly directs the light received from the sun toward the moon. Almost every new moon appearing in its last or earliest phases (if it is observed in a clear sky) on the side turned away from the sun would appear like a glowing, shining white ember. Indeed, its light would be nothing other than a reflection of the earth's light which appears to be so much brighter, the more pointed the sickle-shape is, because at this time the illuminated part of the earth is turned to the moon in an especially extensive area, while conversely, the greater part of the moon on which the light from the earth is reflected more strongly, is darkened to our view.

The earth also transmits light to the moon.

The source of redness in a lunar eclipse.

For just as that reddish light which one observes in the moon at the time of a lunar eclipse is nothing other than the brightness of the sun refracted in the air surrounding the sphere of earth and results in colors that vary according to the thickness or transparency of the vapors around the earth through which the rays of the sun pass, thus, because of the mixture of these vapors with light, we see diverse colors at dawn and in clouds lighted by the sun.

CHAPTER 18

THE APPEARANCE OF THE MOON AND ITS SPOTS

No one will deny that the darker the surroundings are, the brighter every lighted body appears. When we consider the immensity of Space which is supported by no external properties and possesses no qualities in itself, one must easily understand that it does not receive light or brightness from the sun but transmits the solar rays freely without any interference. For this reason, it is completely dark in itself and is imperceptible to us. For this reason every body embraced in this darkened Space shines because of the solar rays it has absorbed and appears bright to us, with the result that the farther away such a body is, the brighter its light seems before our eyes, as can be seen in Book IV, Chapters 13 and 14, so that it finally arpears like a star.

In empty Space there is no reflection of light or brightness.

If, therefore, in this immense and imperceptible Space we were to recede 1,000, 10,000, or 20,000 miles from our earth, it is reasonable to surmise that the earth would appear not only progressively smaller in our eyes, but would present a continually varied appearance to us as the distance changed, being at times wholly lighted and at others, only half lighted, etc. Indeed, if we were to reach as great a distance away as the moon, one can readily see that the earth would appear to us not dissimilar to the moon in appearance because our eyes would be subject to the same conditions on the moon as they are here on earth.

If we were on the moon, we would find the earth to look like the moon.

Furthermore, it has also been shown in the aforementioned Chapter 14 that the farther away an illuminated body is in darkened Space like this, the brighter it seems to gleam and shine, just as we see that clouds which are lighted by the rays of the sun at dusk, although they are dark in themselves, shine like the sun. This is also true for the moon and earth, for if we progressively withdraw farther from them, they appear to us as stars. They lose their yellow or fiery reddish color as the distance becomes very great because all colors vanlsh at a great distance.

Illuminated as well as light-giving bodies appear brighter the farther they are removed.

As for the spots on the moon, they are either waters or seas. For as transparent bodies absorb light and do not reflect it in the same way as rough bodies, so, too, the waters on the moon do not reflect light in the same way as its mountains, caverns, and other rough areas, for these are composed of innumerable tiny surfaces disposed at an infinite variety of angles causing many reflected rays to be dispersed in every direction. Thus the moon's body upon which the solar rays fall, appears completely illuminated from whatever place it is viewed and reflects or diffuses the light received from the sun in every direction.

The spots on the moon are undoubtedly seas.

Now although this may seem unbelievable to us, because we see that mirrors and other bodies of bright surface as well as water, reflect the light of the sun very well, nevertheless this reflection is made only from the outermost surface and indeed occurs within certain limits and angles and if we are located outside of these, we

Transparent bodies do not reflect light except from a certain angle.

see no reflection, but rather a shadow. An example of this is evident in the case of seas, ponds, and large lakes when we look down from a tower or hill, Then they appear to us to be of a bluish color, which is very near to darkness (See Book IV, Chapter 12) and is bereft of all brightness.

If the surface of the moon, therefore, were smooth like a mirror or water, it would not reflect the rays of the sun and consequently, because of its great distance from us, would be completely invisible.

The fact that these dark spots on the moon, moreover, are nothing other than water, cannot be disputed because telescopes reveal that there is a difference between those numerous brightly lighted areas thrust out in rugged and jagged points from what we see in the luminous surface regions.

CHAPTER 19

THE FORWARD MOTION OF THE MOON

Just as satellites pass around Jupiter, so the moon revolves about the earth.

Galileo in his dialogue *De Systemate Mundi*, page 173 says "We do not know the cause which moves the moon in a circle around the earth, etc." Just as the earth and all the planets, however, are carried around by the virtues of the sun and the companions or satellites of Saturn and Jupiter in turn revolve around these planets, so also must we conclude that the moon rotates as an attendant of the earth. For inasmuch as these secondary planets, because of their limited size, are not of a proper and suitable mass relative to the sun's virtue, they are not carried around by the sun, but rather by their principal planet. For this reason the periodic forward motion of the moon is communicated to it through the daily motion of the earth's sphere. Indeed, in a distance of the kind which separates the moon from the sun, the sphere of the sun's virtue is not so strong that it can carry around a small body like the moon and pull it away from the earth.

That this virtue which carries the moon around with it is inherent in the earth is confirmed by the sulphur globe experiment which we descrihed in Book IV, Chapter 15. Thls globe can carry a small feather all around a room when it is floating freely in the air in the sphere of its virtue. Then it can easily be concluded from this experiment that if the moon is carried about, this forward motion has its origin in the impulsive virtue by which the planets are similarly carried forward around the tropics. We have discussed this in the preceding Chapter 16 and in Book VI, Chapter 11.

Therefore, the moon is carried in its monthly forward course by the sphere of our earth in $29\frac{1}{2}$ days. In its revolution, the moon receives the light of the sun in the same way as the earth which rejoices and is renewed by it each day, inasmuch as it does not always present the same side to the source of light.

The moon is carried about by the earth.

Any given day on the moon, however, lasts the space of a half month or almost 15 of our days and the same can be equally true for the night of the moon. Furthermore, even though the moon is subjected unremittingly to the sun's rays on one side for several days, still it does not catch fire because it is outside of the effective sphere of the sun's virtue. For, as we have said, the sun cannot make its influence or heat felt on so small a body as the moon because of its relatively great distance away, while the earth, on the other hand, which has a larger body and is in proper proportion and mass to the sun's distance, could catch fire under these circumstances. (ADDENDUM) Let us not discuss here that vital inner flame or subterranean fire in which our earthly sphere rejoices and of which the moon is deprived since the frlction of the solar rays cannot kindle the heat in this body that it provides in the earth. We discussed this question in Book IV, Chapter 11.

A day on the moon lasts 15 days.

CHAPTER 20

THE MOON'S MOTION

The moon always shows the same side turned toward the earth, and since it does not spin or rotate, it consequently does not possess motion because we do not term its forward movement precisely motion. This was remarked upon previously in Chapter 15, *De Motu*.

The moon possesses no motion.

Since, then, the moon always turns the same face toward us, it is impossible that it be a part of an epicycle, as the Peripatetics affirm, for if it were, it would, of course, change its face when it is carried about the center of its epicycle, and that part of the lunar body which at one time is on the right, at another would be on the left when it has been moved a half turn. The experiment with the sulphur globe, described in Book IV, Chapter 15 shows that the source of this constancy in the moon's face is quite natural. This globe always holds the small feather in its sphere of influence with the same side turned toward it even after it has once been repelled from it, and wherever it is carried about and no matter whether the globe is held above or below it, the feather always turns the same side toward it.

The moon is not a part of an epicycle.

The source of constancy in the moon's face.

If the feather is turned over, however, and directed toward the globe, it can no longer function properly and in the usual way.

The oscillation of the moon.

Furthermore, telescopes reveal that the sphere of the moon oscillates slightly about its own center. In using these instruments for our observations, we find that the spots which are on the rim of the moon's disc now move forward a bit and similarly move backwards, from the north, as it were, to the south and from the south to the north or even in another direction. This subject is further discussed in Father Riccioli's *Almagestum*.

The source of this phenomenon, indeed, is not found in the moon itself but in the earth in whose sphere of virtue the moon is carried. For when the earth holds the moon in its sphere of virtue, with the same side turned toward it, and even carries it around at the same time in a monthly revolution by its own rotation, there arises a slight oscillation in the moon's body. This occurs because it is impossible for a freely hanging body that is round and disposed to move, to be carried around without any oscillation, although the earth's virtue always serves to steady it and maintains the moon's face turned toward it. As a matter of fact, it is entirely reasonable to assume that all planets, the earth included, partake of an oscillation of this kind which is contained by the directive or magnetic virtue.

CHAPTER 21

THE CONJUNCTION AND COMPARISON OF THE MOON WITH OUR SPHERE

The moon is similar to the earth:
(1) In its shape which necessarily must be that of a sphere in view of its perfectly round appearance and its manner of receiving the light of the sun. For if the surface of the moon were a plane, at one and the same moment, it would be lighted in its entirety and at another, would be absolutely without light, for the parts turned directly towards the sun would not be successively illuminated.
(2) In that the moon is no less dark in itself than the earth and is adapted to receive and reflect light.
(3) In its matter which is dense and solid as well as
(4) Its surface which is irregular like that of the earth, as is evidenced by its many heights and cavities which are revealed by the telescope. We have discussed this aspect of the moon in the preceding Chapters 17 and 18.

(5) Just as the regions of the land and sea on our earth are distinct from one another, so we see a varied surface on the moon's disc, some parts of which reflect more light than others. The appearance of the earth would undoubtedly be similar to that of the moon if anyone could look down upon it from the moon or any other body at a comparable distance when it is lighted by the sun. For the surfaces of the watery areas would appear darker because they absorb the light while those of the land would appear brighter because they reflect it.

The regions of the earth and the moon appear to be of the same kind.

(6) Just as, from our position on the earth we at one time see the moon shining with a full orb, at another as a half moon etc., and again being completely removed from our sight when it is hidden from the rays of the sun, so necessarily must the earth appear to us from the moon, provided that there be a distinction in the periods of illumination, for what takes place on the surface of the moon in one month occurs on the earth within 24 hours.

Just as the moon appears from the earth, so, other things being equal, must the earth appear from the moon.

(7) From this it follows that that which happens in one full day to those dwelling on the earth would befall, in the space of a month, the inhabitants of the moon (if there were any), but with this difference, however, that whoever was on the moon, specifically, in the lower hemisphere, (for if a person were located on the other side of the moon, invisible to us, or on its upper hemisphere, he would never be able to see the earth), he would observe the whole surface of the earth revolve every day in the space of twenty-four hours. We here on earth, however, always see only one half of the moon since it does not turn about on itself, a condition which must necessarily prevail if it were to be observed in its entirety.

If anyone were on the moon he would see the earth.

(8) Just as the moon acts as substitute for the absent light of the sun a good deal of the time and by its reflection of this light makes dark nights days for us, so does the earth send out light to the moon by reflecting the rays of the sun, as we have described in the preceding Chapter 17. Furthermore, because the earth is much larger than the moon, the earth's illumination of the moon is much more effective than that of the moon on earth.

The earth is to the moon as the moon is to the earth.

(9) Just as these bodies mutually light one another, so do they occasionally bring darkness to one another, when, for example, the moon becomes eclipsed, being deprived of light because the earth comes between it and the sun. In the same way the moon from time to time comes between the earth and sun and deprives the former of the sun's light. Although frequently and for a longer time the moon remains completely immersed in the earth's shadow, never is the earth in its entirety darkened for as long a period of time by the

One body is sometimes deprived of light by another.

moon, although it does occasionally cause an eclipse on the earth as one might expect when one body deprives another of the light of the sun.

(10) The moon is an eternal and inseparable companion of the earth, for the latter always carries the moon with it in its ecliptic. In the same way as a woman accompanies a man, so does the moon accompany the earth and they proceed around the sun in the same annual forward motion. We do not regard the moon's presence relative to the earth any differently from that of Jupiter's satellites relative to Jupiter, for just as Jupiter has four satellites or companions, so has the earth one.

The moon is a constant companion of the earth.

The difference between the marriage of these two earth bodies and one that occurs among human beings is that the earth and moon affect one another not through intercourse but through emenations of virtues outside of their own bodies. This is a much more satisfactory condition than that existing among men because the attraction of their bodies is not excited for only a short time but continues forever in the eternal sight and influence of one another.

CHAPTER 22

THE DISTANCE AND SIZE OF THE MOON

Since, according to the preceding Chapter 11, the time to calculate the moon's distance is uncertain, whether at its apogee or perigee, as astronomers term it, and since the observations which are taken when the stars are in the northern signs of the zodiac (for then they are more elevated and are less affected by the refraction of the air) are more exact at this time, then the parallax of the moon is usually calculated to be of 53′ (one cannot discuss the number of seconds because of changes in the air, as is to be seen in Chapters 10 and 11). This is a more accurate calculation than those recorded in Book I, Chapter 21.

The most generally accepted lunar parallax is of 53′.

If the parallax of the moon is assumed to be 53′, it follows that the distance of the moon from the center of the earth is 64 radii of the earth. For as the 10,000 parts of a radius (according to the Table of Sines) are to the radius of the earth, 1, so are the parts of a secant of 89 degrees 7 minutes (the complement, as it were, of the parallax of 53 minutes) to the distance of the moon. The following is an illustration.

To find the distance of the moon.

Parts of a radius	Radius of the earth	Parts of the secant of an angle of 89°7′
10,000	1	64.8657

The moon, therefore, is 64.8 diameters of the earth distant from the earth's center.

The moon is distant 64 diameters of the earth.

When the distance of the moon from the earth is computed as 64 radii of the earth, the relative size of the moon to the earth is also calculated from the radii of the earth and moon, according to Book I, Chapter 22.

To find the size of the moon.

The radius of the earth is 860 German miles. The distance of the moon is 64 radii of the earth or 55,040 German miles.

To find the distance of the moon.

The apparent diameter of the moon (which is usually observed with a measuring device) as well as its distance, is found by astronomers to vary significantly in their computations of its hypothetical apogee and perigee, as well as in changes in air conditions, as we have already mentioned.

Since that which is usually observed is to be preferred to that which is found only occasionally, there can be little doubt that one should calculate the apparent diameter of the moon as being 30′. This is evident from the aforementioned Table in Book I, Chapter 22 where astronomers found the diameter of the moon to vary from 25′ to 36′ with 30′ being the average number.

The apparent diameter of the moon is found to be 30′ on the average.

Therefore, inasmuch as the middle course is the safest to follow, we must postulate that the apparent diameter of the moon is 30′ and the radius as 15 minutes. The true radius, indeed, as observed from the center of the earth (acording to Ptolemy) would be less than 2/3 of a second, making it 14′20″. But since the diameter of the moon otherwise cannot be calculated with such precision in seconds, there is no need to begin so laborious a computation.

Furthermore, as the 10,000 parts of a radius are to the distance of the moon, namely 55,040 German miles, so the 434 parts of the tangent (of an angle of 15 minutes) are to the radius of the moon, namely 240 German miles.

Inasmuch as (according to Euclid, Book 12, the last theorem) spheres are the cubes of their diameters, thus when the diameters of the earth and moon are cubed, or (according to Euclid, Book 5, theorem 19) their radii, their dimensions are found. Take the following as an example.

The cube of the moon's radius of 240 miles is 13,824,000 and the cube of the earth's radius of 860 miles is 636,056,000. When this cube is divided by the cube of the moon's radius, the result is 46 152/13824.

The earth is 46 times larger than the moon.

Therefore the earth is 46 times larger than the moon.

CHAPTER 23

ARE THERE OR ARE THERE NOT ANIMALS ON THE MOON?

The opinions of various philosophers.

According to Book IV of Cicero's *Quaestiones Academicae*, Xenophanes was of the opinion that "there are inhabitants on the moon and that it is a land of many cities and mountains etc." The Pythagoreans also believed this for they stated that "the moon appears to be an earthlike body, and because it is like our earth, it is inhabited by very large and fine animal specimens." Plutarch, furthermore, in his treatise dealing with the appearance of the moon's sphere, says, "I think that those who inhabit the moon must find it remarkable that the earth can produce and feed animals, when they see our earth as the lowest dregs of the world, appearing through so much moisture, clouds, and vapors, a place dark and low and without motion, etc.". Macrobius, moreover, in Book I, Chapter 13 of *In Somnium Scipionis* places people and the purgatory of souls on the moon. Achilles Tatius in his *Isagoge de Quibusdam Existimatibus* declares, moreover, "that there is a certain region on the moon which is habitable and is endowed with rivers and other natural phenomena which we see here on earth."

The author's opinion.

As a matter of fact, even though we can easily postulate that areas on the moon are not moribund or unable to produce, still we cannot conclude with certainty that animals are found there which can move from place to place. Should some such be there, however, they would be completely different and of a much lower order and very far removed from our imaginings. Because the moon is devoid of animated life and thus does not participate in the motion of turning as our earth does, then neither will its so-called animals have animated life, which we regard as characteristic of an animal. I am persuaded that this point of view is all the more credible because a variety of sunlight is extremely necessary for the generation of animals. Now the quality of sunshine which gives light to the earth is completely different from that which reaches the moon since (1) in single periods of 24 hours we experience the shifting of night into day, an event which occurs on the moon once a month. (2) The annual ascent and descent of the earth in the zodiac which produces winter and summer is irregular on the moon and is only partially completed there in a single month. (3) Telescopes reveal that at no time is the moon surrounded by clouds, for if clouds gathered there anywhere we would find some of those things obscured which we see on the moon with the help of a telescope. Because there are no clouds, moreover, no rain can fall there nor cause things to grow. (4) The moon's sphere is not at an appropriate distance from the sun so that it can be warmed by it, for relative to its size it is too far distant. Indeed, the distance of earthlike bodies from the sun should be proportionate to their size, otherwise they are beyond the natural reaches of the sun's virtue and do not benefit as they might from the sun's heat, as we have said in Chapter 19 above. From this there follows (5) from which one must assume that on the moon, coldness is all-enveloping (a fact which even visual perception through telescopes seems to confirm to a certain degree) and that this virtue, the cold-making virtue, as it were, arises from the moon and flows down to our earth, as we described in Book IV, Chapters 1 and 6 and likewise at the end of the foregoing Chapter 9.

The moon is beyond the natural reaches of hte sun's influences.

Chilling cold undoubtedly flows down to our earth from the moon.

CHAPTER 24

ECLIPSES

There is no eclipse of the sun, but only one of the earth and moon. For the sun experiences nothing at all, much less an eclipse. Although it seems to be obscured by the moon, its radiation alone is intercepted; that is, the sun retains its light and allows it to stream out without interruption, but through the intervention of the moon, it is blocked from those regions of the earth from which one has a view of the visible conjunction of the two bodies. The earth and the moon, nonetheless, do experience eclipses, some of which are total, others partial. From these eclipses, however, it is not possible to prove that the sun and moon have apogees and perigees.

The sun never experiences an eclipse.

Some astronomers who declare that there is an apogee and perigee of the sun and moon believe that eclipses also differ as a result of their being at varying distances from the earth (the lesser the moon's distance, the greater the solar eclipse in size and duration; the greater the distance, the smaller the part of the sun (there may even be none at all) would be covered, etc.) Empirical science does not support these findings as Father Riccioli is forced to admit in his *Almagestum*, page 233, "because a change in the sun's distance and its apparent diameter cannot be demonstrated by eclipses; for if a person postulates the sun's diameter to be a constant 16′10″, not

Because the time when the moon is at its greatest and smallest distance is uncertain, the variation of total eclipses, then, is uncertain.

"A change in the sun's distance cannot be demonstratd by eclipses."

only will the other eclipses remain the same, but even, a fact which seems more significant, so will the total and annular eclipses of the sun, provided one maintains the radius of the moon to be of the same dimensions as he has already postulated."

Night cannot be made from day by means of a so-called eclipse of the sun.

Since, therefore, the sun greatly exceeeds the earth in size and earth the moon, the lunar body cannot rob the sight of the entire sun from the whole expanse of the earth and thus allow only a portion of the earth to perceive this loss of light. Thus not only lunar but also solar eclipses are not equally seen by all, even when there is a total solar eclipse, that is, even if a certain eclipse is said to be total, nevertheless it does not extend the diameter of its shadow more than 70 German miles over the earth.

As a matter of fact, no eclipse can so obscure the sun in its entirety that a certain thin ring surrounding the moon is not left from it, an observation supported by Tycho Brahe in Volume I, page 134, of his *Progymnasmata*. Thus, the writings of the old historians, Herodotus and Pliny, are not credible on the subject of the eclipse predicted by Thales in the year 585 B.C. "The sun disappeared so completely that suddenly night was made from day because the sun was in apogee and the moon in perigee."

Neither should we accept Clavius' statement in Chapter 4 of his *Super Sphaeram Sacrobosci* regarding the eclipse of 1560 which occured about noon on August 11: "The entire sun was hidden for such a long time in Coimbra in Portugual that the darkness exceeded that of night and a person could not see where to set down his foot. The stars appeared in the heavens in all their brightness and the birds all descended from the sky to the earth in their fear of such great darkness."

In the same vein, the scholarly writers engaged in the revision of the Calendar, made predictions about the solar eclipse that was to occur on the 2/12 day of August in 1654 at 11 o'clock. They sought to persuade the ignorant "that there had not been an eclipse of this kind in the past 2,239 years wherein the sun was to appear to have been lost from the sky, as it were, and the day of the Last Judgement to have been at hand, etc." This, to be sure, did not occur because the sun was not so completely hidden behind the moon that none of its light was visible. As a matter of fact, no one witnessed any darkening of the day at all.

From these and other eclipses, one can see that actual observation disproves the theory that a greater solar eclipse results from the sun in apogee and the moon in perigee or that a greater lunar eclipse and one of longer duration occurs when the sun is in perigee and the moon in apogee. It is also evident that the so-called apogee and perigee of the sun is non-existent and those of the moon appear to be either insignificant or at best, open to doubt.

That one cannot prove the existence of the moon's apogee and perigee by eclipses is clearly shown by the investigations of Ptolemy and Albattani who followed him, both of whom assigned different sizes to the angle subtending the moon in the apogee of its epicycle and the angle subtending it in perigee which was 4′ according to Ptolemy and 5′50″ according to Albattani. In other words, the angle suhtending in apogee was 31′20″ according to Ptolemy and to Albattani 29′30″ etc. Müller in his *Epitome*, Book V, proposition 21 should be consulted on this subject. Since, indeed, these conclusions are reached from the same source of observation, they ought to agree. Differences, similar to these can be cited from observations of contemporary scholars, but they derive from a difference in the height of the air at different times, a subject we will pass over for the sake of brevity.

CHAPTER 25

THE DESTRUCTION OF THE EARTH

Just as the earth has had its birth, so also will it have its destruction – and certainly this by fire, a theory which has come to be supported by many philosophical arguments.

It is the voice of the impious who say, according to the apostle Peter, "Where is the promise of his coming? For since the Fathers fell asleep, all things continue as they were from the beginning of the Creation. For this they willingly are ignorant, that by the word of God the heavens were of old, and the earth standing out of the water and in the water: Whereby the world that then was, being overflowed with water, perished: But the heavens and the earth, which are now, by the same word are kept in store, reserved unto fire against the Day of Judgement and Perdition of ungodly men. But, beloved, be not ignorant of this one thing, that one day is with the Lord as a thousand years, and a thousand years as one day. The Lord is not slack concerning his promlse, as some men count slackness; but is long-suffering to us-ward, not willing that any should perish, but that all should come to repentance. But the day of the Lord will come as a thief in the night; in the which the heavens shall pass away with a great noise, and the elements shall melt with fervent heat, the earth also and the works that are therein shall be burned up. Seeing then that all these things shall be dissolved, what manner of persons ought ye to be in all holy conversation and godliness, looking for and hasting unto the coming of the day of God, wherein the heavens being on fire shall be dissolved, and the elements shall melt with fervent heat? Nevertheless, according to his promise, look for new heavens and a new earth, wherein dwelleth righteousness." (Pet. II, 3.4–11)

With these words Peter sets in comparison the earth and heavens which were in existence earlier with the earth and heavens which are now. Just as they were destroyed by a flood of water when that world perished, so will these heavens and this earth perish in a flood of fire, so that God may avenge the wrongs of men. In the flood, however, the world with the heavens or spheres of other earthlike bodies (that is, the planets, sun, and fixed stars) will not perish. Rather, when the cataracts have burst, it will be the airy heavens above and associated with the earth, that will be destroyed by flood together with all things constituting a whole with the earth. Therefore he who wishes to follow the words of Peter and accepts their validity, understands that the term heavens, here refers to the airy heavens surrounding this whole body of earth, which is to perish by fire. Just as he interprets the destruction of the world through flood as bringing about the restoration and change of the earth for the better, so also does he not regard its destruction as its extermination or disappearance into nothingness but rather as its conversion into a better state.

The words of Isaiah, Chapters 64, 65, and 66 (ERRATUM, read 64 for 46) as well as the Apocalypse seem to support this interpretation: "Behold I create new heavens and a new earth: and the former shall not be remembered, nor come into mind."

Furthermore, theologians who interpret this to signify the renewal of the earth and a change for the better rather than a destruction of the elements, cite Psalm 102 (ERRATUM, read 102 for 101) as an example, "Of old hast thou laid the foundatlon of the earth: and the heavens are the work of thy hands. They shall perish, but thou shalt endure: yes, all of them shall wax old like a garment; as a vesture shalt thou change them, and they shall be changed."

Herein it is maintained that there will not be complete destruction in the future but a change for the better, regardless of the opinion of the Holy Fathers. See also Book I, Chapter 31 in this regard.

If, then, the earth is destroyed, it is to be assumed from these words that the moon must also perish.

CHAPTER 26

APPENDIX TO THIS FIFTH BOOK COMETS

Although it was not my intent to write anything about comets, still an opportunity was presented when one or another comet of significant size appeared about the end of 1664 and the beginning of 1665, at which time this treatise had already been finished for a year. It so happened that my opinion as to their nature was sought by several men, among whom was the most noble gentleman, Stanislaus Lubienietzki, a Polish knight. In view of our subsequent correspondence, he included and published the letters he had sent to me and my answers in return in the order in which they were written in his *Theatrum Cometicum*, pages 453 to 465, which was prlnted in Amsterdam in 1668.

The author never intended to write anything about comets.

From Hamburg to Magdeburg on the 4/14 March, 1665.

Most Respected and Esteemed Sir:

I hope that you will excuse this liberty which I use in addressing you by letter and in interrupting your serious pursuits and studies. I have indeed discovered that you possess the boundless understanding which ornaments those men who are eminent in government service and conspicuous in scientific pursuits. I hold in great honor the bonds of my singular friendship with the most noble gentleman, esteemed advisor and chargé d'affaires of His Most Serene Highness, the Elector of Brandenburg, who bears the name of your dear son. Since I have sought your counsel several times, thanks to him, about the comet that was recently seen, while you, in turn, by virtue of your boundless knowledge, have deigned to set forth your opinion time and again, I confess that I owe both of you a great debt of gratitude. Indeed, your son not only bears your name, but also participates in your remarkable knowledge of natural phenomena, your humanity, and other virtues. I communicated your opinions regarding the origin of comets to learned men with the consent and help of your son, for in view of the newness of these investigations, it seemed best that they be taken under consideration by others. Upon my request, your most distinguished son had the opinion of my friend in Copenhagen sent to you. You disagreed with his assessment and in turn presented your own point of view which I sent back to Copenhagen a while ago. As yet I have not received a reply to this communication. I also await a reply from other learned scholars on this subject. In the

Letter number 1 of the Honorable Lubienietzke to the author.

He uses his friend, the author's son.

meantime, I wish you would set forth your opinion more clearly and more fully, especially those difficult and much disputed principles related to the motion of the earth and empty Space. These assumptions are the bases of your opinion about the origin of comets which I wish you would substantiate by your rational investigations as well as by specific experimental data. In this way you will satisfy the keen anticipation of the learned men who have been your vigorous opponents and who must be won over by your inventive observations. With God's help I have already begun a work for scholars and members of the Christian community containing the observations, communications, and judgments of very learned and outstanding scholars which I have in hand now and which, I hope, with the passage of time, will increase in number. I am confident that the sincerity of my endeavor will have the approval of people everywhere. Indeed I shall expend every effort so that not only recondite information in the fields of natural philosophy and astronomy but even of politics and ethics and other studies of value to humanity may be included therein, making the work both entertaining to read and useful at the same time. I disclosed this project in a letter to the renowned Johann Raue, member of the consular staff and librarian to His Most Serene Highness, your Elector, which was sent a short while ago. I am attaching a copy of this communication to my letter. In this you will see that I rely upon your opinion to no small extent in my eager search for truth. Furthermore, I have done the same thing in a detailed letter written in French and communicated to my friend in Copenhagen which I had sent, along with many other observations, in the care of your most esteemed son whom I regard as a very near and trusted friend. I have followed and continue to follow this course of action, relying more upon inquiry than upon actual demonstration of truth, to which I admit I yield, using my own judgment as the sole criterion. It will be your task as a distinguished scholar to inform me more fully of your belief and to support and defend this with cogent reasoning and indisputable proofs. Do you want to inform me whether I should substantiate your theory by those of famous scholars like Epigenes, Cardano, and Scaliger whom I mentioned in the attached letter? In this investigation and elsewhere I have introduced the opinion of Theophrastus Paracelsus who holds that storms, winds, lightning, hailstones, and earthquakes are caused by an evil spirit. The history of the sons of Job can be called upon to confirm this, they who were beset by a storm stirred up by Satan with God's permission. I have further summoned the authority of the Scriptures which calls Satan the ruler who presides over the skies. I have also the opinion of Kaspar Peucer and Jean

Is the author's opinion substantiated by epigenes, Cardano, Bienewitz, Scaliger, Theophrastus, etc.?

Bodin who contend that evil spirits and genii exercise great power, with divine permission, in the heavens and over the earth. Again I repeat, I should like you to tell me whether I can properly furnish such proofs in support of your hypothesis. I, indeed, believe I should because, to be sure, you postulate that storms wrench and carry off part of the heavens, or to use Pliny's words, Book II, Chapter 48, snatch it away with it and absorb it into the sky, where the tail of a comet is visible, illuminated by the rays of the sun. In conclusion, I bid you, most generous Sir, to rest confident that I will do nothing about publishing your observations without the counsel and help of your most esteemed son, the best of my friends, and that the dependability and sincerity of my pursuits and service will be evident to you throughout. As for the rest, I hope with all my heart that you will be engaged in the affairs of mankind for its own good and for a very long time. We also hope that you and yours will prosper so far as is possible in this vale of tears, and that you will ornament and enrich the realm of literature with many monuments of your rich intellect, as the Supreme Designer of All proposes.

Percer and Bodin believe that evil spirits and genii with divine permission are the source of many phenomena.

To Otto Gericke
Mayor of Magdeburg

Most devotedly, to your
most generous Excellency
for your zeal, cultivation,
and service
Sanislaus Lubienietzki

From Magdeburg to Hamburg on the 29 of May, 1665.

Most Distinguished Sir and well-cherished Friend,

Letter number 2 of the author in reply to that of the Honorable Lubienietzki.

Clearly there is a great deal of material which you treat of in your letter recently sent to me and to Johann Raue, member of the consular staff and first librarian to His Most Serene Majesty, the Elector of Brandenburg. Indeed, those questions which you have raised should be treated in a fundemental way, with all errors removed, if one wishes to skirt the rocks they conceal and to sail into safe harbor under full sail. This, however, requires a great deal of time and can scarcely be encompassed in one letter.

Those principles and fundamentals on which I believe my opinions are based, however, are discussed completely and at great length in my work which was published not long ago and which I recently abstracted and sent through my son, advisor and chargé d'affaires in Hamburg of His Most Serene Highness, the Elector, whom I mentioned before. This, I believe, can sustain the violent attacks of adversaries and can serve as an answer to them. Here empty Space is discussed wherein our world system and all world bodies exist.

The author's treatise De Spatio Vacuo *can be used as a reply.*

This Space, indeed, is limitless, immense, or, as I term it, all-embracing. Here are arranged in a definite order all creatures of God (among which the most important ones are the earthlike bodies, namely the suns, planets, and satellites), and thus our sun, together with its visible planets, namely Mercury, Venus, Earth, Mars, Jupiter, and Saturn constitute our world. These planets have their own virtues or corporeal and incorporeal effluvia. The corporeal effluvium of our earth is air, which is nothing other than its odor and it presses upon the earth with a definite weight (which is equivalent to the weight of water 19 Magdeburg cubits high). Furthermore, air follows the earth and through the conserving virtue of the earth, is not only carried about in its daily motion but is even transported along with the earth in its annual revolution about the sun. In addition, this air can be divided into regions, the middle or the highest of which the comets, or the storms that have been pulled away, claim for themselves where they execute their undefined and irregular movement, forming a ball of fire (a so-called "pyrobolus") until they cannot rise any higher because pure Space begins there. This area is empty of all matter and has no means for moving a body on higher.

Air follows the earth and is carried around with it.

Comets exist in the middle or the upper air and are parts of the air, pulled away by a storm.

The author is not concerned with the opinions of Epigenes and others.

As for the opinions of Epigenes, Bienewitz, Cardano, Scaliger, let these pass because they are not apposite to the subject and I do not have at hand right now a copy of their work. As for whether storms are stirred up by demons, as Paracelsus holds, this should be discussed at some length and a distinction should be made between the demons. At any rate, these questions cannot be treated in a short letter because some principles, based upon established, reasonable explanations, produce one opinion, assenting to the reality of sensible appearances while, on the other hand, a contrary explanation to the system of the world and fundamental beliefs about it brings about dissent. Whatever true reason and experiment commend, however, this is solid, firm, and eternal. But so as not to take up too much time and to appear to be expatiating in too fulsome a manner, I am going to break off the thread of my writing. Commending you to divine care and myself to your favor, I remain,

Most ready to serve your
most Noble Lordship,
Otto Gericke

Letter number 3 of the Honorable Lubienietzki.

From Hamburg to Magdeburg on the 3rd of June, 1665

I learned with certainty that you had received my letter from the page attached to the communication which you sent on March 26 to the most distinguished Brandenburg chargé d'affaires, your son and my dearest friend. From that time I discussed the question of comets with various learned men and introduced your opinion into the discussion. I must confess that I accepted their judgment which found little to commend in your hypothesis (the enclosed excerpts from the letters of the renowned Honorable Leuneschloss, Professor at Heidelberg, and the Reverend Father Riccioli bear witness to this). Thus I thought that it was my duty to warn and ask you in this letter entrusted, as before, to your most esteemed son, to defend your opinion about the vacuum and the generation of comets by employing the strongest possible supports of reason and experiment. For one thing, I should like you to present something in support of this opinion of yours concerning the earth's motion. I admit, indeed, that this has been sustained by the authority of hoary antiquity. I know, to be sure, that according to the statements of some scholars, Plato and indeed Pythagoras and his followers and, in later times, Cardinal Nikolaus von Kues, Copernicus, and others of their adherents had proposed and had wished to prove this theory by logic. I see, nonetheless, that this has been opposed by the artillery of great and strong arguments drawn from the Holy Scriptures and philosophy and has been shaken time and again by their battering. Allow me, distinguished and most venerated Sir, in view of the innate and singular sincerity I feel toward you and your most esteemed son, to say these things to you in the name of friendship and mutual regard, for in asking for a reply from you, I do it not so much for myself (although I confess most freely and willingly that much renown would accrue to my work from your most famous name) but rather so as to have a fair picture of your reasoning. There certainly is no reason why you should have any suspicion that it would be in any way detrimental to your work *De Vacuo* if I happened to include any part of this in my treatise. These excerpts would be published under your name and would bear the title of *Communicatio Magdeburgo-Gerichiana*. With this end in view I have sought to commend the names of other renowned men, God willing, to the everlasting memory of posterity. Among the several objections which attack your theory, there are the following: It is abundantly clear, and has even been recently proved that comets exist in the ether. Furthermore, it seems evident that their matter is not to be found in the bowels of the earth nor in the air, and that the earth cannot possibly lose and thereby contribute as much material as is needed for the generation of a comet body of great size even with the passage of many years. What is more, reliable evidence shows that no winds blow in the highest mountains and no storms are felt there. You

The Honorable Lubienietzki communicated with learned men about the author's opinion. Leuneschloss and Father Riccioli do not agree with the author's opinion.

Authors who have maintained that the motion of the earth exists.

Lubienietzki warns that the author's opinion is not confirmed by reason and experiment.

The contrary opinions of Leuneschloss and Riccioli.

must bolster your theory against these objections. Farewell, and remember to send whatever buttresses you have for your position to your most esteemed son, and continue to regard me with favor.

Stanislaus Lubienietzki

Letter number 4 *of the author in reply.*

From Magdeburg to Hamburg on the 22nd of July, 1665.

The author never had it in mind to write anyting about comets.

I certainly did receive the letter you sent me not long ago wherein you sought to find out my opinion about comets. Although I do not deny that I was not greatly inclined to investigate these questions which involve a more careful study of comets, because in my work, *De Spatio Vacuo* (which treats of the system of the world and all earth bodies and which, God willing, is to be published at some later date) I had no opportunity to deal with this subject, still, inasmuch as you are so persistent in learning my opinion, most revered Sir, I cannot refuse my help in so small a request.

Certain of my experiments must be presented first.

In order to understand my theory better, it will be necessary to select a few experiments from the aforementioned work and present them at the beginning of this discussion.

1. The fact that air presses down upon the earth with a definite weight.

1. The fact that air is an exhalation of this sphere of earth and water, or, that air is a corporeal virtue which flows out from the water and earth into surrounding space: that otherwise this space would be completely empty of all bodies or void, if there were no air. The earth, therefore, is surrounded by air like some soft covering and holds this to itself by the so-called conserving virtue with a weight equivalent to that of water 19 or 20 Magdeburg cubits high and carries it around with it in its annual revolution.

2. The fact that air is compressed in this lowest region of ours.

2. The fact that because of this weight, air is very greatly compressed in itself and, being of this nature, it can be increasingly compressed and contrariwise, can expand itself and stretch out farther where it finds more space. Thus, when two glass vessels, which pharmacists call receivers, (from one of these all air has been extracted, and its opening has been closed by means of a special type of stop-cock) are so joined together that no air can enter or escape from either one and then, when the stop-cock on the one has been opened, the air from the other glass, being filled with air, expands. Thereupon it rushes into the evacuated vessel with such great force that small stones that have been placed in the glass, larger than hazelnuts in size, can be scattered by the blast.

Further data on this experiment can be found in the Reverend Father Schott's work *Technica Curiosa de Mirabilibus Magdeburgicis*, Book I.

3. Visual inspection of such experiments reveals that those very tiny bubbles, as I shall call them, of air which are present in the water but which, because of their minute size cannot be seen, so expand in evacuated space or in evacuated glass vessels that they grow more than a hundred times larger, making them of a greater size than cherry pits. From this it can be assumed (since lower air is very much compressed by upper air and indeed presses upon us here on the earth's surface with the same weight as a column of water 20 cubits high) that lower air is much heavier and denser, and the higher it is, the lighter it is and the more tenuous and rarefied. For this reason air should be divided into three regions, a subject considered in Book V, Chapter 9, of my book.

3. The fact that air can expand greatly.

Lower air is much heavier.

4. Experiments bear witness to this fact, for the air on mountain tops and in towers and high buildings is found to be lighter and more rarefied than that nearer the earth. From all these relevant experiments as well as many others that were carried out, one can certainly assume that air not only becomes more and more rarefied and lighter the higher and higher it goes, but even, if one could go to a distance of 1,000, or say even 2,000 German miles, absolutely nothing more would be left of it and pure Space, void of all matter would be found.

4. The higher air goes, the more it expands until nothing of it is left.

5. Furthermore, experiments prove that the weight of a sphere of air changes from time to time, especially when rain falls (for when much water is released from the air, then it becomes lighter, but once again is made heavier when the air absorbs water. In order that I might anticipate changes in weather, I made a small instrument, namely a long and narrow glass receptacle like a pipe, wherein a small figure of a man was so fashioned from wood that it could be supported in the air and freely moved by it, thereby indicating with his finger the change in the weight of the air at different times. When, as occasionally happens, thisfiigure slips down farther than usual, this is a certain indication that the air at this time is lighter than usual. Experience has shown that at this time there are great and terrible storms in the vicinity, even though they are 100 or 200 miles away. It is to be gathered from this that storms (which undoubtedly have their origin in the clefts of mountains and subsequently rise upwards) attract particles of air, thereby making the weight of air pressing down upon these lower regions lighter. Because of this decrease in the weight of air, great storms arise which move now toward the east, now towards the west, and again in some other direction, indeed even straight downwards causing trees to be wrenched out by their roots.

5. The weight of a sphere of air changes from time to time.

The instrument enclosing a small figure made of wood which indicates storms.

Air is sometimes lighter than it normally is.

Storms, arising from the earth thereby draw up particles of air, a motion which produces gales, hailstones, lightning, whirlwinds, etc.

From all these foregoing experiments, then, I believe it can be

If storms can traverse a quarter part of this earth in this heavy and dense air around us, they can also rise through calm and light air as far as a hundred miles.

Telescopes make it clear that a comet is air that has been compressed.

Its tail varies in appearance according to the disposition of the sun and the eye.

The size and distance of comets.

Refraction and irregular motion make the parallax of comets uncertain.

concluded that if storms of this kind can rise up to the clouds and direct their course here and there into any part whatsoever of the world and according to the violence of their force can blow 100, 200, or 300 miles, and indeed also fall away with terrible force, why cannot these same storms, together with the air to which they are attached, arise from time to time up to the second or third region of air and wander here and there like a ball of fire?

Since, moreover, telescopes make it quite clear to us that a comet is nothing other than a cloud that has been compressed and illuminated by solar rays, I believe, for this reason, that such a storm together with the moister air it has assimilated, is carried to a dark height of rarer and dry air and there is lighted and penetrated by the sun's rays (because it is outside of the earth's shadow). The nearer it approaches the earth, the greater its tail always appears, and the farther it is removed, the smaller it seems because the upper air is always more rarefied. As for the fact that the tail occasionally does not seem long at first, this is caused by the thickness of the head through which the rays of the sun are not able to pass.

Since there are no grounds for objection, some attribute a size to comets greater than that of our whole earth and say that they are located beyond the moon, inasmuch as it is impossible for man to measure their distances, heights, or masses since their true locus (as astronomers term it) cannot be found by reason of their irregular motion and the short period of time they are visible. It is necessary that their "true locus" be determined by observations of more of their movements and that the parallax be found from this information with the "visible locus." Since, however, this information cannot be had, no one can measure the distance or height of a comet or even of a new star (which, however, does not have the comet's irregular motion, nor does it die so quickly).

If, therefore, their height as well as their size is unknown, then there is little reason to believe that the height of comets is very great (for they are not visible like the moon from all points on the earth's sphere) nor is their size of the proportions commonly attributed to them. Thus the scholars of Coimbra in *De Caelo*, Book III, question 3, article 2 declare that the greatest distance at which comets are seen does not exceed 52 miles. Nonetheless one must admit that there are such storms accompanied by clouds and moist air, which can expand themselves to a height of one hundred and more miles (where air does not have so much weight, and where, moreover this massed material has space to extend itself freely), being more than 30, 40, or 50 miles farther than it was before when it existed at that height where the clouds are now.

One can easily conclude that tailed comets do not rise higher than the regions of air extend, from the fact that their tails are nothing other than luminous shadows or refraction of the sun's rays which penetrate the clear head of the comet in this dark and rarefied air or atmosphere. For if they existed in pure Space that is empty of all matter (commonly known as the ether) they would not be tailed comets, but rather hairy or rose-like stars. Were they the latter, they would be at rest, but might seem to us to be in motion because of the earth's annual revolution, unless the attraction of the sun at the same time should be the source of another motion in them. Since it can be visually demonstrated by experiment that fire attracts air and consumes it, so also can such a comet be attracted by the sun (since I am of the opinion that it is a fiery body) and can ultimately even be consumed by it.

The tail of a comet is a cone radiating the light which the sun sends through its head whereby it is extended out at the end.

Hairy stars do not wander like those with tails. If, however, they have motion, perhaps this is caused by the sun or the annual motion of the earth.

These are the facts which I seek briefly to set forth about comets and to transmit to you, most noble Sir, as you asked. I am prepared to accept further information from others who have more exact knowledge than I. Indeed, I frankly confess that up until now more specific research into comets has not engaged me either professionally nor in my leisure. For this reason I urge you to reflect upon whatever good there is in this letter and to interpret what has been presented as you wish.

Farewell and good wishes,

Gericke

From Hamburg to Magdeburg on the 5/15 of August, 1665.

Letter number 5 of the Honorable Lubienietzki.

Shortly after I sent you my letter of June 3, yours of May 29 was delivered to me. I did not wish to answer this correspondence until I had received an answer from you to mine, which, I am happy to say, arrived eight days ago. In each letter, I see a remarkable affability, well suited to a magistrate, which Seneca with true reason sought in such a man. As for myself, I am aware of an uncommon kindness toward me which is due, I am fully aware, to the most esteemed and respected chargé d'affaires of Brandenburg, your son and my own well-beloved friend. Finally, I see in your letters an unusual knowledge of natural phenomena. Both of your letters demand an answer that is consonant with their geniality and friendliness, not only because they were most welcome to me for so many important reasons, but also merit this in their own right. Because there was no excuse for the delay in your reply, therefore, there is no place for indulgence, as I see it, nor should excuses for my

The Honorable Lubienietzki uses his friend, the author's son.

writing be presented timidly. You are indeed a man of great dignity, erudition, and judgment, and for this reason it is fitting that you should set forth your ideas. It is even more appropriate for me to present my opinions, however, who yield to no one in my eagerness for learning and the demands of my close friendship with your most noble son. As for the fact that I asked you and other men of learning so many and such difficult questions about comets, the perceptive and unwearying concern for learning which pervades all your work as well as your curiosity to seek the truth which arises from it and which is always praiseworthy, has effectively shown that divine wisdom, properly applied and directed should be eagerly pursued. In an assessment of studies of this kind, let there be freedom for each man to judge and to publish his own ideas which he believes to be true, or perhaps only probable, and so far as is possible, to defend them. Another person's opinions are not odious to me and just as no one is able to take my opinions away from me, so should he not harbor this thought regarding himself. If he allows this to enter his mind, the care of guarding his just and due freedom of thought falls upon me. That excerpt from your work which I returned to you by your most honorable son, if you wish this to be included in my Book, please return it to me. This holds for the rest of the material as well, especially those passages you wrote to your friend in Leipzig. If you do not wish to do this, however, I will rest content with your letters written on this subject. Each scholar, relying on his right to a free opinion, will examine your theses on the same basis as that of any other scholar, writing on the subject of empty Space, the motion of the earth (which you discuss along with the planets),as well as the generation of comets. It is necessary that these questions as well as many others be submitted to the diverse judgments of various scholars. You are indeed aware of how great is the power of man's freedom; you are aware by what laws it lives in the realm of scholarship. As for what particularly obtains with regard to your opinion about comets, which you hold are storms that have been pulled away, I shall have to speak briefly as befits my intent. I am sure that you will receive all my comments in a friendly, well-disposed way, acting as you asked me to do and, as you are certainly aware, as I have always done. I, indeed, beg you now to do this in even greater measure. There are those who combine the opinion of the Byzantine philosopher Epigenes with that of the Peripatetics, as Libau does on page 44 of his *Tractatus Problematis Cometarum*. He says "Aristotle first described the matter of comets as originating in exhalations of the earth, set afire by the motion or the rays of the sun and burning in the elementary region below the moon". And

The Honorable Lubienietzki asks for the author's opinion on empty space and the motion of the earth.

He combines Epigenes' opinion with that of the Peripatetics.

further on he says together with the Chaldaeans and Aristotle "that hot exhalations are the source of comets which are nurtured into flame in the upper air by suitable nutrients and ultimately are extinguished when all their matter has been consumed." He contradicts Bienwitz and other scholars who hold "that comets are not flaming exhalations but have received their light from the sun's rays." Clearly Plutarch supports Epigenes' opinion in Book III, Chapter 2 *De Placitis Philosophiae* because he holds that a comet is an exhalation of the earth which is borne aloft and is ignited there. Pliny does not give his opinion regarding comets. Seneca, however, devotes serious thought to Epigenes' theory, accepting and rejecting it at the same time. He records in Book VII, Chapter 3 of *Naturales Quaestiones* that comets seem to him to be set afire by a kind of whirlwind of excited air twisting upon itself, and further along in Chapters 6 and 9 he also mentions the whirlwind and the fragment of twisted air. His theory about the power of a whirlwind by which, according to Epigenes, comets are wrenched away and set into motion differs somewhat from Aristotle's thesis which attributes this power to the sun which draws out these exhalations from the land and sea and raises them into the highest region of the air. According to these scholars, each of these bodies lights up those fires in the air. You yourself deviate from Epigenes' theory, however, and in fact come more nearly to agree with the opinion which holds that a comet is a sub-lunar meteor or an exhalation lighted by the sun but is not ignited by it. Not only are Bienwitz, Scaliger, and Telesio, more recent writers, of this opinion, but also Heraclides Ponticus and Metrodorus, both of whom Plutarch, in the reference previously cited, recognizes as defenders of these beliefs. Other theories of Epigenes conform with those of yours, namely that there are two types of comets. One group of these diffuses brightness all around, remains in the same place, and bears a resemblance to hairy stars. (You also call them rose-like.) The other comets stretch out in one direction and draw along their streaming hair. These, you say, wander here and there in a wondrous way with the help of the air and show their tails. I discuss this all the more willingly upon the suggestion of your most esteemed son who advised me to make use of Seneca's testimony which is found in the aforementioned Book VII, Chapter 6 of his work. In this Chapter what Seneca objects to in Epigenes, you will also reject, inasmuch as you very firmly state that "comets, to be sure, appear in very calm air and do not belong to the wind, and since they do not have their origin in wind, neither do they grow with the wind. Furthermore, they appear in one particular place and are not driven here and there like the wind. Rather,

Is a comet a flaming exhalation or is it lighted by the sun? Epigenes said that a comet develops from an exhalation of the earth that is borne aloft.

Epigenes agrees with the opinion as regards the two kinds of comets.

The author has said that tailed comets wander here and there, by means of or with the help of air from an inner drive of their own which is independent of the wind.

Seneca's opinion is not contrary to that of the author.

they are seen above where the winds are permitted to go. What is more they take varied, curved paths, some to the east and others to the west" (which the wind does not direct). They move, moreover, with a regular motion, not wavering, jerking, and eddying, as whirlwinds do. What is more, they last longer than whirlwinds and other similar phenomena which are deprived of a longer existence. We see that Bienwitz favors your opinion as does Cardano in whose judgment "a comet is a sphere in the heavens lighted by the sun whose transmitted rays make a tail for it behind." Scaliger also agrees with you and makes the following comments: "A comet is a body so dense that it can be illuminated, yet so rare that it can transmit rays. We see something similar to this in a glass sphere." Furthermore he says, "I believe that a comet consists of vapor drawn by force from one of the stars into the highest region of the air. This vapor in itself is not glowing but is of a misty composition like smoke or soot that absorbs the sun's rays and in transmitting them is thereby made visible. For this reason it has a tail." (*Exercitio* 79). I have not rashly brought to mind Paracelsus and his spirits which excite storms. Here good and evil spirits come under discussion. Although we support philosophical freedom, still I have relied more upon the authority of the Sacred Scriptures than this hypothesis. For this reason, then, I follow the example of Job and the authority of the Scriptures. For without this I am unable to have wisdom. Now I shall speak of the good demons or the angels. We read very often in the Sacred Scriptures that from the beginning God has used their services in carrying out many tasks. Christoph Rothmann who did not withdraw from entering into the so-called field of combat with Scaliger on the subject of comets, declared in a certain letter he wrote to Tycho "that the lighter exhalations of the earth are carried on high with the help of good angels and are lifted above the moon, with the result that, being lighted by the penetrating rays of the sun there, they appear in the form of comets with tails or beards. These are directed and borne by angel's help to that region where Divine Providence wishes them to be carried, so as to strike terror into the hearts of men." This opinion was not unacceptable to Fromond, pleased Francesco Resta, and was defended by Cabei as Riccioli tells us in his *Almagestum*, Chapter 6, page 38, number 12. In support of this theory, he cites in the same place opinions of Damascenus, Tanner, and others. Indeed, I must confess that theologically and ethically this opinion is the most satisfactory, provided, of course, that it is philosophically acceptable. As to what pertains to the substance and moving source of a comet, this demands more investigation because the aforementioned philosophers do battle with the

Bienwitz and Cardano favor the author's opinion

as does Scaliger.

The Honorable Lubienietzki writes that Paracelsus had held that storms are excited by spirits.

Christoph Rothman thought that exhalations and comets are carried on high with the help of angles.

Fromond, Resta, Cabei, Damascenus, Tanner all agree.

Peripatetics as to what it is that carries exhalations above the moon. They also question the statement that comets are not lighted in themselves, but are illuminated by the sun. Rothmann, then, favors Scaliger's opinion insofar as he teaches that the substance of a comet is illuminated rather than being set on fire, as do Libau, Liceti, and others who say that comets are composed of very dense celestial illuminated matter and give these bodies place in the sky. This is also the opinion not only of Descartes and Ludovico Colombe and others, but also of Tycho, Longberg, Kepler, and many other very famous men ennumerated by Riccioli. These, to be sure, hold that comets are not glowing, but clear and transparent bodies, as did Artemidorus, one of the ancients, when he said that the tail of a comet is composed of vapor illuminated by the sun's rays. Others who are in particular agreement with Rothmann are Guiducci, Galileo, Aresi, and Praetorius who maintain that comets are exhalations of the earth, raised above the moon and illuminated there. They do not agree, however, that they reach this height through the help of angels.

Libau and Liceti and others say that comets are made of very dense, illuminated matter and give them place in the sky.

Descartes, Columbe, Tycho, Longberg, Kepler, etc. follow these.

These are the opinions of which I thought I should I inform you at this time on the subject of comets and I trust this will not cause you annoyance. Although you say that you are not greatly concerned with investigating these observations in depth, I hope, nonetheless, that you will send me some comment about them. In the meantime, I thank you sincerely for those excerpts from your most celebrated work which you sent me. And, that I may present you with something in return, I will forward to you those observations which I recently received from one of my friends who had conversed with a learned man on the subject of Space. The latter held that Space must be corporeal since it consists of three dimensions, which is a property of a body. From this we concluded that Space cannot be empty of a body, for if this were taken away from it, Space would perish and its encompassing bounds would come together.

The author is not concerned about the opinions of others.

A certain man said that Space must be corporeal since it consists of three dimensions.

But, my friend replied to him, (1) If Space itself is a body, then a body would be contained within a body since a body then would be in Space. He thought this supposition was absurd. (2) When the question about a vacuum arises one does not ask whether Space itself is a body or not, but whether (no matter if it is a body or not) it can have existence apart from the other body which is contained in it.

The other made some sort of reply to these objections but my friend said that these answers had not given him satisfaction. Thus, then, both had divergent opinions. It is possible, however, that they will discuss this question further either in person or by letter. If this

My friend contradicted him. For if Space is a body, then a body would be contained within a body, etc.

correspondence should come into my hands, I will see to it that you get it. You may be sure that I am speaking the truth when I say that I am eager to make use of all scientific information I can assemble and that all my zeal and effort will be directed to this end.

The Honorable Lubienietzki has often seen the little figure which indicates the weather at the author's home.

At the home of your most respected son, I have often seen the small artfully carved wooden figure of a man in an evacuated glass tube which indicates a change in weather. I learned from him that when this figure drops down low, then great storms are raging in places one or two hundred miles away from us. Since, indeed, I am going to mention this invention frequently in my work and since its reproduction and description must be furnished from communications I receive from your most respected son, I should like you to tell me how I can free myself of the doubts that beset my mind.

But he does not know how to free himself of doubts.

Experience shows that weather changes according to dissimilarities in land and sky conditions. We perceive that not only the high ridges of mountains and neighboring areas but also lands along the seashore are buffeted and frequently struck by rather violent winds. We see that those ponds are watered by frequent rains which exude exhalations that tend to generate rain. As for fertile Egypt, however, the granary of the world, we have learned that it has no rainfall at all and is watered solely by the Nile, even though it has an abundance of ponds. Are these examples of the alternation of phenomena? For different kinds of weather, like other events happy and sad all over the earth, succeed one another. One may be sure that although we look upon a serene and cloudless sky from either this place or that, still in some other spot, mighty winds and tempests hold sway. This is borne out by the comic writer who says, in agreement, "Here Favonius is calm, here Auster is rainy." How then can we be certain that the little man, as he goes down before our eyes to a lower position, presages storms and lighter air within a radius of 100 to 200 miles at a time when we are enjoying and experiencing clear weather? If only we could know what the position of this small figure would be, could he be both where we are now and also at a distance, 100 to 200 miles away! Then we would know what the state of the air and weather conditions would be in both places and interlying regions. Here recourse should be made to the most trustworthy empirical knowledge and the most reliable teacher available. Pardon the fact, most gracious Sir, that in placing my confidence in your generosity of spirit and my friendship with your most esteemed son, I seem in a sense to abuse both of them. This is quite customary, however, with those who are curious to learn and are desireous of furthering the public good. For when one has the appearance of being unkind, or so to say, churlish, he should not go without praise

The Honorable Lubienietzki asks how we can be certain that the small figure indicates storms and lighter air.

He likewise wants to know the position of the little man if he were to be both with us here and at a distance of more than 200 miles away.

because of his usefulness. You may be sure I am most anxious to look upon your letter with favor and to interpret it rightly so that in advance I may tender you great thanks for it and at the same time, I beg you regard this verbose letter of mine with your well-known sense of equity and kindness. May you fare will and prosperously, dear Sir, most worthy of Nestor's years, and may you look upon me with favor.

Lubienietzki

From Hamburg to Magdeburg on the 17 of March, 1666.

Letter number 6 of the Honorable Lubienietzki.

Although you have not yet replied to the letter which I sent to you on August 8 of last year, nevertheless, I should like to address you in this short note. I do not wish to cause you any trouble, nor have I ever wished to do so, since I am aware that your duties are onerous and that the state of your health is in conflict with illnesses as well as old age which we have found to be generally regarded as a disease in itself. If I have caused you any offense at all, will you please ascribe this, in view of your generosity of spirit, to my constant and unremitting desire for learning? At present I ask and seek nothing from you other that that you answer my letter if you can, despite the pressure of your affairs, and more particularly, your poor state of health and advanced age, wherein I set before you with great, but as I feel, permissible liberty certain trifling questions. Indeed there is not much time left, since I have waited for your answer at great length, with the result that I am unable to attach the correspondence between the two of us to that between your most respected son and myself. Now, however, I can wait no longer, for, God willing, I plan to have this correspondence go to press within the month. Indeed, your answer can be added as a footnote, but I should prefer its being attached to this letter. As I write this, however, I am in no way dictating what you should do. In my last letter, I mentioned a private conversation held between one of my friends and a very learned gentleman concerning empty Space. Since that time I have heard nothing further from my friend regarding this discussion. An associate of his, however, recalled this conversation in essence. He said that a very learned scholar sought to persuade us that Space itself is corporeal since it has three dimensions, but he has not yet convinced us of this fact. For although mathematicians commonly define a body in this way, still it may be that every dimension and quantity is corporeal, solid, and material. So much for his statements. I expect an elaboration of this discussion from

Lubienietzki urges a response.

He repeats his question as to whether Space is a body.

each of them. When I receive these letters I will disclose their contents to you and the public. I am enclosing a letter from the renowned Johannes Müller, Professor of Mathematics in the Gymnasium of this city, which he wrote to me when we were both discussing comets. In this correspondence he made reference to your belief that a comet's movement has its source in the earth's motion and he presented a number of opinions, both pro and con to it. I do not believe I have made his opinions available to you or your most respected son before this time. I hope you will also see fit to make comment upon his remarks. Here is another letter for you which the Reverend Father Schott recently wrote to me about you and which I answered. Still another, written by the most illustrious Fieldmarshal Würtzuber, giving his opinion on the subject of the vacuum, asks me to write to Schott in your support. I have no doubt that these letters will give you pleasure. I am sure everyone must see clearly that the regard and respect shown by a man so outstanding in his intellectual acuity and wisdom should carry weight with you. I, therefore, congratulate you for the recognition of your work as well as for the title of nobility conferred by His Most Serene and Erudite Highness, the Emperor Leopold, upon you and your son, the reward of wo thiness and merit on both your parts. I congratulate you, especially, however, on the recovery of your good health from illness and I hope from my heart that you may enjoy prosperity in good measure for the longest possible time.

According to the author's opinion, the motion of the earth is a consideration in the case of hairy comets, but not of tailed ones.

Fieldmarshal Würtzuber agrees with the author about the vacuum.

His regard and respect.

The Honorable Lubienietzki congratulates the author on the title of nobility conferred by the Emperor as well as on the regaining of his health.

Lubienietzki

From Magdeburg to Hamburg on the 2 of April 1666.

Letter number 7 from the author.

Although I do not in fact wish to refuse you my opinion about comets, still I repeat (as I have already done before) my admission that I have not been greatly concerned with looking into those questions which involve a closer consideration of comets, for in my work, *De Spatio Vacuo*, I had no occasion to treat of this subject. I will not object, nevertheless, to review certain points mentioned in your esteemed letter and to reveal to you, as you desire, what I feel about them. You mention in your letter the various theories of the old philosophers about comets, espectially Epigenes and Aristotle. At the same time you set forth how Libau had tried to reconcile Epigenes with the Peripatetics and in what respects my opinion agrees with Epigenes and wherein it seems to deviate. All of this is unproductive to dwell upon further here.

The author admits that he had not been particularly concerned with those questions which are involved with a closer consideration of comets.

The opinions of the various philosophers cannot easily be reconciled.

I answer, however, that it is not my opinion that the substance of comets originates from exhalations of the earth and, through the agency of the sun's rays, rises. Rather I believe that their matter has its source in a kind of whirlwind generated in the bowels of the earth, which sailors call Orcan, and in bursting forth from the caverns of the earth and mountains, it draws to itself a particle of air (as all whirlwinds tend to do). Enveloped by this whirlwind, the air is carried farther and farther upwards, and does not, as one might expect, descend to the earth, until it reaches that region where air is greatly rarefied and there is no wind. Furthermore, because of its innate whirling nature, this body wanders here and there like a ball of fire, being illuminated by the sun (when it comes out from the shadow of the earth) in the manner of the clouds, and is ultimately consumed by the sun. In my letter to you of July 22 of the past year I explained this in detail in essentially identical terms.

The author's opinion concerning the origin and substance of the comets.

As to whether or not whirlwinds of this kind are excited or produced in the bowels of the earth by spirits, it is difficult to say, nor would I readily dare to give an answer. Rather I should say that there are some living essences, or even animated spirits, which are generated there for the production of storms of this kind and then die out again.

Storms seem rather to be generated by individual living essences than by spirits.

I further learned from your letter that one of your friends had spoken with some other learned man on the subject of empty Space. He does not hesitate to assert that "Space is of necessity corporeal since it has three dimensions. Thus it can be concluded that Space cannot be empty of a body, for if this were taken away from it, Space would perish and its encompassing bounds would come together." I reply to this: Space in itself certainly does not admit of any dimension.

A reply to the objections set forth against empty Space by a certain friend.

Space is that which characteristically contains or holds. In Space, indeed, everything has its being or existence, or can have. It gives way to nothing, not even so much as the body it contains, is eternal, unmoveable, indivisible, and penetrates all things everywhere. It is immaterial whether it is full or empty. Thus, the common man does not believe that Space has existence because it is neither this or that which can be seen, touched, etc., but is something from which all other things are profoundly different, removed, as it is, from everything defined as having a material nature. Thus there is and can be nothing which comes near to being like it in any way, etc.

A description of Space.

Now whatever is measured in terms of a third dimension is measured in terms of body and not Space. Although a person might wish to conceive of a room's having space, for example, insofar as it seems to be encompassed by something material in terms of that which is material, still all that is thought of as a body is not in

Space has no dimension.

actuality a body, nor can that mental concept add or detract anything from Space itself. For indeed, this measurement is made in terms of the position and size of this body, not, however, according to the space itself, because Space is as much inside as outside of this body, etc.

Another objection that Space cannot be empty and its answer.

Furthermore, this man believed that "Space cannot be empty of a body, for if this were taken away from it, Space would perish and its encompassing bounds would come together." Now, as a matter of fact, that conclusion is contrary to my experiments which clearly demonstrate, even to the eye, that all air can be extracted from a round glass vessel, and when this has been evacuated, nothing is left in the glass other than pure Space, empty of all matter. Thus, it would not follow that if a body were removed from Space, Space itself would perish. For when a tower, for example, is destroyed, certainly that Space which it had previously occupied will remain, enabling a new tower or some other building to be erected in its place. Neither does it follow that "if a body were taken away from it, the encompassing bounds of Space would come together," nor as Descartes writes in his book *De Principiis Philosophiae*, part 2, number 16 "would the sides of a container become mutually contiguous." Experiment shows that although this does not occur in the case of a round glass container, the result would be otherwise with a container that is not round, whose sides are broken by the weight of air pressing upon them externally. My volume *De Spatio Vacuo* deals with such questions as these more fully and brings to light results which were heretofore unknown.

Information about the small figure which is an index of storms.

Finally, as for the small figure, or (as I have termed it), the little mannikin, made of wood which was enclosed in a long glass container, evacuated of air, I have indicated that it can show a change in the air by its movement. As a matter of fact, I had not made him in the first place with the intention that he might predict the presence of great storms; indeed, I had not even thought of such a thing. Since I had observed, however, through many clearly convincing experiments, that when it rains or even before the rain reaches us, the mantle of air bounding our field of vision always becomes a little lighter, (the reason for this being that much water falls out of the air at this time), I got the idea that if a kind of device were made wherein this mannikin's figure, made of wood, could be raised by the weight of air or lowered by its lightness, and would indicate the difference in air pressure with his finger, precisely and accurately, then rain that was imminent could also be foretold and expected. I was unaware, however, at the time that if this small figure, contrary to its usual motion, should decline 4, 5, or 6 points, this rather large

When it rains the air is lighter.

When the figure sinks far down, it indicates storms.

descent could indicate the presence of gales and storms (even if they were one to two hundred German miles distant from us). When such a descent occurred, experience subsequently revealed through written accounts that at the same time terrible catastrophes had resulted from storms on land and sea, although where I was, the wind was not nearly this violent. As to what is the real or most obvious cause of this phenomenon, I have not been able to ascertain up to now, but still I can always foretell the presence of storms of this magnitude by means of this small instrument, although I cannot precisely ascertain where these storms may arise. This information I subsequently find disclosed in contemporary periodicals. Yet I have no doubt that in regions where it is calm and without wind, the air presses down with a constant weight upon the earth. A change in our sphere of air is detected by this small figure, both when it drifts downward at the time of rain or storms but also when it rises 3, 4, or even 5 points above the usual midpoint. At this time, there can be little doubt that the air necessarily is absorbing into itself much water, particularly from the sea (for regions of small lakes cannot be sufficient to provide this). Indeed, I should like to see this little instrument made use of on the ocean where men often voyage 100, 200, or 300 miles in a short time. I have no doubt that many investigations and discoveries could be made with it. But what benefit would accrue to me if I should communicate this secret to someone gratis, inasmuch as I have expended a great deal of personal effort in carrying out these experiments? These are the answers, most noble Sir, I am able to give at this time, insofar as my meager resources allow, to your last two letters. May I hope that these ideas are presented in such a way as to satisfy your curiosity and your perceptive judgment. I have confidence, however, that with your intelligence being of the kind it is, you may come to understand for the most part opinions on this ambiguous and intricate subject. As for myself, I can promise you nothing more certain than that you will always find me ready and prepared to carry out whatever favors of friendship you may require. May God keep you safe. Farewell and warmest greetings.

It cannot be ascertained, however, when these storms arise.

The air, or sphere of air in the regions where it is tranquil, surrounds the earth with weight.

The figure occasionally rises even above the midpoint which provides a true indication that the air becomes heavier.

Undoubtedly, just as the air releases water, so does it also absorb it.

The author excuses his limited abilities.

P.S. As for the opinion of the renowned Müller, Professor of Mathematics in the Gymnasium in Hamburg, of my theory of the movements of comets, I remember that I had written in my letter of July 22, 1665, that comets with tails are moved through movement of air, but in the case of hairy or rose-like comets, the situation is quite otherwise. These are at rest, but because of the annual rotation of the earth, they appear to us to move, unless perchance the sun also attributes to them another movement at the same time through its

The comets which move or are at rest.

attraction. For, indeed, it can be visually demonstrated that fire attracts and consumes air, and in the same way, it may be supposed that the sun can attract to itself such a comet (which I believe is, in all probability, a fiery body) and finally even consume it. I should like these observations to be taken in consideration for a proper and true understanding of my opinion, etc.

Von Guericke

The eighth and last letter from the Honorable Lubienietski.

From Hamburg to Magdeburg on the 19/29 of April, 1666.

I confess I should have answered with more dispatch your welcomed letter, as learned as it was friendly, which was written on April 2 of the Julian calendar. Various tasks, which came my way through friends who were passing through the city and visited me and were commended to my care, as well as numerous other duties, prevented my answering in response. Now that the Easter holidays are over, however, I am directing my attention to you. I do give you particular thanks for your kind and understanding reply which was all the more welcome because of the explanation of your opinion about comets. I am aware that even before this exchange, you had said that comets were generated from a particle of air that had been wrenched away, then carried upwards by the force of a gust of wind, and subsequently illuminated in the upper air. I also rightly remember that you had held that there are two kinds of comets, those with tails which wander in the upper region of the air and those that are hairy or of a rose-like appearance which exist in essentially pure air and are at rest. Indeed I did not find this unusual since I realize that natural scientists have learned that comets are both mobile and immobile (a term which may be loosely used) because the motion of the latter is less evident. Those statements which I made to you at one time about the power of demons to disturb the air and to cause storms have sacred as well as profane authority, as I said at the time. As for these opinions about comets, nevertheless, I have not accepted them as necessarily valid, but regard them only within the bounds of possibility. Thus I have submitted for the most part to your own judgment and that of other learned men engaged in experimentation. Recently an individual came to my attention who endeavored to convince his readers in a published writing that a comet is an evil spirit. I admit that he draws upon testimonies and witnesses from the Sacred Scriptures. He does not persuade me to his way of thinking (to speak openly of my feelings), however, and since I will deal with this hypothesis later on, I do not judge it necessary to include

Comets are of two kinds.

The power of demons.

A certain individual wished to teach that a comet is an evil spirit.

these arguments in this letter. Why, indeed, should I continue to trouble you with these remarks? Furthermore, I should now spend my time as judiciously as possible while I am in the process of bringing out my work. Those phenomena relative to empty Space and the change of air which you say have been indicated by your little man shut up in a glass, have been proved valid by experiment and are regarded as cogent. I, indeed, not only regard your efforts and kindness on my behalf and diligence and frankness in discussing your opinion to be that of a good friend, but I also confess that I owe you a great deal in this regard and thus offer you the fruits of my efforts and the studies I have pursued in common with you. Although the publication of my extensive work demands all my time for itself and extracts from me whatever is left after all other things have been attended to, nevertheless, it is possible that when I get a little more leisure time, I will present my thoughts to you on these matters. But if such is not the case (for truly my time is hardly sufficient for collecting and correcting those opinions I extracted from friends on the subject of comets, or for completing what remains on the history of comets), this letter will close our correspondence and conveys the thanks owed you for your consideration and kindness. Farewell, dear friend. and may you have good health and a long life and always keep your feelings of goodwill toward me.

Lubienietzki plans to publish an extensive work.

Lubienietzki

BOOK VI

OUR PLANETARY SYSTEM

CHAPTER 1

THE WORLD, WHAT IT IS, AND WHAT IS INCLUDED IN THE TERM, IN THIS TREATISE

In Infinite Immensity, the Container of all things, where there is nothing below, nothing above, nothing up, nothing down, to the right or to the left, where there is no center, nor middle, nor end and where nothing is removed nor is near at hand, nothing here, nothing there etc., the world exists. Consequently it is undoubtedly the case that the world is nowhere (if no other worlds exist with respect to which this world's position can be fixed).

What philosophers commonly understand the world to be and what is usually meant by this term, this treatise records in Book I, Chapter 1 at the outset. Indeed people tend to associate the term, world, with this sphere of our of earth and water, sometimes even regarding it as a section of our earth, as when, for example, they call America the New World. At other times it can refer to mankind in general because men are the most important animals dwelling upon the earth's surface and are rulers before the rest.

What can be understood under the term, world, in this treatise.

We, however, following the middle course in this treatise, include under the term, world, the sun, together with its spots, as well as the planets (among which we number the earth) and their satellites (in which group we count the moon), together with that enormous intermediary Space which encompasses the whole world.

The world, therefore, is a gathering of stars, a subject which we will subsequently discuss.

Concerning the disposition of the stars.

If this is posited, then the principal question which arises concerns the order and disposition of these earthlike bodies.

In this context, indeed, (1) There is no doubt that Saturn is the most distant of the planets and is followed by Jupiter and Mars. (2)

There is also no doubt that the sun spots are nearer to the sun than all the other bodies. (3) Furthermore, there is little doubt that the moon is nearest of all to the earth. But should the position of the sun with its spots as well as Mercury and Venus and the stars nearest them, be fixed beyond Mars or should the earth, together with the moon, be placed there? Here one finds grounds for dispute. For if one positions the sun immediately beyond Mars, then the earth will be the center of this world; but if one locates the earth along with the moon beyond Mars, then the sun will be the center of this world.

See Plate IV.

Without a preference for either theory, let us draw a sketch on a piece of paper which represents the immensity of Space which embraces this world; herein we will distribute and arrange these earthlike bodies as reason dictates.

First we will assign the earth, a, roughly to the middle of this chart, Then let us fix the position of the sun, b, in another place, removed at any distance you want from the earth. The spots, c, c, are nearest the sun, then Mercury, d, and Venus, e, whose planetary orbits circle the sun. Thus necessarily these orbits, c, c, d, and e, must pass around the sun and describe their courses within the orbits, as it were, of Mars, Jupiter, and Saturn.

Because, however, Mars appears to us much larger when it is opposite the sun than when it is nearer in conjunction with it, (the same observations hold for Jupiter and Saturn, although with a much smaller difference in Jupiter than in Mars and accordingly an even smaller one in Saturn than in Jupiter), it follows that these three planets move in an orbit, not around the earth but properly around the sun. Consequently their orbits are more naturally orientated around the sun as their focal point than the earth. Indeed, this difference in their appearance clearly indicates that the earth is not in the center of the orbital paths of these planets (unless we wish to postulate eccentrics which we described in Book I, Chapter 4) and to attribute irregularities to these essentially perfect earthlike bodies, which is contrary to all reason.

The following bodies, therefore, must also revolve around the sun: the orbit of Mars, fg, which of necessity encompasses the earth together with the moon, where Mars at point, f, is nearest the earth and opposite the sun, while at point, g, it is in conjunction with the sun. The same may be said of the orbit of Jupiter, hi, where Jupiter at point, h, is nearest the earth and opposite the sun, and is in conjunction with the sun at point, i. Finally, the orbit of Saturn, kl, where Saturn is nearest the earth at point, k, and is opposite the sun and at point, 1, is in conjunction with the sun.

It now remains for us to assign a place to the moon which can only be very near to the earth, with its orbit embracing the earth, or its path encircling it, as is represented by mo.

Finally, the orbit, mn, must of necessity be assigned to the earth together with the moon. Thus (unless we wish to accept Tycho Brahe's concept of the solar system which was sufficiently well disproved in Book I, Chapters 9, 16, and elsewhere) we attribute forward motion to the planets (and the earth as well) about the sun and place a stationary sun in their midst.

The sun naturally is situated in the center of this world.

If this is the true system of our planetary world, then all the fixed stars are stationary, and if they are stationary, there is no reasonable basis for assuming that they are situated in some spherical kind of shell at a distance from our sun. It is much more reasonable to assume that they exist throughout the immense abysses of Infinite Space at different distances from some determined point or other.

The fixed stars are stationary and are not in a kind of spherical shell.

Thus, we must assume that our world will be a gathering of individual earthlike bodies (wandering stars, as it were), arranged in a prescribed and definite order, of which the king, so to speak, is the sun. It is situated in the center, as on a royal throne, and rules, cares for, and governs the planets. To such a point as the forces and virtues of this ruler extend, so far is the world thought to stretch out. Where they cease, moreover, there also one naturally supposes that the outermost bounds of the world lie.

The world is a gathering of wandering stars.

As for the earthlike bodies themselves (with the exception of their king which is endowed with individual virtues of its own), they are composed of a mixture of the four so-called elements arranged in such a way that is appropriate to their nature. Moreover, the more important individual planetary bodies revolve about their own centers and are mutually related to one another through certain similarities as are Venus, Mercury, and the earth. It is manifestly clear, moreover, that individually these bodies discharge their own corporeal and incorporeal virtues and are susceptible to different kinds of influences etc. Still they are not all designed in exactly the same way and according to the same plan by the Supreme Maker of the world, but actually differ considerably from one another.

Earthlike bodies are composed of the elements.

They differ, nonetheless, from one another.

CHAPTER 2

IN WHAT SENSE THE WORLD IS ONE BODY AND WHY IT SHOULD BE TERMED A UNITY

Philosophers generally agree that this world consists of diverse bodies

Philosophers hold that earthlike bodies have their existence in some matter.

and substances which, nonetheless, are united by a special bond into one body or unit.

According to Peripatetic doctrine, the whole world is composed of various substances and thus all the earthlike bodies have their existence in an intermediate material Space or in some sort of celestial fiery matter, a kind of substance which binds them one to the other. In this way they mutually form one body or one world.

Tycho Brahe, indeed, believes that some sort of rarefied essence exists.

According to Tycho whose opinion has been accepted by all astronomers in years past, there is a kind of matter filling the heavenly Space between earth bodies, a sort of rarefied air or fifth essence, which travels with the sun and its huge planetary system each day and circles the earth in twenty-four hours, leaving the earth immobile in the center.

The author holds that earthlike bodies have their existence in pure matterless Space.

In our opinion, however, the intermediate Space between the earthlike bodies is not a material substance, but is immaterial (as we have said in Book II, Chapter 4 and elsewhere in discussing Space) through which the sun's virtues are dispersed, not indeed into infinity, but to the bounds of the sun's influence, or simply to the point where the sun's rays do and can effectively extend. In such a sphere of virtues, then,the planets are contained and move, thus being united and joined together into one body or one world. This world has its existence in a certain natural, rightful, and harmonious arrangement, disposition, and collection of its bodies which, by the virtue of the sun, their king, as it were, are so arranged with respect to one another, they could not be otherwise according to nature. In just such a way every stable government has its establishment in a king and the hierarchy of its distinguished leaders.

CHAPTER 3

THE FORWARD MOTION OF THE WORLD

Inasmuch as there is a great difference between motion and forward motion (translation), one should reconsider at this juncture what we said about motion in Book V, Chapter 15 and forward motion in Chapter 16.

The world as a whole cannot partake of forward motion.

The world as a whole cannot partake of forward motion. For in Infinite, Immense Space there is no forward motion in itself and nothing in existence can move itself from one region of Space to another (or to put it another way), from place to place because in Space there is nothing up or down, nothing on the right or on the left, and absolutely no respect by which forward motion can be directed to one thing toward another. Where there is no place, then a body cannot draw near to or leave anything or he carried here or there. Just as nothing can rise in Immense Space, so nothing can fall in it or have weight or lightness. When something is heavy or light, however, then it is capable of forward motion for this is produced through the agency of some second body or material substance. Here one may ask if there are not, in fact, many other worlds (as is believed) to which this world would tend to be drawn. Should this be the case and even if this world of ours should fall or rise for 10,000 years or countless 10,000's of miles, it would nonetheless be in the same place and would remain there because there is no respect by which one thing is related to another in Immense Space and there is no distance so great that can be compared with it. If, however, it is posited that there are many other worlds, still no one of them can partake of forward motion (1) because one cannot be carried to another (2) because forward motion, as we have said, cannot be produced without some external source or without the help of a second body. Any world whatever, being solitary in itself and joined with no other, would be incapable of supplying support to another, but would remain unmoving in its own space and not participate in forward motion.

In pure Space or a complete vacuum, nothing can move itself from one place to another.

Just as in a living body which consists of different parts or members participating in varied motion, so also in the parts or members of the world, there are diverse motions, some being relative to the members themselves and others involving forward motion. As for the center of the of the world, it necessarily must be deprived of all forward motion because if the whole is not moved from place to place, then its center is not so moved. Bodies, however, can participate in some sort of motion, particularly internal rotating motion without the presence of forward motion, as we see with spheres which are rotated by lathe turners, no matter how quickly they are turned, but which still remain immobile in themselves, insofar as forward motion is concerned, and remain in place. Thus it must be concluded that the center of the world is not moved from place to place, although it does revolve about itself, and that its motion is the initial mover in the world.

The center of the world cannot participate in forward motion.

The center of the world can naturally revolve about itself.

Since, therefore, there is nothing outside which initiates movement in the world, the world in itself is deprived of all forward motion. It follows, then, that all motion, including forward motion, (of members or parts, as it were, of the world, namely the earthlike bodies),

The parts of the world can be moved by an internal force.

which is observable in the stars, does not originate in an external, but rather in an internal force.

CHAPTER 4

THE PERIPATETIC CONCEPT OF THE WORLD

The Peripatetics believed that the motion of the world has its origin in a First Mobile, the outermost or Tenth so-called Heaven. They assert that this completes a revolution within a period of twenty-four hours (consequently, as can be seen in Chapter 17 of this Book, a point on its equator at any given moment always moves forward more than 130,720 German miles, a concept which is contrary to all reason, and carries along with it all the stars, the countless fixed stars as well as the rest of the planets, moving from east to west). How, then, do they explain the periodic motion of some individual planet, if a person asks about this? They postulate a Ninth Heaven to which they attribute a contrary motion, one from west to east, dependent upon other poles of the zodiac. We may as well pass over in silence the many and varied eccentric motions of individual planets, for whose approach toward and drawing back from our earth they find excuses, all of which they cannot understand or explain rationally, as is to be seen in their theories of planetary motion.

Astronomers commonly attribute many motions to the stars originating from without as well as within.

For indeed, it is impossible and contrary to nature that motion should be produced in any body (as we see in the preceding Chapter 2) by external bodies, unless it is initiated internally, for all motion, as was said in Book V, Chapter 15 in this regard, has its origin in itself from an internal first beginning. Since, however, according to the Peripatetics' thesis, the center of the world is deprived of all turning motion, then the movements or turnings of these so-called Heavens should not be called movements, but forward movements. Forward motion, however, cannot take place unless some external source is present, which, however, is not there (because, according to Aristotle, there is nothing beyond the last revolution). Therefore there can be neither motion nor forward motion in these two Heavens, much less since the motion of the Ninth Heaven is said to be opposite that of the Tenth Heaven.

Motion cannot be created from something outside.

There can be no forward motion of the First or Second Mobile.

As for conscious elements (to which certain writers attribute the revolution of these imaginary Heavens), Michael Havemann states in his "Astraea", "They seem to make no more sense than the action of the country man who brought along a wagon to bring home some bread he bought for a copper. For why should they prefer to draw

There are no conscious elements.

the immense world system around with great difficulty when they can explain the turning of the earth in a much easier way?" Indeed, if there were conscious elements,they could not draw anything around outside of the world in that Infinite and Immaterial Space because nothing is there on which they could place their feet or with whose help they could bring about this rotation.

God himself, moreover, does not draw the Heavens around because Immensity is not self-moving, nor does it have anywhere to move but the "Everywhere" at hand; much less should anything be attached to Him to lead around, since He functions in the midst of everything.

God himself does not draw the Heavens around.

It follows, therefore, that the course of the planets or their forward motion does not originate in some source external to the world, but from something internal, namely the center of the world, the sun, which continuously turns about itself and the earthlike bodies surrounding it through its own virtues and influences.

The forward motion of the heavens has its origin in an inner source.

If, therefore, the course of the planets is governed by a center, then the postulation of spherical heavens is false and it follows that the world is not a body united by some sort of matter, because a material body whose center rotates also turns its whole body together with all its parts at the same time. We see that the planets do not operate in this way, one of which is carried around and completes its forward motion in a space of 30, a second in 12, a third in 2 years. This rotation from the center cannot be produced by any material substance but through virtues emenating from the center and operating upon these widely dispersed bodies.

The world is not a body united by some sort of matter.

CHAPTER 5

THE MOTION OF THE WORLD ACCORDING TO THE FOLLOWERS OF TYCHO

Although the followers of Tycho declare that the intermediate space between the earth bodies is some sort of subtle matter, an aura or fifth essence, as it were, they contend, a nevertheless, that whatever the matter present there, because of it the world forms one material body. Furthermore, the whole planetary system, together with its center, the sun, completes its course around some other point in this system, namely the earth.

The followers of Tycho say that the world body is filled by an aura.

Likewise that the whole planetary system revolves each day about the earth.

One asks with justification how a body can turn itself when all the parts contained in it do not rotate at the same time and one part remains immobile.

One may further ask with regard to this world described as a united and unified body, how so vast a planetary system (whose diameter contains more than 76,000 diameters of the earth, in comparison with which the earth is only a point) can run through so huge an orbit in 24 hours between the sun and earth, together with its own center, the sun, and at the same time can turn itself, together with all its parts, namely the planets, about this, with so many diverse motions, and still follow a course about a certain center, one indeed contained in this system, namely the earth? Surely this is contrary to all reason.

The orbit of Tycho's system is contrary to reason.

If this system of Tycho's lay outside of some point or other which it does not contain within itself, such a revolution could be conceived of (this could not actually occur, however, because in etherial Space, no forward motion can be produced without the help of some other body), but since, in fact, it contains that point, namely the earth, it is impossible that this system rotate without drawing along the earth with it.

CHAPTER 6

THE MOVEMENT OF THE WORLD AS SET FORTH BY THE AUTHOR AND IN PART BY CERTAIN OTHER WRITERS

The sun rotating with extreme speed about its own axis carries planets in its sphere of activity at the same time.

In Infinite, Immense Space where no place can be described or defined with respect to something else of the same kind (in the eventuality that there are other suns, for example), our sun has its place, rotating very swiftly and without cease about its own axis and center toward the east, and emits its virtues far and wide into surrounding Space. These virtues touch the planets contained in its sphere of influence (which is the source of their heat, as is shown in Book IV, Chapter 11) and make them turn. This source of motion makes all these stars of our world move and carries them around without difficulty and with the greatest ease (because their bodies are free, existing in themselves in that Space which is otherwise occupied by no other matter and is in its entirety most ready to receive all things created).

The very rapid rotation of the sun can even be seen directly.

In clear weather especially when the sun is near the time of its rising or setting, we can see that it revolves very rapidly (an observation that becomes even clearer if one uses a telescope, as Father Anton Maria de Rheita confirms in his *Oculus Enoch et Eliae*, Book IV, at the end of Chapter 1, where he writes, "The sun, observed through a very good 15 foot telescope with colored glass (see, however, that you do not make this observation without at least a double colored convex lens unless you willingly subject yourself to the danger of being blinded) appears as a continually seething and glowing bronze sphere of pure flame, ceaselessly rotating.") One must conclude, therefore that there is a valid reason for this and that the sun, as the center of the moving world, with its own virtue or so-called rays has a very great capacity for moving and driving the surrounding planets and indeed carries the nearer ones about more rapidly than the more distant ones, as observation itself demonstrates, when

Astronomers observe and calculate that

		Years	Days	Hours	′	″	‴
Mercury	complete their		87	23	15	16	–
Venus	periodic forward		224	17	53	2	–
Earth	motion in		365	5	49	15	48
Mars		1	321	23	31	56	49
Jupiter		11	317	14	49	31	56
Saturn		29	174	4	58	28	30

The sun carries the planets around in their natural order of rank.

The sun, however, is actively moved from within because it is composed of fire and fire cannot exist without motion, as can be seen in Book III, Chapter 12 which discusses fire in a vacuum. We see, moreover, that gold, the heaviest material of all, shows movement in its center when it is being liquified by the heat of fire in a container; we must assume that this motion also occurs to a much greater degree in the sun.

Thus, the sun's internal motion and its virtues spreading out spherically in every direction produce forward motion in the planets by means of the sun's friction acting upon the planetary bodies from a distance.

The sun's virtues radiate through unbelievably vast Space.

Although the sun's virtues permeate such vast and enormous Space and its rays are dissipated far and wide to such an extent that human intelligence cannot comprehend it, (indeed, its scope is completely unbelievable to us mortals), nevertheless, one must not suppose that these virtues extend out infinitely, for just as the strength of no created being extends itself infinitely, so do the sun's virtues have their limits, and when they have become exhausted by great intervening distances, they disappear. Clear reason dictates, however, that these virtues must extend even beyond the sphere of Saturn, for although we cannot define their bounds with certainty, nonetheless, we must assume that they end in the spheres of virtues of the nearest

Finally, however, they become weaker and their influences cease at the bounds of the world.

fixed stars, whichever ones they may be, perhaps Sirius, Vega, etc.). At this point, then, one surmises that our planetary system ends.

CHAPTER 7

THE EARTH BODIES

There are three kinds of earth bodies.

In the world, or even let us say, in the universe, there are three principal types of primary bodies which we also designate by the term, stars, through direct observation. Otherwise they are known as suns, planets, and moons.

Solar bodies.

Solar bodies are those which have light in themselves and for this reason they are called light-producing bodies, as are the sun and fixed stars. From a point of observation on each of these, the sun would not appear greater nor different from them if they were viewed from the sun.

Planetary bodies.

Planetary bodies are those which derive light from another source, namely the sun (or other fixed stars if bodies of this kind happen to be near to them), on which account they are said to be illuminated. From each of these, the earth would not appear different from them if they were viewed from the earth. These bodies are transported in ethereal Space within the sun's sphere of virtue. They are also endowed with a turning and directing virtue and are enveloped by their own atmospheres.

Moon bodies.

Moon bodies are satellites of the planets and are of a lower order. They are not carried around by the sun but by the planets themselves because, by reason of their smallness, they are outside of an appropriate and natural distance from the sun's sphere of influence. They do not embody the turning and directing virtue.

The planetary and moon bodies are different from one another.

The planetary and moon bodies are opaque and are composed of rudimentary elements and do not embody identical properties but are as different as they are numerous, with the result that they reveal the unbelievable majesty of the Creator to his creatures.

They hang freely in empty Space.

The location of these bodies.

All earth bodies exist in ethereal Space without any weight; the solar bodies where the divine Hand has placed them; the planets in their own orbits where they are set in motion and carried around in the sphere of the sun's virtue. The moon bodies move around the sphere of the planet with which they are associated and are drawn with it and by it in the direction of its forward movement.

Thus it happens that there is no point to which a natural force would carry them; rather they are freely suspended without support. They have centers other than the center of the earth upon which they depend and they maintain themselves relative to these points and strive to conserve this relationship. If they did not do this, all the bodies of world would necessarily fall to ruin, whether up or down, in complete disorder. Indeed, if you should transfer a part of the moon that had been torn away to some other earth body, the earth, for example, it could not remain there but would be repelled.

Earthlike bodies do not have weight.

A part of one body would not be tolerated in another.

Even though the sun appears to us who dwell on earth as the largest as well as the brightest and the hottest of all the stars, still the study of Optics teaches us that if a person should observe some other star from a point close at hand, Sirius, for example, it would appear to him in the same way as the sun appears to us from our position on the earth. Father Athanasius Kircher in Rome also confirms this in his *Itinerarium Ecstaticum*, Dialoge I, Chapter 9 where he describes a journey into the region of the fixed stars. Here he remarks among other things that "the star, Sirius, as well as other fixed stars, are solar bodies or suns which are illuminated in themselves and produce light in the stars orbiting them,(in this number, as is apparent from various other references, he includes the wandering stars, or planets, which belong to the system of fixed stars), and fill these planets with heat to the limits set by nature." This point is further discussed in our subsequent Book VII, Chapter 4.

If a person could come sufficiently close, the fixed stars would seem like suns.

CHAPTER 8

THE SUN

Reference should be made to Book I, Chapter 12 on the subject of the sun.

The sun is a solid body and a fiery image of God. It is the heart of this world and its primary moving force, ruling the whole band of planets like a king.

What the sun is.

The sun's motion is its rotation and spinning around on its own axis. This motion appears to us as twinkling, (especially in the fixed stars), but can be seen with the naked eye when the sun is setting or rising at which time it is clouded by vapors and a greater thickness of surrounding air.

The sun's motion.

Although some scholars wish to set this revolution only once every twenty-four hours, others every.month, and still others every year to accord with the motion of the sun spots, still visual observation reveals that this motion is extremely rapid and of very brief duration. Furthermore these philosophers very often contradict themselves, as is evident in the previously cited Chapter 6 of the aforementioned *Oculus Enoch et Eliae* where the author writes, "Viewed through a telescope, the sun appears as a body of continuously seething air and flame which is forever turning around on itself." He contradicts this statement in Book IV, Chapter 2, section 7, page 232 of the same work where he seems to be convinced that "the sun turns about on its axis only once a year," when he writes, "To be sure if that enormous flaming sphere of the sun revolves with as great speed as many believe in the space of a month, and not a year, there is no doubt that all the space near the sun would be burned with the terrific motion and heat of the fire. For since rapid motion tends to create fire and burn even those bodies which are naturally very cold, what do you think would happen in the immense region of the sun as a result of the very rapid revolution of this enormous fiery sphere? It would necessarily result that a point on the surface of the sun's equator would pass through about 6,285 miles every day, etc."

Turning about in empty Space, the sun has no effect on its vacuity.

What effect, indeed, can the sun have or what forces can it disperse in the ether which is nothing other than pure Space, empty of all matter? Where there is empty Space, to be sure, no effects nor traces of heat can be produced. Father Rheita, indeed, marvels at the speed of the sun's revolution when he, in agreement with Tycho, postulates that the sun revolves each day about the earth, and in Book IV, Chapter 1, section 6, page 194 sets the sun's distance as 2,073 earth radii, and consequently calculates that it travels 13,205,800 German miles each day. Furthermore, more significantly, since according to Tycho's system, the sun draws the whole planetary system with it around the earth every day, why is not the ether in the entire world set afire by this motion?

The speed of the sun's rotation is incomprehensible.

The sun, therefore, must revolve about its own axis rather than follow a so-called course around the earth (since motion is as natural to it as to all living bodies). Thus it turns about its own center or axis, not in the space of a year or a month, but in a moment of time and with the greatest speed, as can be visually observed. This rotation is so rapid that there can be nothing more worthy of wonder in the world. It is, moreover, a natural phenomenon and the omnipotence of God is made manifest by it. Indeed, this rotation is also necessary, for when the sun, in the course of its continuous turning, disperses its forces through pervious Ether or pure Space, the planets are influenced and are made to turn in the same direction as the sun, namely toward the east. Those that are more distant from the sun complete their periods more slowly and those that are nearer are carried around more rapidly.

The sun's rotation makes the planets revolve.

We have further observed through the telescope which the most eminent Elector of Mainz, Johann Philipp very generously gave us as a gift, that the sun, in addition to its usual turning motion, very often bubbles violently, as it were, and on one side or the other bulges outwards in the shape of a mountain. Suddenly, however, it will return again to its usual roundness, a phenomenon which I believe Father Scheiner refers to in his *Rosa Ursina* as follows: "There is a certain solar phenomenon which I discovered empirically but which heretofore has always been apparent. Aside from the wavering on its rim, the sun often flashes, as it were, with certain sudden lightning, just as if it were shattered internally by some light produced within and shines out like a lightning flash, dancing with momentary splendor etc." Scheiner had shown Father Kircher this same phenomenon in Rome and writes in his *Itinerarium Ecstaticum*, page 22, "and when the sun, not unlike the sea, rolls up mighty surges, I cannot help but conclude that it is nothing other than what it in fact is, namely a very stormy surging of the solar ocean." Simon Mayr in his *Mundus Suus Jovialis* also identifies this as the flashing and boiling up of the sun's matter. He says, "I say the surface of the sun surges together in the same way as gold that has been liquified by extreme heat. Here, too, the rolling motion and flashing is similar to that found on the surface of gold." All subscribe to this theory who make careful observations of solar phenomena of this kind.

The sun not infrequently begins to seethe.

Observations of certain authors.

When Riccioli, moreover, observed the image of the sun through an opening in a plank, himself being more than 170 feet away in the Church of San Petronio in Bologna, he discovered that "the sun moves with a very rapid motion, but also with a great and clearly perceptible tremor, and that its edges seethe with remarkable surgings, etc." Riccioli, indeed, ascribed this tremor to the continuous fluctuation of the air, a condition, however, which prevailed, not in the air but in the sun itself. Nonetheless, he is forced to admit that the rotation of the sun is extremely rapid.

We, however, attribute the source of this seething to a certain remarkable force in the sun which perhaps attracts the air of the surrounding planets to itself, for everyone is aware that fire attracts our air, in fact, it is evident from the experiments in Book III, Chapter 13, that fire is fed by air. There is little doubt that the sun is a fiery body, admittedly not a simple fire like ours, for when we direct our eyes towards it as it moves, we perceive the flames on its outside, while in the inside we are aware of a certain darkness, the source of its life, as it were). It must be presumed, therefore, that when the sun receives fresh air from the planets, the exterior flames which surround its perimeter, surge up and spread.

The reason for this seething.

The sun is fed by air.

According to Diogenes Laertius, Zeno also thought that "the sun is fire, nurtured by the sea, by taking to itself its most subtle vapors." Cicero, moreover, in Book II of *De Natura Deorum* says "the sun is fed by the moisture of the ocean." This agrees with our experiments wherein it is demonstrated that the greatest part of the sphere of air is generated from water. In this way, nothing in the world perishes but is finally returned into the same essence.

Stars, being the armies of the heavens, undoubtedly perceive one another.

Furthermore, according to Chapter 1 of Book IV, because homogeneous virtues feel or intuitively sense one another, there is little doubt that light-giving bodies (since they are essential to all vision), mutually perceive and sense one another. It must be presumed, therefore, that the sun perceives the stars. For these reasons, philosophers are moved to postulate that stars are animate and possess feeling and intellect, as did Cicero in Book II of "*De Natura Deorum*"; Pliny also attributes a mind to the sun and Origen, moreover, in Volume I of his *Commentarius in Johannem* and Book I, Chapter 7 of the "*Periarchon*" states that stars are not only animate but are even capable of good and evil. Cajetan supported him by explaining the words of Psalm 135, "He who made the heavens with the faculty of perception, etc." We, however, leave everyone to his own judgment in this regard.

CHAPTER 9

THE SIZE OF THE SUN AND ITS DISTANCE FROM THE EARTH

The sun's distance is found from its parallax.

It is more than clear from Book I, Chapters 23 and 24 how great the variation and uncertainty in astronomers' calculations are as to the distance and size of the sun. Inasmuch as this entire calculation is based on the true parallax of the sun, this phenomenon will have to be of prime consideration here. The size of this parallax, however, is found by astronomers to vary greatly, the extent of which variation can be seen in the 3rd Table of the aforementioned Chapter 24.

One reason for these divergent computations lies in the fact (1) that astronomers torture themselves constantly about the apogees and perigees of the two luminaries, the sun and moon, when, as a matter of fact, the sun itself has no apogee or perigee and it is doubtful whether the moon has any or not, as is evident from Book V, Chapter 11. Many astronomers are forced to this same conclusion since some find the apparent diameter of the sun to be 30 minutes in perigee, a number which others find it to be in apogee, as is evident from Table 1 of the aforementioned Chapter 24.

(2) This variation also arises from an imperfection in astronomical instruments which cannot be divided accurately enough into minutes and seconds.

(3) The state of the air, moreover, and its variable temperature as well as the changing height of a star on the horizon also contribute to these differences. See Book V, Chapter 10 on this subject. With an alteration in the size of the diameters of the sun's, moon's and earth's shadows, moreover, the distances and sizes deduced therefrom are also diverse.

No mortal man, therefore, will ever measure the distance of the sun so exactly that he will not err 50 to 100 radii of the earth. It is indeed the case that observations even made on the same day and with identical instruments are at variance. Thus the diameters of the shadows of the sun, moon, and earth which some individual or other has observed, should not be accepted as valid, but rather those diameters which astronomers as a group have found.

The apparent diameter of the sun at apogee, however, is commonly found to be 30′ (for we do not see the apparent diameter at the time when many hold the sun is in perigee, because at this time the sun is in the southern signs of the zodiac and is much lower on the horizon and consequently more subject to the air's refraction) and the apparent diameter of the earth's shadow is 80′ and the apparent diameter of the moon is 30′, as can be seen from the preceding Chapters 22 and 24 of Book I.

If, then, one rejects the apogees and perigees of these two luminaries and accepts those four measurements which Ptolemy supplied for ascertaining the distance of the sun, one finds this by calculating the greatest height of the star above the horizon on an average at all times of the year. According to this calculation the apparent radii of the sun and moon is 15 minutes, the distance of the moon from the center of the earth is 64 radii of the earth, and the apparent radius of the earth's shadow as seen on the moon at the time of a lunar eclipse is 40 minutes.

The distance of the sun can be found from four measurements.

From these four measurements which no one should reject nor can deny that they have been found with greatest frequency by others, shall we seek to discover the distance and size of the sun.

First, in the right triangle, ibc, is the angle, b, the radius of the earth's shadow at 40′ and the side, bc, the distance of the moon from the earth at 64 radii of the earth or 55,040 German miles, if one calculates the radius of the earth as being 860 German miles.

See Plate XX in Book V.

Thus as 10,000 parts of the earth's radius are to the distance of the moon	so are the parts of the tangent for the radius of the earth's shadow of 40′
55,040	

116361
55040
4654440
5818050
581805
6404509440

to the radius of the earth's shadow, namely 640 German miles.

Subtracting these results from the radius of the earth, there remains for kg, 219, 549 German miles.

Secondly, in the right triangle, gik, the two sides gk and ki are known to be equal to ic and bc. Together with the right angle at k, the angle at i is sought, to which the angle at di is equal because the lines ki and bd are parallel.

Side ki	Parts of the radius	Side kg
55,040 miles	10,000,000	219.55 German miles
		10,000,000
		2195500000.00

39889.17 parts of the tangent of the angle i or d which amounts to 13′42″

How large the parallax of the sun would be.

If these 13′42″ of the angle at i or di are subtracted from the angle, age, or the apparent radius of the sun, which amounts to 15 minutes, there remains 1′18″ for the angle of the solar parallax, bag, (because the external angle, age, is always equal to the two opposite internal angles, according to Book I, proposition 32 of Euclid). Likewise if one subtracts 13′42″ from 90°, the complementary angle, dgb, of 89°46′18″ remains.

Thirdly, in the triangle, dbg, the axis of the cone's shadow or the line, db, is sought.

As the sine of the angle, d, 13′42″	is to the sine of its complement, that is of the angle, dgb, 89°46′18″	so is the radius of the earth to the axis of the cone's shadow, db.
namely 39,889	that is, at 9,999,920.3	1

The computation for the axis of the shadow of the cone, which is the line db, and is 250.69 radii of the earth.

Fourthly, take the triangle from the radius of the earth to the middle point of the sun; this is the triangle, bag, in which the radius of the earth corresponds to the angle of parallax, bag, 1′18″. When this is subtracted from the 90° of the right angle, an angle horizontal to the center of the sun, that is, angle agb, of 89°58′42″, remains.

As, therefore, the sine of the angle of parallax, bag, =

1′18″	is to its complement, namely the sine of the angle, agb, which equals	so the radius of the earth is to the distance of the sun, ba.
namely,	89°58′42″	
3,781.5	9,999,999.3	1

This will give 2644.4 radii of the earth for the distance of the sun which is the equivalent of 2,274, 184 German miles

The sun is distant 2,644 *radii of the earth.*

Furthermore to find the size of the sun

2,644,4	the distance of the sun
plus 250.7	the length of the shadow of of the cone, db,
gives 2,895.1 for the line, da.	

As the length of the line, db,	is to the radius of the earth	so the length of the line, da, is to the radius of the sun.
250.7	1	2895.1

The result is 11.5 radii of the earth for the sun's radius, or to put it another way:

The radius of the earth is to the radius of the sun as 1 *is to* $11\frac{1}{2}$.

As the parts of the line, ba, (which are posited as being 10,000,000)	are to the distance of the sun, ba, namely 2644.4	so the true measure of the sun's radius is to the tangent of our angle for the apparent radius, abe, which equals 15′ or 43,633

Thus the sun's radius is 11,538 radii of the earth which is equivalent to 9,923 German miles.

Because, however, every body which is seen from a visual angle of 15′ is distant 229,18 of its own radii from the eye (as we have shown in Book IV, Chapter 13), it follows that if the radius of the sun, 11,538 is multiplied by 229,18, the result, in terms of the sun's distance away, is 2644,28 radii of the earth, as is proved above.

The sun is 1,521 *times larger than the earth.*

Since, therefore, the radius of the sun is 11.5 or $11\frac{1}{2}$ radii of the earth, the cube of the sun will be 1,521 and according to Book V, proposition 19 of Euclid (as we have said in the preceding Book, Chapter 22 concerning the moon), the proportion of the sun to the earth will be as 1521 is to 1. Consequently, the sun is 1,521 times larger than the earth.

From the numbers that had originally been given and have now been verified, we can be reasonably certain of the distance and size of the sun, for no one can deny (1) that the radii of the sun and moon are found to be 15′ and although this is not always the case, still they usually appear to be of this size when they are higher above the horizon and freer from the surrounding impure air, (2) that the distance of the moon from the center of the earth is 64 radii of the earth and although astronomers practically never agree upon the same observed distance, because of the so-called apogee and perigee as well as the changes in atmosphere, still they find that the distance most commonly observed is 64 radii of the earth. (3) It is also true that the radius of the earth appears to us at the time of a lunar eclipse to be of 40 minutes, and even if it is not always of this size for the reasons already mentioned, still when the moon is at the highest point in the heavens and thus is freer from the refractions of the air, it is generally found to be of this size. Thus we compute:

The angle of the solar parallax is 1′18″.

The distance of the sun is 2,644 radii of the earth.

The size of the sun is 1,521 earth bodies.

Indeed, as long as astronomers adhere to the belief in the existence of apogees and perigees, they do not take into account the relationship of different heights and depths of air to changes in the height of a star nor do they reduce it to a more general concept. Their discrepancies are of such magnitude that one may say the distance of the sun is 1,000 and another 14,000 radii of the earth. Furthermore one may declare that the sun is 166 times larger than the earth and another say it is 262,144 times greater.

If the sun travels in its course about the earth, at any moment it would cover 166 *miles.*

Furthermore, since the sun is at least 2,644 radii of the earth or 2,274,000 German miles distant from the earth, it follows (if it is true that the sun within the space of 24 hours travels in a course about the earth) that in one hour it would complete 595,571 German miles and in one minute 9,926 miles and in one second or moment of time, 166 miles.For if the diameter of the sun's sphere is 4,548,000 German miles, then according to the proportion of the diameter, 7, to the circumference, 22, the daily orbit of the sun must be 14,293,714 miles, which, divided by 24 hours, is 595,571 miles per hour etc.

But since, according to the Copernican hypothesis, the earth is situated where the Ptolemaic system locates the sun, then the earth within a year travels through these 14,293,714 miles. If this figure is divided into the 365 days in one year, the earth moves forward about the sun each day 39,160 German miles and in one hour, 1,631 miles in a minute, 27 miles and in a second, almost a half mile.

On the other hand, if it is the earth which travels over this course within a year, it moves forward at any moment almost a half German mile.

Finally, it should be remarked here that those who intend to begin this calculation require the Tables of Sines, Tangents, and Secants carried out not only in degrees and minutes, but also in seconds. Since, in fact, our Tables were carried out only to minutes, the reader may ascribe any error in calculation there may be to the imperfection of these Tables.

CHAPTER 10

THE WANDERING STARS OR PLANETS

Chapters 12 to 19 of Book I should be reread on the subject of the individual planets and the end of the aforementioned Chapter 19 concerning planets in general.

Six wandering stars or planets are visible, namely Saturn, Jupiter, Mars, Earth, which is counted as a planet, Venus, and Mercury. Of the rest, those lying between Mercury and the sun appear to us as spots when they pass with their satellites between our vision and the sun although they are invisible in themselves. These planets are very numerous but because they are quite small, one cannot ascertain their number accurately.

Six wandering stars are visible to us.

The fact that each planet carries with it one or even more secondary planets or companions is in part visually ascertainable and in part to be concluded from reason.

Planets have companions and are opaque bodies.

The planets are in themselves opaque bodies whose surfaces are not illuminated in their entirety, but only in that part which is turned toward the sun. Thus they borrow all their light from the sun and reflect back its rays to our eyes. Ethereal Space, indeed, receives no light and that which is without light is invisible and therefore appears as black as shadowy night. An earth body found here, however, receives the light of the sun upon it. Such a planet, which is reduced in size because of its being at a greater distance from us, appears much clearer and brighter to our eyes because of the light's being constricted and results in its looking like a star, as we said in Book IV, Chapter 14. Just as the planets, therefore, appear as stars to the earth, so does the earth appear as a star to the planets, small though

Empty Space does not receive light and thus seems black to us.

Just as planets appear as stars to us, so does the earth appear as a star to the planets.

it may be. For indeed were there a man on any one of the planets, he would perceive the planets and their motions and the earth, as well, in the same way as a man looking out from the earth.

The spheres of the planets are surrounded by air.

The individual spheres of all the planets are surrounded by an atmosphere which is nothing other than an odor-bearing or vapor-bearing effluvium – air, in other words. Rheita confirms this statement in his *Oculus Enoch et Eliae* as does Kircher in the *Itinerarium Ecstaticum*.

They are not precisely round.

The bodies of the planets are indeed spherical, although they are covered with many blemishes like the moon or our earth. These bodies are not smoothly rounded, geometrically speaking, but only appear so visually. This uneveness of Jupiter's and Venus' surface can be seen through a powerful telescope. See *Itinerarium Ecstaticum*, at the end of page 267.

The planets are material and opaque bodies.

That the rest of the planets like the earth are composed of a hard material, solid and opaque, which absorbs the rays of the sun, or are made of the four so-called elements, the aforementioned Rheita, Copernicus' opponent, declared in Book IV, Chapter 1, page 475 and Kircher in his *Itinerarium Ecstaticum*. Furthermore, because of the appearance and aspect of the lunar star and a contemplation of the other stars which are at a greater distance from us, we are led to the conclusion that the moon is composed of material and substance similar to that of the planets and that the rest of the stars are made of the same matter as that of the lunar sphere.

Planets turn themselves about their own axes.

Planets were believed by certain philosophers to be animals in the ether.

Each one of the planetary spheres turns itself about its own axis for the sake of its own preservation (See the subsequent Chapter 12 on this subject) and for this reason they are believed by some to be earth creatures. They say that it does not follow that only those things are animals which have flesh and blood, eyes, hands, and feet. Rather it is an internal component which contributes motion that makes a living body. For although planets are without eyes, hands, feet, and other endowments common to the animals of the earth, nonetheless, they have that which contributes to their own preservation and makes them move with more freedom and ease than the animals of the earth. Furthermore, just as these owe their existence to two sources of heat, namely their own natural heat and an external heat having its origin in the sun, so also do the planets embody heat like great creatures of the earth. Thus, Cicero in Book II of *De Natura Deorum* says, "Upon consideration of the divinity of this world of ours, one should attribute the same divinity to the stars which have their origin in the most noble and pure part of the ether, so that they may be called, with complete justice, animate beings which have perception and understanding.". And shortly thereafter

Cicero's opinion.

he says: "To Aristotle it seemed absurd not to believe that an animal is born in that part of the world which is most adapted to bearing animals. Stars, indeed, dwell in the ether which is the seat of an animal's birth since it is extremely tenuous and is always moving and vital and likewise is also of clear sensibility and mobility. Therefore, since stars are born in the ether, one must agree that they embrace perception and intelllgence." If, however, planets are animals of the ether, they will also have a vegetative and growing spirit, a subject which is discussed in Book V, Chapter 4. We leave this discussion to men of wisdom and those who are born researchers, for it is no interest to us to deny or affirm this speculation.

CHAPTER 11

THE FORWARD MOTION OF THE PLANETS

Just as God created the sun to be the principal source of light as well as heat for the planets and gave it the power of life and death over them, so also does this body, in the course of its very rapid revolution, carry all the planets into the same ethereal region in which it itself turns through its own force in that most fluid Space. (See also Book I, Chapter 12.)

The sun carries the planets around in a circle.

For the sun, while it is continually turning with great rapidity about its own axis, carries, at the same time, the whole sphere of its virtue in a circular course (in which all the planets are embraced). This sphere, in turn, carries with it the planets contained therein, among which number is the earth, in an order commensurate with the mass or weight of the body of each. Whichever planet has a body more solid or larger in size, the greater the impulsive virtue of which it is capable and by which it is carried around in a greater circumference: whichever has a smaller body, however, this turns in a smaller circle. Thus the forward motion of the planets depends upon an harmonious relationship in the distances as well as the periods of motion of each, nor can any irregularity befall these bodies whatever individual course they each may have.

Just as the planets are made to revolve in a circle through the impulsive virtue (which, in these solid bodies is produced to the greatest degree possible by the continuous revolution of the sun), thus through this virtue they are driven to a still more distant point by the sun insofar as this virtue can extend itself in any body proportionate with its bulk or mass. This point is actually in the middle of the sphere of virtue of each, or to put it another way, on its

How the forward motion of the planets is carried out.

equator. The planets always extend out a bit beyond the equator because of an innate impetus they have until they reach that point in the sphere's activity where the circle begins to contract, in other words, the Tropics. (See also Book IV, Chapter 3.) Thereupon they are forced to turn back and again to slip obliquely to the point of greater circumference, the equator, as it were, in accordance with the succession of the signs as we have discussed rather completely in Book V, Chapter 16.

The nearer planets are carried about in a circle more quickly.

Consequently the more distant any one of the planets is from the sun, the greater the circular course it makes and the more slowly it completes its period, for because the sphere of the sun's virtue is more extended here and the planet is more distant from the sun itself, it is weaker and of less force. To such a degree as a planet is smaller than the sun, by so much does it revolve in a smaller circumference and so much more quickly is it carried around in its course.

Since Mercury is nearest the sun, it must necessarily complete more periodic revolutions in its course than the rest of the planets. As for the sun spots or the invisible planets, the nearer they are to the sun, the more quickly they are carried about the sun (on the 10th, 20th, 30th, 40th, or 50th day, as it were). It follows from this that Saturn, the farthest removed from the sun, completes its periodic orbit within $29\frac{1}{2}$ years; Jupiter within 12 years; Mars within 2; the Earth within 1; Venus within $7\frac{1}{2}$ months; and Mercury within 3 months; as can be seen in Book I, Chapters 14 and 15 etc. and Chapter 6 of the preceding Book.

The periodic revolution of each planet.

Finally, we mentioned above in Book V, Chapter 6 that in the case of such rapid motion, planets sometimes leave something of their atmosphere or air behind them in the ether or empty Space. This matter, then, being gathered together and collected into a ball (under the continuous turning of the sun), and illuminated by the sun, is made visible to us, especially when it is present in great abundance from the larger planets, and appears as a new shining star because of the greater distance. The brightness of this body, indeed, should not be ascribed to its turning or rotation, as is the case in the sun or fixed stars, but only to the oscillation or vibration which is caused by the shining light received from the sun and fixed stars on bodies of this kind that are separated at a distance from our eyes. In this regard, Book IV, Chapters 13 and 14 as well as the Appendix of Book V should be read. See also the end of Book VII, Chapter 2.

What is the origin of a new star?

CHAPTER 12

PLANETARY MOTION

All planets participate in the turning motion.

In the same way as we have described the turning motion of the earth in Book V, Chapter 15, so does each one of the planetary spheres revolve around its own axis for the sake of its preservation. One can see this very clearly from the diverse phenomena associated with their different spheres as observed through telescopes.

Thus Father Rheita said in *Oculus Enoch et Eliae*, Book IV, Chapter 2, page 204, "I do not question that planets turn about their own individual axes continuously, for indeed, the phenomena attendant upon Jupiter absolutely prove that Jupiter turns in this way." Francesco Fontana and Kircher in his *Itinerarium Ecstaticum* confirm this.

But since the bodies of the planets are of diverse size and therefore carry out their periodic forward motion in different ways, one can easily conclude that their motions of revolution are also varied and require different lengths of time. Indeed, the larger their bodies are, the slower they turn and vice versa. Thus in Saturn and Jupiter one observes no shimmering is apparent (which is nothing other than the turning of a star about its own axis), because the turning motion of these planets is slow. Contrariwise, we see that Mercury and Venus are reddish in color and shimmer to some degree by reason of their more rapid motion. They do not move as quickly as the fixed stars, however, whose turning is the most rapid of all.

From the foregoing, it is quite apparent that Jupiter turns itself in a shorter length of time than Saturn, and Mars than Jupiter, and Earth than Mars, etc.

It is otherwise with the secondary planets, however, which we also call planetary satellites. These do not revolve, as we said in Book V, Chapter, 20.

CHAPTER 13

THE DISTANCE OF THE PLANETS

In Infinite and Immense Space there is nothing, as we have said, that is far or very far away, for the greatest distance imaginable which escapes our comprehension, recognition, and power of expression is as nothing in comparison with Infinite Immensity.

In God, therefore, stars have enough space, even though they may appear to be much more distant from one another than they actually are: for indeed, in order to fill Immense Space, what can there be so far away that may bound it?

The distance of the planets is not sufficiently certain.

A way to measure or to prove with certainty the distances of these bodies is precluded to mortals as is evident in Book I, Chapters 20 and 25. Here the reader will find so many and such great discrepancies in the results of astronomers that they cannot be satisfactorily reconciled. For example, some say that Saturn in the middle distance is 10,000 radii, while others say 140,000 radii of the earth distant from here. The same discrepancies can be found with the rest of the planets.

Since, however, the height of the smaller of two perpendicular bodies is related to the larger as the square of the former's motions is to the square of the latter's, there is no doubt that, from the most insignificant to the largest, these earthlike bodies are so arranged and ordered relative to one another that they cannot be disposed more suitably or appropriately. Accordingly, their periodic revolutions (See Book IV, Chapters 3 and 4) conform exactly with their position in the heavens. It must be concluded, therefore, that as the periodic motion of one body of this kind is to the square of its distance from the sun, so also is the periodic motion of another body of this kind to the square of its distance from the sun. The square root of this number gives the distance of such a body from the sun.

The method of measuring.

Now the periodic forward motion of the earth around the sun is completed in one year. The radius of this period (according to the previous Chapter 9) is 2,644 radii of the earth, the square of which is 6,99O,736. Thus, according to the Golden Rule, the square of Saturn's distance (which completes its periodic forward motion in the space of $29\frac{1}{2}$ years) will be 206,226,712. When the square root of this is taken, Saturn's distance from the sun is found to be 14,369 radii of the earth.

This is the calculation:

As the earth's periodic forward motion in one year	is to the square of the earth's distance from the sun	so Saturn's periodic forward motion in $29\frac{1}{2}$ years is to the square of its distance from the sun
1	6,990,736 29.5 34953680 62916624 13981472 206,226,712	$29\frac{1}{2}$
		This is the square of Saturn's distance from the sun. When the square root of this is taken, the result is 14,360 radii of the earth, which is its distance from the sun.

The distance of the planets from the sun according to the author's calculation.

Jupiter, moreover, completes its period in 12 years less 50 days. If a similar calculation is made, the square of Jupiter's distance from the sun is 83,888,832. The square root of this is 9,159 which is Jupiter's distance in terms of earth radii.

Mars completes its period in almost two years or 98 weeks. The square of its distance is found to be 13,981,472, the square root of which is 3,739 or Mars' distance from the sun.

Venus completes its periodic forward motion around the sun in 32 weeks. As 52 weeks, the earth's period is to the square of its distance from the sun, namely 6,990,736, so the 32 week period is to the square 223,703,552, the square root of which, 2,049, is the distance from the sun.

Finally, Mercury completes its period in $12\frac{1}{2}$ weeks. The square of its distance from the sun is 87,384,200 and the square root of this number is 1,290 which is Mercury's distance from the sun in terms of earth radii.

The distance of the planets from the earth.

As for the distance of the planets from our earth, this can also easily be found. The middle distance of Saturn from the earth is taken to be the same as the former's distance from the sun. This also holds for the rest of the planets. Moreover, the nearest distance of Saturn, when it is in perigee, that is, is found by subtraction. When the distance of the earth from the sun, namely 2,644 radii of the earth is subtracted from Saturn's distance, 14,360, the remaining distance is that of Saturn in perigee, namely 11,716 (ERRATUM read 11,716 for 11,696.) (See Table below.) radii of the earth. On the other hand, when these numbers are added together, namely

Saturn's and the earth's distance from the sun measured in earth's radii, the distance of Saturn in apogee is found which is 17,004 radii of the earth. The following Table shows this clearly.

The Distance of the Planets from the Sun as well as the Earth in Terms of Earth Radii

	Distance from the sun	Distance from the earth when it is in perigee or nearest to the earth	Distance from the earth when it is the middle distance	Distance from the earth when it is in apogee or farthest from the earth
Sature	14,360	11,716	14,360	17,004
Jupiter	9,159	6,515	9,159	11,803
Mars	3,739	1,005	3,739	6,383
Earth	2,644	–	–	–
Venus	2,049	595	3,020	4,693
Mercury	1,296	1,348	2,644	3,940

CHAPTER 14

THE SIZE OF THE PLANETS

Just as astronomers disagree with one another as to the distance of the planets, so also are they at odds in determining their sizes, as is evident from Book I, Chapter 25. Thus Tycho Brahe asserted that Saturn is at least twenty-two times larger than the earth while Riccioli said it was 891 times larger. Others invert the order of nature, as it were, and declare that the planet, Mars, is farther removed from the sun than the earth but nonetheless is twice, three times or even ten times smaller than the earth. It is impossible, however, for a body which revolves in a larger orbit to be of less mass and weight than one which turns in a smaller orbit.

The sources of these discrepancies spring (1) from an uncertainty regarding distances, a subject touched upon in the preceding Chapter. (2) from a difference in the apparent diameters which astronomers calculate in such different ways that the whole problem is beset by uncertainty, as is clear from the Table of Riccioli, page 708. We have indicated the causes for this in Book V, Chapter 10. A small variation of one second, more or less, which is found in the apparent diameter of these stars, results in an unbelievable difference in size, not to mention the one, two, or three minutes which more often than not is assigned to a certain star. (3) from the singular nature and quality of light and brightness because of which one finds it makes little difference whether a light or bright body is nearer by many thousands of miles (or even earth radii), or, in fact, is farther distant. No change in its size can be perceived, as is to be seen in Book IV, Chapter 14.

For these reasons it is clear that there is no way to define a body or sphere of this kind with certainty, be it large or small, from the size which our eyes perceive it to be.

However much astronomers disagree in this matter, nonetheless, we are very sure that the sizes of these bodies can be calculated from their periodic revolutions around the sun.

For as the period of one year is related to the body of the Earth, 1, so will Saturn's period of $29\frac{1}{2}$ years be essentially related to the body of Saturn, $29\frac{1}{2}$.

From this it follows according to the Golden Rule and the periods of the planets, which were noted at the end of our preceding Chapter 13. (ERRATUM, read 13 for 11) that:

Saturn	is larger than	$29\frac{1}{2}$	
Jupiter	the earth	12	
Mars		2	times.
Venus	is smaller than	2/3	
Mercury	the earth	$\frac{1}{4}$	

CHAPTER 15

ARE THERE ANIMALS ON THE PLANETS?

Certain ancient philosophers have asserted that there are animals on the moon and Heraclides, the Pythagoreans, and the followers of Orpheus have extended this theory to the rest of the planets. For as the Space in which the planets have their existence and being is one and the same, so Cardinal von Kues writes in Book II of *De Docta Ignorantia* (as recorded by Rheita in *Oculus Enoch et Eliae*, Book IV, Chapter I, folio 179), "A man cannot know whether the kingdom of our earth is in a more exalted position or one of lesser consequence than the regions of the other stars. For although God is the center of all stars, encompassing them entirely, and is the source of the different characteristics in beings dwelling throughout the universe, having created them so that the many regions of the heavens and the stars may not be empty and that this earth of ours may perhaps not be alone of the lesser stars to be inhabited, nevertheless, the

Cardinal von Kues' opinion.

intellectual and spiritual nature of those dwelling here on the surface of our earth does not seem capable of being, more exalted or perfect in accordance with God's plan, although there may be inhabitants of another kind on the other stars, etc."

Kepler's opinion.

Kepler inclines toward this opinion when he says in his *Astronomia Optica*, page 250, and the *Dissertatio Cum Nuncio Sidereo*, page 10, "It does not seem unlikely that there are inhabitants, not only on the moon, but even on Jupiter itself, etc."

Father Rheita's opinion.

Father Rheita makes this observation in his work referred to above "Indeed since it seems to be tacitly assumed and appears quite clear that there are many bodies in the heavens which up until now have lain hidden from us because it has been only during our lifetime that we have found so many hitherto unknown phenomena around Saturn, Jupiter, and other planets, now we seem to have an even greater reason to theorize and philosophize a little about the other worlds that have been only partially known to us before this time. The ancient astronomers as well as Nicolaus von Kues were of this same opinion. In view of the likeness found in Jupiter's splendid system to the moon's revolution around our earth in so orderly and precise a fashion, this phenomenon must in some way imply a kind of similarity between Jupiter and this lower world of ours. Thus we can reasonably assume that there is something hidden away in Jupiter's sphere which causes its satellites to travel so consistently around it in an unceasing and ordered course. That the rotations and turnings of the aforementioned satellites are caused only by the action of our earth, seems to me as far from the truth as to suppose that the monthly revolutions of our moon around its center, the earth, occur because of Jupiter or Saturn. The fact that these satellites move within such bounds, therefore, is sufficient proof that they properly look to the sphere of Jupiter and it certainly does not seem believable that they attend a Jupiter that is completely bare and devoid of all life.

"We have, indeed, frequently heard of kings who are very secretive and whose hidden resources and treasures are not at all accessible. Why, then, should we poor creatures wear ourselves out in vain in pursuit of the hidden and eternal treasures of the wealthiest king of the ages? It is certainly better to regard reverently phenomena of this kind as proofs of the omnipotence and wisdom of God rather than to make foolish conjectures about them.

"There can be no doubt that many things are hidden in the heavens that have not yet been revealed and perhaps shall never be known to us in this life, inasmuch as for reasons understood by God alone, they have been reserved for the great joy of the life to come. Thus they should not be said to be without existence or being because they are not known to any of the inhabitants of this mortal sphere of ours. For since the reality of the existence of things is dependent upon a creature endowed with reason but is independent of his knowledge and cognition, there can be no perception of this reality without knowledge, although a rational creature may be capable of perceiving it. Thus, by intellectual projection we have learned that there have always been antipodes, although no one was aware of them two hundred years ago. Who of sane mind can doubt that Jupiter's satellites were created at the same time of its birth and that the rings of Saturn are coeval with it, although before the telescope was first invented, no mortal man – not even in a dream – had foreseen either phenomenon to exist in nature nor to have been there since the first beginnings of things.

"Cardinal von Kues, therefore, who has already been mentioned, seems clearly to feel that many secondary worlds, actually integral parts of the whole universe, are encompassed within the highest vault of creation. Had he seen the satellites of Jupiter, Saturn's rings, the lunar phases of Venus and Mercury, now scythe-shaped, now half full, now gibbous and now full as well as the mountains, valleys, and cavities of the moon, what then do you think he would have concluded? Undoubtedly he would have been much strengthened in his own opinion, for certainly we can see no lower species encompassed by and imprisoned in one individual alone (unless perchance you accredit the veracity of the tale of the phoenix bird), but all are reproduced and multiplied individually and in infinite numbers. Thus if one theorizes that the greatest and best Creator had followed this design in making a secondary world or earth, I do not believe that such a conjecture would be contrary to reason, nor perhaps even contrary to faith, since we have no special knowledge, indeed, from which we can directly conclude that there are not a number of worlds, for if there were such knowledge, it would be found in the Sacred Scriptures or from the Church, something that is evident in neither place. Indeed nothing actually found in the Scriptures which seems to favor one opinion more than another unless it is that all things are said to have been made by God.

"Although the Holy Scripture does not mention another world, inasmuch as it does not expressly teach that another such world exists, still neither does it deny its existence; thus it apparently does not favor the one hypothesis more than the other. For this reason, it does not follow (to quote the words of the learned Father Mersenne in his *Commentarius in Genesin*, Chapter 2, page 1080) that there are not several worlds, particularly subordinate ones which are actually integral parts of the whole universe. The Sacred Scripture passes over this possibility in silence, but indeed, there are many things which it does not mention but which we nonetheless are aware were undoubtedly created. Perhaps it sought to record nothing other than those things which God judged necessary either for his own glory or for our safety. The author we have just cited says that now it is generally agreed that an argument taken from negative authority is not cogent: that is, when the Sacred Scripture does not mention this or that thing, ergo it does not exist etc.

"But whatever may be the case, let this be held as certain that many of God's creations are hidden from us and more are cloaked in ignorance than are are known, while many things that have existence, are not believed to exist. Furthermore, I believe that among natural phenomena there are many things which have frequently been described as having existence which perhaps do not exist and never will.Contrarily, there are many things which are generally regarded as non-existent which, nonetheless, may, in fact, exist. I believe, however, that it is safer to agree with opinions that are generally accepted and that only with caution should one apply oneself to new theories."

Reasons why a plurality of worlds is credible.

The arguments by which it is adduced that animals exist on the planets are to be found in the preceding Chapter 10. More evidence can be added to this list, for (1) because all planets revolve and have forward motion, movements similar to those of our earth; (2) and because it is incredible that so many bodies of such enormous dimensions which exceed this earthly sphere of ours many times in size, are empty of animals in which this sphere of our abounds; (3) and because these bodies do not seem to have been created for the purpose of giving light to the earth since they have no light in themselves (for no one lights a torch for domestic use which is greater than the actual dwelling); (4) and because the omnipotence of God is not restricted only to us mortals who dwell on a point in Space, as it were; (5) Thus, it is contrary to reason to suppose, in view of the vastness of immense Space and the Ether, that the earth alone with its inhabitants, lies concealed as one spike of wheat in the broadest of fields.

No men exist on the other planets.

But if a person proposes that there are men on the other planets, he is as far from the truth as a star of this kind is distant from us, for no animal which is on this sphere of earth also exists on the other planets since there is a diversity in all things causing everything which is there to differ basically from the bodies we know. Everywhere in all the regions of this earth, we see that there is a difference in the water, herbs, planets, trees, stones, etc., as well as in the animals of the earth and later, etc. and that no genus is completely like another. As a matter of fact, no man is absolutely like another, nor are his actions, abilities, arts, words, beliefs, knowledge, etc. in all respects similar to another's, since no man has the innate nature, ethics, and behavior of any other man. "Nature rejoices in variety and herein lies the world's perfection. Thus, too, the desire and passion for different and new things is never wanting, for if it were, no man would possess anything about which he would be concerned or which he desired very much, and death itself would follow.

Variety gives pleasure.

As for the person who is unwilling to admit the existence of animals other than those he sees on earth, he should know that no brute animal exists in America that is similar to any of those in Europe, Asia, or Africa. Should he speak, then, about the characteristics, appearance, and stature of those animals which he has never seen or about which he has never heard? Surely he cannot, and much less can he describe or imagine those beasts which exist on so many different and widely separated earth bodies.

Variety is found in everything.

Galileo replies to me as follows in the first Dalogue, page 40 of his *Systema Cosmicum*, "Whether grasses or plants grow or animals similar to ours exist on the moon or any other planet, or whether there are rains, winds, lightning, etc. such as prevail on earth, neither do I know nor believe and am much less cognizant whether men live there. Nevertheless, I do not see how it can necessarily be inferred from this, because there is nothing similar to the things existing here on earth, that no change or variety is found there and that other things cannot have being which are altered, brought to birth, and dissolved, which are not only different from those we know but indeed are completely removed from our imaginings and are incapable to the highest degree of being intellectually conceived. I feel certain that it is just as if a person were born and brought up in a vast forest among wild beasts and birds and had never known anything about the element of water. Such a person would never conceive in his imagination that there was a world in nature different from the earth, filled with animals which move about swiftly without limbs or wings, not only on the surface like the wild animals of the earth, but also far down in the depths. Not only would these animals move forward, but they could also remain immobile in whatever place they might wish, something that birds cannot do in the air. Furthermore, men could even live there and build palaces and kingdoms, and would enjoy such ease in traveling that without any labor and accompanied by their families, homes, and even their entire communities, could they transport themselves to far distant regions. For this reason I say, and indeed I know, that such a man, although

Galileo's opinion.

he were endowed with most perceptive powers of imagination, would never have conceived of fish, the ocean, ships and fleets: and thus it may be that on the moon, removed at so great a distance from us and composed of matter perhaps much different from the earth's, certain substances may exist which produce effects that not only are different from our imaginings but actually are completely foreign to us. For these materials may have no similarity to things we know and thus may be completely alien to our thoughts.

Whatever can be encompassed by our imagination, this must either have been seen before or is made of things or parts of things already seen before, like sphinxes, chimeras, centaurs etc."

Inasmuch as variety is present in all things, there is nothing on this earth that is present on the other planets.

If, therefore, the men or animals of this earth are not on the other stars, then neither will the acts and deeds of God, His laws and regulations, etc. be the same as those here on earth. Consequently the manner of life, as arranged by the leaders in charge,(perhaps rational animals which may exist there), will not be similar to that found upon this earth of ours, but will be designed in accordance with their notions and intellects, particularly since we see that God does not operate in all regions of this earth of ours with identical and equal reason, inasmuch as He gave certain laws to the Jews which He did not give to other people and also made certain concessions and granted lenient regulations to others which were not accorded to everyone else. Indeed, He has always operated in different ways with different men.

God's omnipotence is manifest in variety.

It is generally believed that nothing exists elsewhere except by the great munificence of God which enables the creatures living there to make manifest His Praise, Omnipotence, and Honor by functioning, moving, etc. in a manner, however, that is completely different from our own.

Rational animals can also exist on the planets.

Our statement that rational animals exist elsewhere should not seem remarkable to anyone, for it is a necessary corollary. If it is true that there are animals there, then without doubt there is variety in these animals for in those beings which are different, one is always more noble or excellent than another. Thus certain animals will undoubtedly be more noble or more outstanding than others, (as for the stars themselves, we will say nothing now), and consequently will hold positions above the rest. As to what kind they are and what their actions are and whether events transpire there similar to those here on earth, etc., there is no one who can guess since the actions of God are infinitely Infinite and are so wondrously and wisely ordered and disposed that we are compelled to admit that we know nothing other than our ignorance. It should not be presumed, however, that God has preserved His infinite Omnipotence in secrecy, but that which He did not give either to these or those creatures, He left to be discovered by others. For it is compatible with infinite omnipotence that the Infinite be manifest in finite things for His infinite Glory.

CHAPTER 16

THE TRUE SYSTEM OF THIS WORLD

Although almost all modern astronomers (among which number is the most renowned astronomer of our time, Father Giovanni Riccioli, a Jesuit and Professor of Philosophy, Theology, and Astronomy at the Academy in Bologna who brought out his famous treatise, the new *Almagestum* in 1651, which embraces the new and old astronomy), reject the Ptolemaic system, which we discussed in Book I, Chapter 4, because of its incurable defects, still there are those who follow the Peripatetic School or the Pontifical Edict issued in the year 1616 by the College of Cardinals wherein it was decreed "that the Pythagorean thesis and that of Copernicus supporting a belief in a mobile earth and an immobile sun in the center of the world should be prohibited and condemned since it was contrary to the Sacred Scripture," they are compelled to recognize no other system than one which involves an immobile earth, and thus can recognize none other than that of Tycho (which I discussed in Book I, Chapter 9). This system, whether it be true or false, they are forced to defend. The question here arises whether they can investigate with any degree of accuracy the science of the heavens in this way or the right and proper relationship and position of the earthlike bodies.

By Pontifical Edict in the year 1616, *the doctrine of the earth's mobility was prohibited.*

In the first place, as for the Peripatetics, they generally accept the precepts of Aristotle as plausible and consequently employ his writings as their bases of argument. Prevented from setting forth precepts of a non-Peripatetic nature, they are not concerned with that which sane reason and experience teaches but rather adhere and submit to authority. Indeed, there is a difference between believing and knowing (for to believe is to submit to a person by reason of the authority of the speaker while to know is to recognize a result through its cause) and for this reason one can easily conclude that a theory based on the recognition of its cause is valid and should be preferred to another which is backed only by the authority of the speaker. As Augustine says, "Why do you bid me believe if I can know?" Since the Copernican system can be understood through causes, while that of Tycho cannot, (as we have already proved in various references

There is a difference between believing and knowing.

and will discuss further later on), the former system, therefore, must be true and preferred to the latter.

Secondly, as for a veneration of the Sacred Scriptures, certainly this should be preserved safe and in tact; thus in no way shall our writings criticize this sacred text. But inasmuch as certain scholars wish to explain the location of earthlike bodies in terms of their own concepts and thus seek to discredit the Copernican system and defend that of Tycho, I now raise the question whether the veneration owed to the Scriptures should be extended beyond the province of the Scriptures, which points out the path to eternal Salvation, and be expanded to embrace philosophical, mathematical and astronomical, indeed, even pedagogical questions, (among which the question of the location of the sun and earth in this world holds priority) and whether those controversies which arise in these studies should be resolved and refuted by means of this text. Or rather should controversies which tend to disturb learned scholars but not ordinary men, for the most part, since thev cannot understand them, be set aside and should the Scriptures be allowed to speak to the subjects of the rising, setting, and journey of the sun and the immoveability of the earth in a popular fashion which is attractive to popular concept, opinion, and manner of expression? For in Kings I, Chapter 7, verse 23 it is written "and He made a molten sea, ten cubits from one brim to the other: it was round all about and a line of 30 cubits did compass it round about." Now inasmuch as common workmen roughly set the proportion of the diameter to the circumference as 10 is to 30 while mathematicians define it as 7 is to 22, thus if this molten sea were completely round, its circumference would consist of $31\frac{3}{7}$ cubits. From this it must be concluded that the Holy Spirit was content to reckon the proportion of the diameter to the circumference as 1 is to 3 in accordance with the usual custom, an approximation which is not accurate. Similarly with other questions of this kind which pertain to secular knowledge and science it generally adapts itself to the common opinion and outlook of the ordinary man. Had it sought to correct all false opinions and to teach men science and the arts through the word of God, there could not have been books enough to encompass this knowledge. "He, therefore, gave the world for their dispute." Ecclesiastes, Chapter 3, verse 10. Meanwhile nothing more can be said today than that the sun rises, the sun sets, and the sun suffers eclipses, etc. Otherwise, if a person should try to express this more exactly he would embark upon a long and wordy discourse.

From the Scriptures one learns the path to Salvation, not solving mathematical problems.

Thus, when all the data has been considered which was written on this subject in Book I, Chapter 8, we cannot grant that Tycho's hypothesis is consonant with reason.

Tycho's system is contrary to reason.

1. Indeed, it is contrary to sane reason and common sense that not only the so-called sphere of so many fixed stars, countless in number, but even the entire world system, including ♄, ♃, ♂, ♀, ☿, ☉ (See Plate IV in Book I, Chapter 9 whose indescribable vastness Father Riccioli himself is more than willing to concede, since, as can be seen above in Book I, Chapters 24 and 25 of our work, he sets the sun, the center of Tycho's world system, as 38,000 times greater than the earth with Saturn 891 times and Jupiter 685 times greater, and calculates the distance of this center of the planetary system, the sun, as 7,000 earth radii distant from the earth, and Jupiter as 47,552 radii distant and Saturn 90,000). He postulates that this system turns itself every day around this tiny earth of ours, comparable in size to the smallest grain of sand. A person should consider the size of this system and then compare the vastness and distance of its bodies with our sphere of earth and water. Nonetheless the followers of Tycho maintain that this entire system revolves around the earth in the space of 24 hours in the manner described and that the sun, the center of this system and the leader, as it were, (when it is in the middle distance from the earth) in a moment or two of time, as much time as is required to say a single word, travels with its whole entourage of planets over a space of 547 German miles (Riccioli, page 127). Of this number, Saturn occupying the outermost position, at any given moment, travels 5,633 German miles: as for the rest of the planets, we will pass over without discussion their very rapid daily movements. It may be remarked, however, that every planet keeps to its own individual motion about the sun as well as observing that along the zodiac.

The speed of the sun's and Saturn's course according to Riccioli.

Who can rationally conceive of such a system? Although the Reverend Father Riccioli tried to give assistance to this theory and devised a semi-Tychonic system wherein he disassociated Saturn and Jupiter from Tycho's system and assigned them to that of Ptolemy. The only visible result from this shift seems merely to be the substitution of old errors for new. (May this most distinguished gentleman accept my apologies.) Indeed, there can be no rational basis to remove these two planets from the rest of the system and assign completely different positions and movements to them, which would be the source of a great diversity of motion in the sky, particularly since those same irreconcilable faults which were the source of the rejection of the Ptolemaic system would recur anew in the case of these two planets.

The semi-Tychonic system of Riccioli is not satisfactory.

When the whole solar system moves, all things contained in it are moved, including the earth itself.

2. It is impossible for something to revolve in a circle about some defined and given point and simultaneously to rotate about itself while still preserving, at the same time, that point contained in it, the earth, for example, immobile. Thus reason dictates, when Tycho's whole system is transported with its center, the sun, from place to place in the great sphere which the sun describes ahout the earth, that everything contained in this system of Tycho's, including the earth, no matter where it may be, is compelled to follow along with it. For a sphere which is moved in its totality with its center from place to place, infallibly takes along with it all things contained within it.

The earth is not in the center of Tycho's world system.

3. Furthermore, the followers of Tycho can assign no precise place to the earth in this world system. It is not in the center of the great orbit of the sun because they say this is irregular, with the sun now in apogee, now in perigee, and now in the middle distance. Thus the earth is not in the center of the sun's orbit, much less in that of any planet because (as is to be seen in Book I, Chapter 25) these are frequently 10, 20, 30 thousand etc. radii of the earth, more or less, removed from the earth. Therefore the earth is in the center of Nothing, when it actually should be in the precise center of the world, inasmuch as all the followers of Tycho seek to base their hypotheses on this assumption.

4. Since, according to the followers of Tycho, our whole planetary system with its intermediate heavenly matter, which they call air, transports itself every day from place to place and since there is no empty space in the nature of things, one may ask how it is possible that a vast system of this kind, filled with its own air, can penetrate with such evident speed any other air, namely that which is believed to fill the space below the so-called starry sphere?

In Tycho's system, any planet can be established as the center of the world.

5. The fallacy inherent in this system becomes even more apparent from the fact that a world system similar to ours can be established by any planet whatsoever, and in this way it, too, can be assumed to be the center of the world with the same attendant phenomena. Thus the inhabitants on Mars or Jupiter or Saturn, if there are any, can declare that the rest of the world revolves around them and that they are situated immoveable in the center of the world system. Indeed of all the Tychonic systems of this kind, none could be established more fittingly than by Saturn, were this planet assumed to be the center of the world with the sun and all the planets including the earth being carried around it.

6. There are, moreover, countless other difficulties which are associated with Tycho's system. For the sake of brevity, we will not discuss them here but direct the reader to Book I, Chapter 9 and

the authors mentioned there as well as others who deal with this subject. From this it will be quite evident that all the difficulties found here can be corrected if the sun is placed in the center of this world.

7. Furthermore, great and difficult problems arise with respect to the fixed stars when the earth is set as the center of the world according to Ptolemy's and Tycho's hypothesis. In the following Chapter we shall say something about these bodies and the diverse and irrational motions, contrary to reason and nature, which would result.

CHAPTER 17

THE BOUNDARIES OF OUR WORLD

It is clearly evident that the fixed stars bound this world, for we see them on every side beyond the sphere of Saturn, which is the outermost body of this world's planetary system. Since, however, there is such variation in their distances which cannot be perceived by the senses or the intellect, as has already been discussed in Book I, Chapters 26, 27, and 28 and is considered in more detail in Book VII, Chapter 1, for this reason Philip Lansberg, although he believed that the stars have a sphere of their own and are subject to motion from west to east or the precession of the equinoxes, nonetheless, in his *Commentationes in Motum Terrae Diurnum et Annuum*, page 32, declares "that the spheres of the planets" (by which he means this entire planetary system of our world) "should not be regarded as part of the Eighth or Starry Sphere etc." Subsequently in his *Uranometria*, page 1235, he concludes with these words, "Astronomers, both ancient and modern, are in error in their concepts of the heavens when they go so far as to deny the existence of the Eighth Sphere, with the result that they make one heaven from two. For indeed, the sphere of the stationary stars cannot be part of the planetary system because the latter in a certain way is its center. Thus Tycho Brahe should be criticized all the more when he calculates the radius of the sphere of the fixed stars to be 14,000 radii of the earth when we, to be sure, have demonstrated in Elementum 6 that the radius of Saturn's sphere is 14,880 radii of the earth, etc."

The fixed stars are not part of this world of ours.

Not only does the extreme smallness of this entire planetary system of the world suggest a separation between it and the imagined sphere of fixed stars but there is also clear and easy to understand proof offered us from the fact that the sphere of virtue of the king or governor of this world, namely the sun, cannot extend into Infinity because it tends to vanish, weakened by the interval of distance which exists in the vast Space lying between the sun's sphere of virtue and those of the nearer stars. It is here that the extent of this planetary system finds its terminal bounds and the spheres of virtue of the stars nearer us have their beginning.

This earth has its bounds where the sphere of the sun's influence ends.

Furthermore, almost all contemporary astronomers agree as to the existence of that vast interval lying between Saturn's sphere and the fixed stars. Riccioli, for one, assumes the distance of the so-called stellar sphere to be 210,000 radii of the earth, although, in fact, the distance is actually much greater.

Contemporary astronomers agree as to the existence of that vast interval lying between the sphere of Saturn and the fixed stars.

Since, therefore, the distance of the first stars is so great that it confounds human intelligence, one must suppose (as will be made manifestly clear in the following Book) that other stars, dispersed here and there throughout the immensity of this Infinite Space, are 5, 10, 20 and more times higher and farther removed than the former. Inasmuch as their number is incalculable and the magnitude of their glowing bodies is incomprehensible, (See Book I, Chapters 29 and 30), it follows that this First Mobile of Ptolemy or Tycho functions contrary to all reason, since these scholars attribute to it the amazing ability not only of turning itself about within 24 hours, but even of carrying along all these stars simultaneously in an identical course and with similar speed, so that any one of the first or nearer stars is made to run through 152 radii of the earth, that is 130,720 German miles (according to Riccioli's computation, page 419) on the equator at any given moment. And if this is the case, what can one say about the more distant stars?

Why the so-called First Mobile is contrary to reason.

Furthermore, it is difficult to understand to what end they would suggest such an unbelievably rapid course, not to mention the unceasing motion of so many countless thousands of enormous bodies which are so different from the earth and of which every one is several times larger than our planet. Some say that this makes manifest God's Omnipotence in which nothing is impossible. But would we be well advised to look for God's Omnipotence in anything so uncertain and contrary to reason as this imagined motion, or rather should we take cognizance of many other examples of His evident authority which appear more reasonable and with which more people are in agreement? We shall leave this question for others to judge, although in this way a person could cleverly devise any sort of paradox and peddle this as the truth under the pretext that there is nothing impossible to God.

That very rapid motion of the First Mobile would be to no purpose.

As a matter of fact, the impossibility that this outermost Mobile could exist is clear from Book V, Chapter 15 which deals with

The rotation of the First Mobile is contrary to the nature of forward motion.

motion, as well as from the preceding Chapter 4. All motion takes its existence from some intrinsic internal moving property. This so-called First Mobile, however, does not have an internal moving impulse and therefore cannot move itself by itself. Neither can it be made to turn by some external medium, intelligence, for example, which we have already mentioned in Chapter 4. And even if someone should say that God himself made this First Mobile turn, one may reply that God acts only indirectly and in no way carries anything from place to place. Because Infinite Immensity is everywhere it does not have a place where it may move or transport itself.

The cause of the First Mobile should be sought on the earth.

By rights, therefore, one should seek the cause of the apparent motion of the fixed stars nowhere else but on earth which creates this appearance in the stars through its daily revolution, at which time we ourselves are turned about with the earth. Because the earth travels 1/16 of a German mile a second (according to Book V at the end of Chapter 15) it seems more reasonable to assume that this is the cause of the stars' apparent motion than to put faith in a dream which involves a distance of so many thousands of miles through which the fixed stars must travel in a single moment, given the assumption that the earth is stable.

Furthermore, the Second Mobile does not exist in the nature of things.

Furthermore, there is a Second Mobile or so-called Ninth Heaven which is of a similar nature and through whose individual motion they say (during the time when the putative sphere of stars is so rotated in 24 hours by the very rapid motion of the First Mobile compelling all stars to be turned simultaneously) that the fixed stars are moved by another motion of their own, contrary to the First Mobile, but in conjunction with the celestial poles moving from west to east, a motion which is termed the precession of the equinoxes. From this in turn arises the wavering motion of "trepidatio" or the movement of rising and falling as was discussed rather fully in Book I at the end of Chapter 2. There surely is no doubt that just as the daily rotation of the stars cannot take place, neither can this secondary motion. Indeed, its existence is even more unlikely than the former's since it opposes the First Mobile with one force operating against the other, and by nature, the stronger force would have to overcome the weaker. Moreover, there is great difficulty in the assumption that the First Mobile carries along all the lower bodies, the fixed stars, as well as the planets, simultaneously within the 24 hour period from rising to setting, when the Second Mobile in a contrary motion in conjunction with the poles of the zodiac from setting to rising, moves the stars with a very slow motion taking from between 25,000 to 36,000 years, while the planets move with a swifter motion in accordance with the individual periods of each one. If this is so, why then does not the Second Mobile (as the First Mobile makes all stars rotate uniformly within 24 hours) move the fixed stars in the same way with retrograde motion as the planets? This Mobile does not admit of such regularity, however, if it turns the nearer sphere (namely that of the stars) slowly while the others (namely the planets) are moved more rapidly the farther away they are. Indeed this is contrary to nature for these bodies should be carried along more slowly, inasmuch as all force or impulsive power exercises greater influence on bodies that are nearer than those that are farther away, etc. One asks, moreover, what is the nature of the solidity of this starry sphere in whose depths are encompassed countless large bodies which enables them to be pulled around all together with the greatest speed in an orbit, without any change of position or being forced apart from one another. Likewise one wonders what is the reason for and usefulness of this motion, especially since "trepidatio", which is a defect of nature, causes so many bright and glowing earthlike bodies to carry out such irregular motion.

If one assumes the motion of the earth, then many of the unreasonable motions suggested by astronomers can be removed.

Many similar difficulties can be brought up ad infinitum relative to the Ptolemaic and Tychonic system all of which may easily be removed by assuming the mobility of the earth.

Copernicus, therefore, in taking away all forward motion from the fixed stars, ascribes this retrograde motion to the earth and transfers it to equinoctial points of the earth's equator, inasmuch as the annual motion of the earth's axis and equator do not exactly coincide with the motion of the center in longitude.

Copernicus does not admit forward motion to the fixed stars but transfers this to equinoctial points of the earth's equator.

What Riccioli opposes.

Still we are not completely satisfied with this theory of Copernicus nor can we understand his reasoning regarding the irregular motion in the earth. This precession of the equinoxes, then, and the anomalies associated with it, should rather be inputed to a slower forward motion in the earth's ecliptic (whereby the fixed stars, immoveable in themselves, seem to be moved toward the east), as we described in Book V, Chapter 16, *De Telluris Latione*.

If this assumption is made, even though this forward motion is subject to some anomaly or unevenness (for with the sphere of earth, it would be unreasonable for one to devise a mathematical rule to account for its increase in a certain year or period of time and then to compute an equal increase thereafter for all time), there is no apparent necessity, in view of this slight irregularity in forward motion, to devise still other motions, like rising and falling, or "trepidatio." For if, by reason of the earth's increase, one grants there may be a slowing in its ecliptic of some 50 seconds yearly, which over a period of 72 years is the equivalent of one degree, all anomalies of this kind as well as the postulation of the stars' forward motion can be removed.

The motion of equinoctial precession should be ascribed to the earth.

Since, moreover, according to Chapter 10 of this Book each planet must increase individually, undoubtedly Mars, Jupiter, and Saturn for the same reason (it is not so clear with Mercury and Venus because they always accompany the sun) never reach the same fixed star in their periods in the zodiac, but remain a little behind, as can be seen in Book I, Chapters 17, 18, and 19.

From all the foregoing, then, it is quite clear that the system of the earth's motion is supported by the most solid mathematical foundation, and if this is accepted, all the phenomena of the planets and fixed stars can be explained much more easily than by the generally accepted Ptolemaic and Tychonic theory, for which no natural cause can be produced to account for either the order or the disposition and movement of these bodies that gives a credible explanation.

BOOK VII

THE FIXED STARS AND THE BOUNDARY WHICH CONFINES THEM

CHAPTER 1

THE DISTANCE OF THE FIXED STARS FROM OUR EARTH OR RATHER FROM THE SUN

We now come to a subject that tends even more to transcend the subtlety and insight of all intelligence. For it is, indeed, in a contemplation of the distances, infinite numbers, and vastness of those bodies existing beyond the world of our planets, as well as the enormous distances between the dispersed luminaries themselves, that the Majesty, Omnipotence of God, and the unfathomable depths of His creativity is more manifest than anywhere else.

Omnipotence is manifest nowhere more clearly than in the infinite number of His creatures.

First of all, however, we must investigate the container of these bodies, namely Space, which is infinite in expanse. In this no size so great nor distance so large can be postulated which is less credible than any other either greater or smaller. Neither can any body be called so huge than which another cannot be found which surpasses it in size, nor can so great a number be mentioned than which some other is not infinitely larger. In Book II, Chapter 12, indeed, we said that no quantity is truly independent in itself for those words, great, small, vast, tiny are not absolute, but relative terms whereby the same thing compared with different bodies can at one time be called vast and at another incomprehensibly small. Indeed who can deny that a person can easily mention a distance in Infinite Immensity so great that the whole sphere of the stars postulated by astronomers, compared with it, may appear as nothing more than a point? Just as so great a distance is incomprehensible intellectually, however, so must it be admitted openly that the numbers, distances, and sizes of stars which all astronomers have generally accepted, are as difficult to comprehend as any others one may mention. For just as our

In Infinite Space, any distance, size, or number is to be judged as small.

Greatness or smallness does not exist in Space.

In Infinite Space, one can easily mention so great a distance that in comparison with it, the whole universe may appear as a point.

intellect is confounded and cannot comprehend numbers beyond myriads of myriads, so also does this happen to us when we contemplate the stars.

Indeed, astronomers have assigned only one sphere to all the fixed stars: these are the followers of Ptolemy, Tycho, and Copernicus, but in measuring their distances in the heavens, however, they are figuratively poles apart from one another.

The height of the stars according to the Ptolemaics.

The followers of Ptolemy commonly attribute a parallax (computed from the earth's center) of some four seconds to the fixed stars. Using this number, they compute the distance away of the starry skies to be 45,225 radii of the earth. Riccioli, however, assigned to the stars a parallax of only one or two seconds, thereby computing the distance to be 100,000 or 210,000 radii of the earth. One should reread Chapter 26, Book I of this treatise on this subject.

According to the followers of Tycho.

Tycho Brahe, without relying upon geometric calculations and perhaps by using that which is known instead of unknown (since he determined that Saturn is 12,340 earth radii distant from the earth) set the distance to the fixed stars as 14,000 earth radii. His follower, Father Anton de Rheita, however, put this distance at 20,000,000 earth radii. See Book I, Chapters 26 and 27.

The height of the stars according to the Copernicans.

Inasmuch as the Copernicans discovered that the earth's radius made absolutely no perceptible parallax to the fixed stars, they proposed another parallax of 30″, 20″, or 10″ for the diameter of the earth's great orbit. From the calculations of these scholars, as well as his own for the diameter of the earth's great orbit, Riccioli has assembled the various heights of the starry sphere as follows

	Copernicus	47,439,800	
according to	Herigone	49,502,400	
the theories	Galileo	49,832,406	radii of the
of	Bouillau	60,227,920	earth distant
	Lansberg	61,616,122	
	Kepler	142,746,428	
	Wendelin	604,589,312	

This is more fully discussed in the aforementioned Chapter 28 of Book I.

The height of stars is uncertain.

Thus no sphere can be assigned to them.

Now, in actual fact, these are distances which cannot be investigated or verified, for if one wished to examine them with greater accuracy, he would discover that these distances are not arrived at by measurement, but only by guess and conjecture. Along with Pliny we hold "that it is the height of frivolity and rashness to wish to determine with certainty the distances of the stars which we cannot measure nor conjecture by what, where, and how they are bounded." For this reason it is impossible to assign to them any defined heaven or sphere.

I do not agree with Galileo's statement in the third Dialogue of his *Systema Cosmicum*, namely that the stars are distributed on the surface of a sphere at an equal distance from the center. Rather I think that their distances are so varied that some are two, three, and four times more distant from us than others.

The opinion of Father Anton de Rheita regarding the different distances of the stars.

Father Anton de Rheita confirms this point of view in his *Oculus Enoch et Eliae*, Book IV, Chapter 1, section 7 when he says, "A simple man, worthy of reverence, of bygone times believed that the fixed stars were inserted as nails in the tablet of the firmament. We, on the other hand, believe that they are spheres of enormous mass which are separated from one another at different intervals in the space between the circumference of Saturn's orbit, the outermost planet, and the concave side of the firmament. Thus we say that many of these very bright spheres are closer to the earth and others farther away, etc.".

Kircher's opinion on the subject.

Father Athanasius Kircher subscribed to this theory in his *Itinerarium Extaticum* where he says, "There is no doubt that the stars are not all on the same surface of the heavens; rather they are at incredible distances in the innermost impenetrable and immense recesses of the same." He likewise says in another place in the same reference, "The position of the stars varies, and so great is their difference in size and brightness that we cannot conceive of it." He likewise says elsewhere, "The stars are removed from one another by as great an interval as that separating the first star from the earth. The rest are distributed some at greater and some at lesser distances from the earth."

It is probable and consonant with reason that the stars do not exist on the same level in the heavens but that some are nearer to us than others.

Reasons why stars do not exist on the same level.

(1) One reason is that they appear to he crowded together. For when we see stars which are near to one another, coalesce, as it were, into a small cloudy mass, it does not follow that they are actually very close together, for they may be, in fact, quite a distance apart. Vision cannot distinguish between nearer and more remote bodies which are at a distance, for they cannot be seen clearly because of the great intervening space. Rather we tend to perceive all bodies as being at one and the same distance. For example, clouds, falling stars, and other such meteors seem to us to be in the heavens. Rising stars and setting stars appear to us to be conterminous with the horizon, for indeed, they do seem to rise up from the ocean and to

sink down into it again. Thus if a person could stand in a space far removed from here, beyond all visible stars, he would see the whole mass of stars at an equal distance from his point of vision.

(2) If the stars were made to turn in the same sphere at the same height, God and Nature surely would have distributed them in specific designs, encompassed, say in four, five, or six-sided figures. Inasmuch as their situations are not the same, their apparent difference in appearance arises as much from their greater or lesser distance from us as from their greater or lesser size.

(3) Using the best telescopes, we very often see that two or even three stars are so near to one another as to appear to be joined together in a two or three-bodied cluster. This observation should not be accepted as valid, however, for the following reason: although these stars appear to fall in a straight line with our eye, in actual fact, each one is greatly removed from the other. For just as our sun has a vast space around it in which no fixed star is present, so individual stars greatly removed from all the others according to their difference in mass, kind, size etc., are at varying distances from us, or from our sun.

(4) Although stars appear to us to be larger or smaller in size, still their distances cannot be determined accurately because we do not know which of them is nearer to us. For if the larger is farther removed, it can appear to us, nonetheless, to be of the same size as a smaller. We do not deny, however, that the brighter the light emitted by this or that star, the greater we can assume its proximity to be.

The stars nearest to us are Sirius and Vega.

Thus there is little doubt that the stars nearer to us, or our sun, will be Sirius in Canis Major and Vega in Lyra, by reason of their very bright light. Their distance from our eye, nevertheless, is so great that it would be completely credible (because no parallax can be perceived in them), that if we were in the neighborhood of a star of this kind, not only the great orbit of the earth, which the latter describes about the sun, but also the circuit which Saturn itself makes, would appear so contracted as to be seen from an angle of 30′ (that is, of a size similar to that of the sun or moon). And just as we, from our point of vantage on the earth, cannot see the wanderings of Sirius or Vega, so neither can the wanderings of our sun be perceived from them.

If we were on one of the stars, our sun would undoubtedly appear as a star.

Thus it is clear, if we were on one of the more distant stars, (say one of those which appear to us as of second or third magnitude) that our sun would seem to be of the same size as such a star. And thus it follows that if one were on stars of sixth magnitude, our sun would likewise appear to be of the same size, and if these stars were still farther away, they would completely disappear, as would our own sun.

The true distance of the stars cannot be determined.

Let us assume, however, that the great orbit of the earth is seen from Sirius under an angle of 30′ (which corresponds with our visual image of the sun and moon), and that the radius of the great orbit of the earth, or the earth's distance from the sun, is (according to Book VI, Chapter 9) 2,644 radii of the earth, it follows (according to Book IV, Chapter 13) that Sirius is 605,956 earth radii distant from the sun or earth.

Calculation

As the radius parts, namely 10,000,000	are to the parts of the tangent of an angle 89°45′ which is the complement to 15′ or half of the aforementioned angle of 30′, 2,291,816,636	So is the value of the radius of the earth's orbit, 2,644

that is, to the distance 605,956 which, in this case lies between the Dog Star and the sphere of earth, or the earth itself or even the sun. The parallax which the great orbit of the earth would make around the sun relative to the nearest star, namely the Dog Star, should be computed as of 15′. But inasmuch as astronomers up until now have been unable to discover a parallax, it follows that because of the exceedingly great distance, as well as the refraction and change of the air, which we discussed in Book V, Chapter 10, a parallax is completely undiscernible. Therefore nothing certain can be determined about the distance of the stars, but one can assume that the distance of the nearest stars would be so great that if a person looked at the great orbit of the earth from one of them, it would appear in an angle of 1′, that is, as Venus or Jupiter appears to us.

If a person were to look out from the Dog Star at the great orbit of the earth, presumably it would appear of the same size as Jupiter.

As for the rest of the more remote stars, one can easily conclude that the parallax of the earth's orbit around the sun would grow so small relative to the very great distances involved that it would disappear and could not be more than a point in size. Thus those distances presented by Riccioli on the basis of the assumptions made by those astronomers who were followers of Copernicus (even though they may seem too great in the case of the nearer stars), relative to the more distant stars, these distances are not overgreat, but are actually even too small.

There is no reason, therefore why the followers of Ptolemy and

The space or sky between Saturn and the fixed stars is not especially great.

Tycho should greatly marvel that the Copernicans believed that a vast, and to their way of thinking, an absurdly expansive empty space for stars lay between Saturn as well as between the fixed stars themselves. Rather they should wonder at the fact that they should concentrate all the fixed stars in one sphere and state that outside of this, there is no vast empty Space, but only Infinity, which the Aristotelians term "the Imaginary."

Which is more reasonable, to bring the stars into one sphere or to scatter them in the space which extends outside?

We shall leave it to others, therefore, to judge whether it would be more absurd to hold that so many bright, enormous bodies are concentrated in one sphere and leave that Space empty which extends outside to Infinity, or to assign to each one of these bodies, in proportion to the radiation of light and heat they emit, roughly as much Space as exists between the sun and one or another star nearer to it? For what use can there possibly be in regarding this Imaginary Space as infinite and in leaving it empty, or in collecting into one sphere an unknown multitude of enormous light-giving bodies, whose nature is to disperse light all around through many tens of thousands of miles?

The effluvia and forces of the stars do not leave intermediate Space useless.

It is the wish of everyone to show himself as more noble, more excellent or more important than others of his peers.

A planet is a neighbor to a planet and a fixed star to a fixed star.

And even if so vast a space is assigned to each one of these bodies by the Copernicans, still this is not to be regarded as a useless vacuum. Indeed it must so please the divine Immensity to place its creatures widely apart for its own greater glory that each may be of an atom's proportions relative to the vast Space surrounding it. And just as it is the nature of all living things in this world always to be seeking greater space, and just as all things which are of the same type have a neighbor and seek to extend themselves up to the point that it itself extends (as, for example, we see no farmer or countryman who does not have a neighbor, nor any district at whose limits another is not found, nor any citizen who is not a neighbor to one or many others, nor any state which does not look upon a neighboring one, nor any governor or king who does not have regard for another neighboring governor or king. In such a way each and every thing always looks upon that which is of its own kind and condition and attends it with friendship, hate, or jealousy and at the same time conducts itself in such a way as to appear more noble or important), in the same way also, one can surmise that each of these heavenly bodies extends itself far and wide by means of its exhalations up to the limits of its own kind and expands its virtues between itself and its neighbors. Thus, in accordance with the condition, nature, mass, quality, etc. of each, these bodies function, come together, influence, and look upon one another, for good or ill.

CHAPTER 2

THE SIZE AND NUMBER OF THE STARS

In general it should be understood that nothing certain can be said about the size of stars. The basic requirements for this information lie in an exact knowledge of their distances as well as precise measurements of their diameters. Unless these two items are mutually consistent, all calculation breaks down, for if we alter their distances, the mass or size of the bodies will be changed.

In determining the size of stars, astronomers differ greatly.

The more common opinions of astronomers on this subject are presented in Book I, Chapter 30 where one can see both how they differ from one another and what sort of errors they commit, with the result that some say stars of the first magnitude (among whom Tycho Brahe is one) are at least 45 times larger than the earth and certain others, like Riccioli, who declare that they are 5,355 times larger than the earth. Riccioli, moreover, following the assumptions of certain Copernicans, finds the size of the stars of first magnitude, their apparent radius being posited as 18″, so great as to exceed the earth in size 15,000,000,000,000 and more times, as is to be seen in the aforemention Book I, Chapter 30. The reason for this excessive figure lies in the assumption that the apparent diameters of the stars are greater than they actually are, for the eye sees stars much larger than it should according to the rules of Optics. The reason for this is that with such far distant light-giving bodies, their sphere of light increases the size of the lighted body (See Book IV, Chapter 12) through the constriction and squinting of the eye and thereby it is made to appear much greater than it actually is. One should turn to Book IV, Chapter 14 in this regard.

The reason for the difference.

Indeed, there is a rather large number of stars which, by reason of their great distances, are virtually out of the range of sight and whose true diameters could not appear to us even as much as 1‴; which is a third of one minute in size. Nonetheless, we see these in an angle of some seconds because of the attendant sphere of virtue of the lighted star which, being unbelievably far away and at the same time contracted and smaller, still appears brightly visible and joined to the star itself. See Book IV, Chapter 14, number 2.

Why are bright bodies not as small as other objects?

At greatly removed distances, therefore, the size of lighted bodies grows, contrary to common belief and optic principles which indicate that it ought to decrease.

Thus, although Riccioli in his *Almagestum*, Book VII, Chapter 11, page 717 had listed these magnitudes specifically as I have recorded in the aforementioned Book I, Chapter 30 of our work,

The reason for exaggeration in the size of the stars.

basing his hypothesis on an assumed parallax of fixed stars and an apparent diameter of 18 seconds, postulated by certain followers of Copernicus' hypothesis, nevertheless he could not prove the validity of such great measurements for the reasons already given. For although stars are rather large heavenly bodies and there are many among them that are much larger than our sun, nonetheless, they should be placed in a definite relationship of some kind and size to one another.

Why Sirius should be regarded as one of the nearer stars.

Since, therefore, Sirius (as well as Vega) is undoubtedly the first or nearer star to us and because of its turning motion, namely its twinkling, it appears brighter and clearer than the others, it follows from our calculation of its distance (which, however, is not sufficiently great, for its distance must clearly be much larger, resulting in the star's being of greater dimensions than our calculation indicates) that with its diameter of 6 seconds, it is 68 times larger than the earth. By way of example:

Its size

As radius parts	are to the distance of Sirius	so the tangent half the apparent diameter, namely 3‴
100,000,000	605,956	1,454

is to the true radius of the star in 8.8 earth radii. If, therefore, the true radius of this star contains 8.8 radii of the earth, according to Book V, 19 of Euclid, its body will contain 681 earth spheres. On the other hand, if a person should seek the apparent diameter of our sun as seen from Sirius (that is, under whatever size angle our sun would appear from there), he could easily find this by employing the Golden Rule inversely, (See Book IV, Chapter 13) because the distance between both stars as well as the true diameter of the sun is known from Book VI, Chapter 9 to be $11\frac{1}{2}$ earth radii. He should proceed, then, as follows

If a person should look at the sun from Sirius, our sun would appear little larger than this star.

As the presumed distance of Sirius	is to the radius parts	so the known radius of the sun
605,956	100,000,000	11.5

is to its tangent 1,896, which when you consult the Tables you will find to be essentially 3″ 55‴, and from this number doubled (that is, under an angle of 7″ 50‴) the sun will be seen from Sirius, should a person be there. Indeed, as we have already said, because Sirius is undoubtedly a great deal farther off than 605,956 earth radii, then our sun must also appear there in a smaller angle than 7″ 50‴!

Galileo found the apparent diameter of stars of the first magnitude to be of at least 5‴!

Further, as for the fact that we have set the aforementioned apparent diameter of Sirius at 6″ the reason is that Galileo generally assigned diameters of 5‴ to stars of the first magnitude, as is to be seen in Book IV, Chapter 14. Sirius, indeed, appears the largest of all the fixed stars and for this reason we add one second to it. For the reasons already given, the Reverend Father Riccioli rightly assumes the apparent diameter of stars of first magnitude to be greater, namely 18‴, and that accordingly their mass necessarily is greater. Indeed, we ourselves do not deny that certain of these stars (which appear to us) are not only 681 times greater than the earth, but also, as Riccioli declared, are two, three, four, five, ten, even one hundred etc. times greater, as well as being even smaller than our sun.

There are stars without a doubt, which are ten and even a hundred times greater than our sun.

Omnipotence consists in variety.

To be sure Omnipotence consists in variety and we find this in all kinds of things and species as much in quantity as in quality, etc., so that one thing surpasses or differs from another, either in these respects or in those; for example, one may be small, another great, one may be in a pitiable, imperfect, sickly, etc. state, while another may be mediocre, and a third glorious, healthy, and relatively perfect in form, etc. All, nevertheless, exist for the manifestation of the Creator's omnipotence; for the blind man (John, Chapter 9) had not sinned nor had his parents, causing him to be born blind, but it was done that the works of God might be made manifest in him. Thus in the human race, one man always supercedes another, either this one or that, for one man is poor, another rich, one simple, another distinguished, one a farmer, another an ordinary citizen, a noble, knight, prince, or a king, etc. with the result that degrees and varieties can be found among men without end and "he who finds himself in a better position should give greater thanks for this to the Divine Majesty and should take care, all the more, not to spurn the one beneath him, for he himself could easily have been granted a more wretched condition by God." In the same way, therefore, among the stars, one gives forth more light than another, is more splendid, brighter, clearer, greater or smaller, etc., and differs either in these or other respects from the rest so that difference is without end, in the same way as Omnipotence has no end.

No end can be assigned to the multitude of stars.

Furthermore, as for the multitude of stars in existence, we have said in Book I, Chapter 29 that ancient scholars had reduced the stars to a certain number, namely those visible to the human eye, and had separated them into constellations as the celestial charts indicate, but that modern astronomers, by using telescopes, had found not only those mentioned by the ancients but others as well, countless and beyond number. And thus Father Anton de Rheita in Book IV, page 197 of his *Oculus Enoch et Eliae* declares, "In the constellation of Orion alone twice as many stars can be counted as the scholars of old found in the universe, 2,000, as a matter of fact, while in the Milky Way, they are actually without number.". Thus Riccioli in Book VI, Chapter 6, number 5, page 413 of his *Almagestum* declares, "If a person asserts that there are more than 2,000,000 stars, he makes no unbelievable claim."

The fact that these stars cannot be observed in Infinite Space does not prove that they do not exist, in the same way as a person cannot assert that a forest comes to an end because no trees can be seen in the distance. Similarly it appears likely that, even if a person could push beyond these throngs of visible stars (even with the help of the best telescopes), he still could not come to a place where all these spheres pass out of sight and no heavenly torch of this kind would be visible to shed its light. Indeed, since we discovered in Books II and III that there is no end to the Space of the world and that the existence of so large a number of illuminated bodies cannot be denied, where can we possibly set a limit to them? This will be further discussed in the subsequent Chapter 4.

The Milky Way.

So as not to pass over the Milky Way and the stars in the nebula in silence, one should realize that there is a mass of very small stars that escape our power of discernment as well as those shrouded in mist. For in whatever region of the heavens you turn a telescope, an enormous mass of stars immediately comes into sight, some of which are quite large while an incalculable multitude of small ones are completely indistinguishable from one another. Even before the invention of the telescope, Democritus (as Plutarch records), Ptolemy, and Albertus Magnus reached this conclusion through rational thought. Now in certain regions of Infinite and Immense Space (or Infinity, as I shall call it, although this terminology pertains only to the regions of stars inasmuch as Space itself cannot be divided into regions), there exist certain masses of very small stars which are heaped together; these are not dispersed at very great distances from one another as the larger bodies are (because it is pleasing to God to show his Omnipotence all the more in the great variety of all things), but are scattered about quite near to one another. Consequently their systems, together with their neighboring planets which are naturally encompassed by their spheres of air, appear greatly reduced and constricted because of their enormous distance from us with the result that they surround and cover a star, as it were, with their own effluvia and thus make them all appear dark and nebulous to us.

Finally, as for those stars which they call novae, all these we relegate to the families of comets, for they take their existence from the ethereal effluvia or atmosphere of the planets, being left by them from their spheres in the course of their periodic motion. For if these new stars were created in the regions of fixed stars and came into our sight, they would not perish as quickly as they do but would continue to be in existence. These stars, therefore, which indeed appear to us very far removed, should not be assumed to be much farther away than Saturn's sphere, even though, when we look at them, we seem to feel that they have their place among the fixed stars. When we consider the matter from a mathematical point of view, however, one easily comes to recognize the fact that the fixed stars are higher than the novae by many thousands of earth radii. See also the end of Chapter 11 in the preceding Book VI.

The so-called novae.

CHAPTER 3

THE FIXED STARS AND WHAT THEY REALLY ARE

There are three kinds of earth bodies according to Book VI, Chapter 7, solar, planetary, and lunar, and of this number, the solar bodies alone have light in themselves, That the fixed stars are self-illuminated, however, and do not receive light from the sun is evident 1) from the great distance between them and our sun, inasmuch as Sirius, the star which is nearest to our sun according to the foregoing Chapter 1, is at least 606,000 radiii of the earth distant. Thus the sun would not appear any different to someone on Sirius than it appears to us here on earth. This is shown in the preceding Chapter 2. Sirius, moreover, does not bestow less light on us than it does on our sun. The sun, moreover cannot give less light to Sirius than to all the other fixed stars. Father Anton de Rheita confirms this statement in Book IV of his "Oculus Enoch et Eliae" when he remarks, "I say, indeed, that it is impossible for stars to receive more light from our sun than our sun can receive from them. For what light does Sirius give our earth when it is enveloped in night's darkness? What brightness does it give to our deep shadowy darkness?" 2) from their twinkling, which is nothing other than the very rapid turning of the stars about their own centers. For just as our sun continually revolves with very great rapidity (we discussed this in

There are three kinds of earth bodies.

Fixed stars are self-iluminated.

The twinkling of fixed stars.

Book VI, Chapter 8), so do all the stars turn about their own centers. This movement, then, is the source of their twinkling. Plato also supports this hypothesis and Father Raphael Aversa in his *De Caelo*, question 34, section 3, agrees. For since the circumferences of the stars appear to us now one way, now another because of their extraordinarily rapid motion and since the outlines of their lighted bodies are seen at one time to be more or less contracted and at another extended, there arises this kind of wavering which is characteristic of fiery stars because they do not participate in forward motion and therefore conserve themselves by means of the turning motion which is incredibly rapid. Although their twinkling appears varied in intensity, for some twinkle more than others, the reason lies as much in the differences in their distances, which in many cases are great, as in the locations of their bodies. All twinkle, however, inasmuch as their individual centers rotate, a fact which has been verified by Tycho Brahe's observation. Biancani adds that "the nearer they are to the horizon, the more they twinkle." Kepler asserts that Sirius projects more defined and longer rays of light than Arcturus, that Aldebaran sends out slow rays and that Antares emits very rapid ones. These differences arise because their positions in the heavens are not the same and for this reason, one may easily conjecture that rays of unequal length are transmitted by a body that directs its polar side to us rather than its equatorial. As for the revolution of the planets, inasmuch as this cannot in any way be compared with that of the stars since one sees little or virtually no twinkling in them, and whatever there is, the higher ones, being farther away, seem to give out less than those nearer to us or lower down in the heavens. (These matters were discussed in Book VI, Chapters 12 and 13) When one takes serious thought, it is clear that a point (which the sun, in fact, represents among so many thousands of widely separated lighted bodies) can in no way be adequate for lighting them. Furthermore, it is reasonable to assume that just as the bodies of stars are so reduced in size when the distance is great enough that they finally disappear into nothingness, so also does the solar body of this world become so small that it finally vanishes.

A third reason that the fixed stars are not lighted by the sun.

Writers who state that the fixed stars are suns.

For these and many other reasons Galileo in his *Systema Cosmica*, Dialogue 3, expressly declares "that the fixed stars are just so many suns, like to our own sun." Kepler subscribes to this opinion in his *Dissertatio Cum Nuncio Sidereo*, page 9, as does Father Anton de Rheita (who, indeed, asserts that the earth is stationary) in the aforementioned Book IV of his *Oculus Enoch et Eliae*, page 177, when he writes to Caramuel von Lobkowitz, Esq., in these words, "Since it is impossible that the fixed stars borrow light from the sun because of their incredible distance, it seems that we should admit and acknowledge one of two possibilities, namely that the fixed stars either have light in themselves and accordingly can be regarded as suns, or that they borrow their light from some higher and very strong illumination. Thus a person would not be foolish in the least to suppose, since the light of the fixed stars is so vivid and shining bright, that this indicates and proves that there is very strong light or some other kind of brightness that is much higher than our light and is found at an extremely great distance which we cannot see."

On page 196, he says, "Therefore, I do not know whether a person should be criticized if he supposes that above and beyond the imagined distance of the great heavens from the earth, there are many fixed stars or suns revolving around their planets or that the planets are revolving about the aforementioned moving bodies." In this statement, he seems to be at a loss to determine whether suns (which he associates with fixed stars) revolve about planets or whether planets revolve about suns. As a matter of fact their nomenclature ought to correspond with the bodies themselves inasmuch as stars (which he admits are suns) are called "fixed" for the reason that they do not partake of forward motion while planets are so named because they wander. It, therefore, must undoubtedly have been Father Rheita's opinion that the wanderers, or planets, turn about the stars or suns themselves.

Aristarchus of Samos agreed with this also. His works were recently revived in France by Marin Mersenne in his *Novarum Observationum Physico- Mathematicarum*, Book III, page 8, where he says "As for the fixed stars, there are some who believe that one of them is the head and occupies a special position in some world system or other, just as the sun occupies a special position as the head of its own system. In the same way as the sun has planets grouped and carried about it, so these scholars believe any one of the fixed stars can have other bodies like planets grouped about it, even though we do not see them from the earth because of their extremely great distance from us. Similarly our planets are not visible from the fixed stars, for it is certain that this distance is so great that compared with it, the distance between the sun and earth is imperceptible."

Some have held that any fixed star can be the center of a world system.

Thus Cardinal von Kues in *De Docta Ignorantia*, Book II, Chapter 12, believed that "each star makes its own individual world."

Cardinal von Kues' opinion about fixed stars.

It is also reported that Giordano Bruno was of the same opinion.

In any case, I can say that before I had heard or read anything about these matters, the following thoughts came into my mind, namely that fixed stars are suns and by and large resemble our own sun with wanderers or planets undoubtedly revolving around them.

Early thoughts of the author about fixed stars.

For, of what other use would this enormous interval and vast Space be wherein stars appear scattered everywhere throughout the enormous abyss at different distances from one another? Neither does this nor the following similar thought which now comes to mind seem strange to me, for who, heretofore, would have believed that there are satellites around Jupiter and Saturn, bodies whose existence was revealed by telescopes only later on? Thus it seems very likely that there are planets carried about fixed stars throughout encompassing Space from which the former receive light so that their brightness may have a purpose. Kircher says in his *Itinerarium Ecstaticum*, page 410, "for even if they cannot be seen by mankind from the earth because of the great distance, reason nonetheless dictates that this must be the case because it seems likely that a certain parallelism exists between all systems of spheres in the world. Who could persuade you that the Earth, Moon, Mercury, Venus, Mars, Jupiter, Saturn, and other stars which revolve about the sun, when viewed from a star in a nebula, appear only in the form of such a star? Although the former do not have their own proper light and shine only with the light of the sun, still one ought not say that they do not exist even though they are not visible. I should like you to think that the same occurs in the vast reaches of the firmament."

Father Kircher's opinion of fixed stars.

Certainly it is not my wish to discuss and to peddle these ideas of mine as being valid or to defend them against the opinions of others. For since these hypotheses are dependent upon conjecture inasmuch as no one has made a trip to the stars to confirm their validity by actual experience or, by contrary reasoning or observation, to give proof that they are false, we leave this discussion so as not to damage those ideas which are true or to give strong support to others whose validity we cannot prove by visual demonstration. Perhaps a time will come when, through the development of better telescopes, those ideas will be confirmed which now are regarded as false. In the meantime, René Descartes declares, "The infinite power of God can never be sufficiently increased, imagined, and praised, and for this reason we must take care that we do not have too much pride in ourselves. This might occur if we try to assign limits, of which we have no reasonable proof, to the world (understand this to mean, not "world" because this has an end, but rather the "creations of God" which are without end),assuming that our mental powers can be carried beyond that which has actually been done by God, or imagining that all things have been created for our sakes alone."

Now let us turn to Father Kircher's opinions in more detail regarding the fixed stars and their satellites.

CHAPTER 4

FATHER KIRCHER'S OPINIONS CONCERNING THE FIXED STARS ALONG WITH OUR OWN REMARKS

Father Athanasius Kircher, a widely celebrated mathematician of Rome who penetrated farther into the domains of the heavens with the help of his excellent telescores than most other astronomers, brought out his book several years ago, *Itinerarium Ecstaticum* with Cosmiel and Theodidactus as interlocutors. The Reverend Father Kaspar Schott furnished the second edition of this text with notes and in great kindness sent this to me last year as a gift.

In Chapter 9, dialogue 1, page 341 ff., there is a discussion about a trip from Saturn into the region of the fixed stars.

From the region of the stars, Theodidactus cannot see the earth nor the planets of this world of ours

Number I. Theodidactus asks, "Where are you taking me, Cosmiel? What is to be the end of my journey? Where, now, are the sun, the earth, the moon, and the other planets? Indeed, all seem to have disappeared." Cosmiel replies, "Why, you have now entered into the region or kingdom of the fixed stars, to employ your earth-bound terminology, that you seemed so desirous of seeing." Theodidactus: "Oh! Cosmiel, I see the whole countenance of the heavens ornamented with new and hitherto unseen stars which are differentin their position, size, disposition, and light in a wondrous way. Where are Aries, Bootes, Cassiopeia, Orion's Belt and the rest of the stars I know so well?" Cosmiel: "Here, within the region of stars, your eye now sees everything dispersed, broken up and changed into another order." Theodidactus: "Are all these masses of stars the same distance from the earth?" Cosmiel: "Not at all, but you must know that these nebulous masses, brought together in the shape of a circle, are as far removed from your eye as even the first star of the firmament is from the center of the earth and are distributed, some at greater, others at lesser distances than this fixed star, from the earth.

nor the constellations.

Our Comment

The stars are separated at enormous intervals from one another.

It is clear from the foregoing that because of the appearance of new stars and their immense distance away, Theodidactus finds this region is totally separated from our planetary system (which has completely disappeared, except for the sun) and that the stars which appear to us in mass in the heavens are separated from one another at as great a distance as the nearest fixed star is separated from the earth.

Number II, page 143. Theodidactus: "I do not know, in truth, where I should tarry. I see huge stars which are hardly smaller or of less brightness than the sun. I see, moreover, some in a variety of phases like the moon. Can it be that there are bodies in the immensity of this Expanse which have a similarity with the sun and moon?" Cosmiel: "Why do you doubt? Did you think that stars of this kind are lighted by the sun? Well, you are hallucinating! Do you see this star diametrically opposite you: it is not particularly different from the Dog Star? What do you think it is?" Theodidactus: "It is some star of the first magnitude." Cosmiel: "How greatly mistaken you are. This is your sun, that admirable body, than which nothing in our world can be equalled in size or is more wondrous. By reason of the brightness and the power of its light all lower beings live; by reason of its indescribable virtue, all creatures of the earth are nourished and kept alive. Now do you think that such a small star as this is powerful enough to light this world?" Theodidactus: "No indeed, quite the contrary. I do not think that this star could give out as much light as Sirius or Vega to illuminate our sphere of earth. Indeed, if we were a hundred times farther from the sun than the sun is from the earth, who would believe that such a very small star could illuminate these enormous spheres of the world as well as those at a much greater distance from us which are as yet undiscerned and undiscovered?"

He sees new planets.

Our sun appears there as Sirius.

Our sun cannot illuminate the stars.

Our Comment

It can be concluded from this that the fixed stars, which we see, are suns and that among them there are some bodies, invisible to us here on earth, which undergo a remarkable variety of phases like our moon. These bodies are lighted by the suns, as our moon and the other planets are, and must be planets of these suns or solar bodies. As for his declaration that our sun is lower than the others, this, I believe, he bases on a long-standing belief, but surely no one can know, in so great a multitude of stars where no end is defined, what is below and what is above.

It can be concluded that the stars are suns and that they have planets.

Number III, page 344. (ERRATUM, read 344 for 145) Cosmiel: "Do you see that star in the east which in aspect is almost as large and bright as our sun? This is the star which your astronomers call the Dog Star." Theodidactus: "Can one believe, then, that in the most distant recesses of the heavens, suns are found?" Cosmiel: "Why not? It is not only fitting that bodies of a solar nature be found there, but it is even necessary lest so many and so great earthlike spheres should be committed to eternal shadow and lie in motionless

Cosmiel sees Sirius almost as large as our sun.

inactivity without solar light. For if the Dog Star were taken away when it was shining alone on an earth-like body, what do you thlnk would happen to this body?" Theodidactus: "Indeed, I believe that it would be encompassed by eternal darkness and everlasting gloom." Cosmiel: "The same would happen to countless systems of heavenly spheres if there were not something to give them light and to warm them by its heat to the limits intended by nature. For all these stars are of a kind which they call first magnitude."

If the fixed stars were not suns, then the rest of the earth-like bodies would be encompassed by eternal shadow.

Our Comment

If suns have their place in the so-called firmament so that the great number of earth spheres may not lie in everlasting shadow, then there must also be earthlike spheres there which have no light of their own. Moreover, inasmuch as earthlike bodies are of a three-fold nature, (as we have said in Book VI, Chapter 7) namely solar, planetary, and lunar, so planets and moons as well as suns must also exist in the most distant reaches of the heavens. These three types of bodies constitute a world system. Therefore all stars of the first magnitude (indeed, not only these, but all fixed stars, constitute individual systems because we do not know whether this or that star is really of the first magnitude when a larger one which is farther away can appear as a star of the sixth magnitude. Just as the Dog Star does not light up our earth when the sun has been taken away, neither can our sun give light to the earthlike bodies on which the Dog Star shines.

It follows that there are also planets and moons in the firmament which are nonetheless invisible to us and consequently that there are individual earth systems.

Number IV, page 347. Cosmiel: "But, that you may understand all this more clearly, let us hasten to the Dog Star sun." Theodidactus: "How far away from it do you think we are?" Cosmiel: "As far as the earth is from your sun." Theodidactus: (with that remark, we hastened on our way.) "Oh! Cosmiel, my friend, I thought you wanted to take me to the sphere of the Dog Star and now I find myself between our earth and sun." Cosmiel: "What makes you think this?" Theodidactus: "I see our moon." Cosmiel: "The moon which you see is not the earth's moon but a dark sphere revolving around the Dog Star. Lighted by the Dog Star sun it necessarily shows phases similar to those of our moon. There are many of these bodies in the immense depths of the heavens and they can never be perceived by the human eye."

The Dog Star sun.

Certain planets revolve around the Dog Star sun.

Our Comment

It is clear from the foregoing (1) that the star which we call the Dog Star is a sun in its own locale and that Theodidactus was only as far away from it at that time as we are from our sun; (2) that Theodidactus believed that he was going to be taken to this sun, when in fact he had been brought to a dark earthlike sphere that revolved around the Dog Star and was lighted by it. This, indeed, appeared to us as the moon (for at such a distance which corresponds to our distance from the moon, it is necessary that the planets also show phases similar to those of the moon except for the difference in their turning motion), but it was not actually a lunar body since it revolved about its own sun. It had to be a planetary body, therefore, because there are only three kinds of earth bodies in existence; (3) It should be remarked that one can conclude from Kircher's hypothesis that there are countless bodies of this kind in the immeasurable depths of the fixed stars that cannot be seen by us because they borrow light from their own suns which is a great deal weaker that that of the suns or solar bodies themselves.

Number IV (likewise), page 348. Cosmiel: "You will see all these phenomena much more clearly, however, once you are settled on this moonlike sphere." Theodidactus: "With that he deposited me on the sphere resembling a moon. It consisted of liquid and solid and was of a similar composition to our own moon except that there was a world of difference, as they say, from its strengths and innate properties. My dear Cosmiel, the light that touches my eyes is too bright and I cannot look any longer on the heavens. Take me, therefore, to the other side of the sphere so that I may examine that whole aspect of the sky more carefully under the shadow of night: and this he did. Here I discovered those sights that words cannot describe with sufficient eloquence. For indeed I saw above me a countless multitude of stars reaching as far as the order of the sixth magnitude."

Theodidactus is deposited on a planet associated with Sirius

and in the sector where it is day.

He wishes for another locale where it is night.

Our Comment

It is to be gathered from the foregoing that (1) Cosmiel or rather the author of the colloquy does not accurately distinguish here between planetary and lunar bodies, but takes one for the other. It was, however, a planetary body that was referred to in the foregoing account and not, in fact, a lunar one. (2) Although lunar as well as planetary bodies for the most part are composed of liquid and solid matter (See Book VI, Chapters 10 and 15 in this regard) they still differ greatly from one another. To be sure it detracts from Omnipotence to create the same thing without any variation for from this show of variety, Omnipotence is duly recognized. (3) Theodidactus, being deposited on the face of the sphere which at that time was lighted by its sun (that is, where it was day), did not have a view of the stars. We saw, however, without a doubt, on the lighted surface of that sphere those sights which are beyond verbal description. In his enthusiasm, the author perhaps neglected to describe anything about them. Thereupon (4) he goes to the opposite side of the sphere, where it is night, and looks at the face of the heavens from that region which he discovers is quite different from that which he saw from the earth.

Number IV (likewise), page 348. Theodidactus: "My dear Cosmiel, how strange everything appears to me! Are these stars of the firmament or are they not? Indeed, I am certain that none of them is seen from the earth." Cosmiel: "Indeed, they belong to the firmament and you correctly remark that one of these is seen from the earth." Theodidactus: "Now, show me, I beg, the sun, moon, earth, and the other spheres turning around the earth." Cosmiel: "See that star of the second magnitude? That is your sun which your philosophers believe lights the whole world." Theodidactus: "But where is the earth, the moon, and the rest of the planets and stars?" Cosmiel: "All are out of sight because of the very great distance, with the exception of Jupiter and Saturn which you see in the region of the sun and which appear as stars of the fourth magnitude."

Theodidactus sees other stars.

Our sun appears as a star of the second magnitude.

Planets associated with our sun cannot be seen.

Our Comment

(1) There certainly is no doubt that the stars present a different appearance when seen from the neighborhood of Sirius, which has been the subject of our discussion. It does not follow, however, that none of them can be seen from the earth. The Reverend Father himself in his first Dialogue admits that stars are not contained in one sphere, but are distributed, some at greater, others at lesser distances from the earth. Therefore, some that are situated even beyond the Dog Star are visible from the earth.

(2) If our sun appears from the Dog Star as a star of second magnitude, it must follow that the sun is much smaller than the Dog Star but this conclusion must be a source of doubt.

(3) In the same way as the author in Dialogue 4 acknowledges that the spheres revolving around Sirius and other fixed stars can never be seen by the human eye, so also should it be conceded that

the planets of our sun, whether they be Jupiter or Saturn, etc., cannot be seen from the Dog Star.

Number V, page 349. Theodidactus: "But what does this mean? For I see that neither the sun nor Jupiter and Saturn move, but that they almost always maintain an essentially constant distance from one another." Cosmiel: "You should know that the distance away these spheres are, as seen from your eye, is so great that the whole circumference of the solar heavens as well as the heavens of the other spheres we have mentioned are contracted into a space essentially equal to that of the solar body. I should also like to persuade you to look at the spheres in the heavens when you are situated on the earth, for all appear to be separated from one another with eternally equal distances and unchangeable intervals. This, nevertheless, is not the case, for all these bodies or stars which you see in the firmament complete their courses not unlike the sun, moon, and other planets lying below, in enormous circles. Do you not see that new stars appearing time and again, now rise above the field of vision of this moonlike sphere and indeed, at another time, descend? All of these complete their own course about the sun and moon in the heavens and have their own laws of ascent, descent, and decline prescribed for them as is the case with the planets lying below."

Because our planetary system is so far away, it is contracted into a space essentially equal to that of the solar body.

Father Kircher believes that the fixed stars together with their satellites move in the same way as the Tychonic system describes with the sun and its planets following their course around the earth.

Our Comment

It is evident from the foregoing that the Reverend Father Kircher in defining diverse planetary systems, equates our solar system with that of Sirius. He says: "These bodies in the firmament complete their courses in circles etc., not unlike the sun, moon, and other planets." Just as this old concept regarding the stability of the earth and the motion of the sun has fixed such deep roots in his mind that he firmly believes that not only the planets, but also the sun is carried about a center, namely the earth, so also does he conclude, insofar as the system of Sirius is concerned that this sphere with its accompanyomg stars (or planets) necessarily is carried about a center point, which is nothing other than some other earthlike body. There is little doubt, furthermore, that the common precept "when there is one inconsistency, many follow" applies here. For this is indeed the case since the Reverend Father at one time ascribes movement to this Dog Star sun (when, as a matter of fact, according to Book VI, Chapter 3, no sun, functioning as the center of its system, can carry out any forward motion) and at another declares it is stationary, when he says, "All these, (that is, the planets moving around the Dog Star) complete their own courses about this sun and moon of the star world as well, would have their own laws of ascent, descent, and decline prescribed for them as is the case with the planets lying below, (which undoubtedly means our own planetary world) etc." If this is so, then it should follow that the Dog Star sun together with the Dog Star moon is the center of a system of this kind. Such a supposition, however, is contrary to reason (since there cannot be two centers), as well as inconsistent with the Reverend Father's hypothesis that all the bodies in the firmament are moved in great circles.Thus there is no doubt that the Dog Star sun must be the center of its system and this moon of the Dog Star one of its planets.

Number V (likewise), page 350. Theodidactus: "If these things that you say are true, Cosmiel, our telescopes undoubtedly would have recorded differences in motion of this kind and would have observed these moons, now full, now half, now horned, and finally completely disappearing from sight." Cosmiel: "As for the fact that phenomena of this kind have never been observed by astronomers, this does not prove that such do not exist. By reason of the extremely great distance of these spheres from the earth, that sphere alone is visible to our eyes which is the largest and brightest among these aforementioned bodies. The Dog Star sun or Sirius, is one of these. Similarly, from this sphere you see nothing of the lower lying spheres except the sun since they have been left behind together with their unobservable orbits in the immensity of their distances away."

Because of their incalculable distances away and the lack of light, the satellites of the fixed stars cannot be seen.

Our Comment

From the foregoing, it is clear (1) that the author, the Reverend Father, includes planets under the designation, moons. For all planets, seen from different angles, appear now as full, now as half, etc., for that which happens in the case of moons every month, occurs with planets every single day; (2) that because of the incalculable distance of these spheres from the earth, we see only that one before our eyes which is the largest and brightest among them, namely the sun, the Ruler and Sovereign of this system. This body shines out with hrightness and underlines the truth of what we have already said at the end of the Comment to Number IV, namely that neither Jupiter nor Saturn can be seen from the Dog Star. As for the fact (3) that the author calls our planetary spheres lower, this undoubtedly arises from a common conception which considers the antipodes as lower, when in actual fact, so far as the sphere of earth is concerned, nothing is below or above. Similarly, I do not understand how anyone in considering our planetary system and that of Sirius can declare that one is lower and the other higher.

Theodidactus comes to the sun which we call the Dog Star.

Number VI, page 351. Theodidactus: "I beg of you, Cosmiel, let us now see this Dog Star sun," – and behold, in the wink of an eye, he set me down on the sun. "Indeed, Cosmiel, I see that this sphere is truly similar to our sun. Here the waters of the flood roll along in a mighty surge." Cosmiel: "There is no reason for us to tarry here for the fluid substance is of a kind similar to that of our sun. You should know, however, that all solar spheres of this kind differ in their virtues and composition in accordance with the location and position they hold in the universe. Since these characteristics are known to God alone, there is no need to set them forth more fully here. If I were to describe individual characteristics to you, you would realize that no matter how many books you might write, they would not apply to the universe as a whole."

Our Comment

Without doubt the Dog Star is similar to our sun, but just as all things are vastly different from others of its kind, so also are these suns different from one another, for the Omnipotence of God is Infinite and all nature is always seeking after variety. The variety of these bodies, therefore, is infinite.

Now Theodidactus is brought into the Innermost parts of the firmament.

He sees Aldebaran accompanied by many satellites as well as Vega.

Number VI, page 352. Now Theodidactus is brought into the innermost parts of the ethereal firmament and in the course of his journey comes upon countless solar spheres together with their moons. Among this number is Aldebaran, remarkably visible with its countless accompanying stars: then the solar body of Arcturus, the largest in size together with its followers; finally the solar bodies of Capella, Vega, and Algol, bright and conspicious, attended by their multitude of spheres that have never been observed from the earth. Furthermore, all, like the countless kinds of natural bodies here on the surface of the earth, are different from one another in the forces, composition, and qualities with which each is naturally endowed. Still, there is none which is not composed of moisture, earth, and others of the elements, and none which does not possess its own center and atmosphere.

Our Comment

I know that all fixed stars, which we see among their throng of satellites, are alone visible to us while the satellites themselves are not, (as the author also stated in Number V, page 350.). Thus any fixed star constitutes a world system of its own, for a single world cannot admit of two or more suns.

Number VIII, page 352. Theodidactus: "My dear Cosmiel, what is the significance of that very large sun I see behind us?" Cosmiel: "Why, that is the Pole Star or Polaris." Theodidactus: "I have learned from my instruction in astronomy that this star, always being stationary in one and the same place and attached, as by a nail, to the pole, describes a circle not more than 3 degrees distant from the world's pole. Now, moreover, I discover that the orbit in which it turns does not exceed that of our own sun and at least can fairly be equated with it. Furthermore I see it accompanied by its own moons." Cosmiel: "Polaris, turning in a very great orbit, is so far distant from the earth that an eye observing from the earth's surface, sees it as being no more than a point, and the diameter of its orbit as no more than 6 degrees. Thus it is no wonder that it always has the appearance of being stationary and of proceeding very slowly in its orbit." Theodidactus: "But why is this not the case with the rest of the stars?" Cosmiel: "Because they are not situated in the same place but in others that are completely different."

Polaris is an immense sun accompanied by its own moon.

Our Comment

One can gather from the foregoing that the Pole Star is a larger sun than ours and is accompanied by its moons which are actually planets, although the Reverend Father commonly calls the bodies which borrow light from their sun, moons. As for the motion of the Pole Star, the comments recorded above are quite contrary to reason, (with apologies to this Reverend Gentleman). First, as to the statement that in its slow motion this star covers an interval of not more than three degrees, a subject greatly disputed in astronomical studies, one should understand that this motion is not actually a characteristic of the star itself but is dependent either upon the First so-called Mobile or the daily motion of the earth causing stars seemingly to revolve in a twenty-four hour period, or because of some change in the magnetic axis in the earth, a subject discussed in Book IV. Furthermore, as to his effort also to attribute an orbit to the fixed stars, such as he believes our sun together with its planets describes around our earth, a hypothesis which is in accord with Tycho's belief, this hypothesis seems to be completely untenable. One can understand, however, how he came to be of this opinion. Perhaps he agreed with Aristarchus who held that each and every star is the head and essential element of some system, much as the sun is the principal part of its own system, and also with Tycho's hypothesis that our sun describes an orbit each day about the earth. Thus he assumed that the fixed stars in the same way describe circuits of this kind together with their satellites about other centers. If one accepts these postulates then it must follow that the systems of the fixed stars correspond with Tycho's hypothesis and consequently that any center whatever of these systems is an individual earth. In the same way as Tycho's hypothesis is false, however, so also are these of the Reverend Father. It seems much more reasonable, therefore, to define their position, together with that of their solar bodies, as consonant with Copernicus' hypothesis and to include these earth centers in the company of the planetary family.

Numbers IX and X, page 353. Theodidactus: "I should like to know and have you point out to me where that very celebrated northern pole of the world is, before we go any farther." Cosmiel: "Everywhere in this vast pervious ether no solid spheres are found; therefore necessarily the concept of the poles must be abandoned. You must know, therefore, that this whole Space which we believe stretches from the earth to the outermost bounds of the earth bodies, and which the Jews call "Rakihangh" and we, "Expanse", is filled with very fluid and tenuous air which the ancients called "Aether" or "Ethereal Air". And advisedly I say "filled" as there is nothing in the nature of things so solid or hard whose pores it does not penetrate. It was, moreover, so created by God's great wisdom to the end that there would be no place for a vacuum to exist that is absolutely abhorrent to nature.

In vast ether there are no solid spheres.

Kircher describes "Expanse" as being filled with very tenuous air or ether which penetrates everything.

It does not admit of a vacuum.

Our Comment

All things have their existence in "Space" or "Expanse" which penetrates all things as Father Kircher himself avows. That it is human nature always to rely upon one's habitual preconceptions, however, is evident here for the Reverend Father, (since he could not admit to the existence of Empty Space), imagines that Space is filled with an ether which nonetheless penetrates everything. Because, indeed, ether is nothing other than very rarefied air and air is a corporeal substance (as we demonstrated throughout almost all of Book III with our various experiments and because reason itself argues that only corporeal matter can fill a vacuum), for this reason it is impossible that ether, being of a corporeal nature, can penetrate bodies: rather it is "Expanse" or "Space" itself which penetrates all things.

Air is something corporeal.

It cannot penetrate solid bodies.

It follows, therefore, that if "Expanse" is filled with air, "Expanse" is one thing and air another. And because they are different, one can be separated from the other, namely the corporeal from the incorporeal. In consequence of Space being rendered empty, that

"Expanse" and air are separate entities.

putative air (in Space) would be completely useless and for this reason would not exist.

Certain Aristotelians will reply, "If Space is not filled with air, Space is empty. Since this is inadmissable, some hypothesis must be devised which does not contradict the authority of Aristotle."

It is very well attested in Books II and III and elsewhere that in nature empty space does not exist; as for ethereal air and its function in empty space, nothing has been established. If, however, a person wishes to regard a vacuum as being absolutely nothing in the Aristotelian sense (in this event it would not be one of the two things that exist) it cannot be so regarded, as we have said in the aforementioned Book II, Chapter 7, because there can be no third thing. Thus it must be the case that Space is empty and that emptiness without space cannot be spoken of or exist.

Number X (again), page 355. Cosmiel: "In this enormous expanse of ethereal air, then, the Creator of the world disposed the mass of countless spheres in such a way, relative to the position of the earth, that all might follow their movements from rising to setting and their courses around the earth. Since, indeed, this Expanse of ether consists of no defined form but permeates the world in an uncertain and indefinite flow, the Supreme Architect disposed these spheres that had their place of birth here, with such harmony and symmetry or order in relation to one another that no one would interfere with another or cross another's paths. And since individual bodies carry out their own motions, with the exception of daily motion, in their own orbits, it was necessary that one occupy a higher place than another or a more distant one, when following identical courses in their daily circuits, so that certain middle, and as it were, secondary, suns might illuminate the individual spheres dependent upon them, when their orbits preclude their having light. Indeed, He placed them near solar spheres but at such a distance from them that they might enjoy their bounty and still not impinge upon one another's sphere of authority.

In Space or Expanse the Creator of the world disposed these countless bodies.

No earth body interferes with another.

The varied motions of the stars.

Suns give the individual spheres dependent upon them light when they have no light.

Our Comment

Stars do not imping upon one another's jurisdiction.

All the foregoing is eminently reasonable. That this Expanse, however, is filled with some ethereal substance and that the earth body is situated in the midst of these countless spheres in such a way that all complete their daily motion from rising to setting about this as a central point, is without support. Even Theodidactus testifies to this in the words that follow.

Number XI, page 356. Theodidactus: "Indeed, I comprehend

Theodidactus cannot understand what force there is which can pull so massive a system of spheres around in twenty-four hours.
It is said that the earth bodies are directed by intelligences.

these statements, but I cannot understand what force there is which can pull around such a massive system of spheres with such incredible speed. Cosmiel: "All spheres which number beyond count, have individual intelligences governing them whose duty it is to direct each and every one of them in accordance with the position designated by nature in the vast reaches of the ethereal universe and to make them all revolve in an appropriate fashion about the sphere of earth in such a way that they cannot deviate a mite from their prescribed course or change the distances between each other."

Our Comment

As for motive intelligences, it was declared in Book VI, Chapter 4 that such do not exist. Nonetheless, that individual spheres can carry out their own motions in such a way that they do not deviate a bit, is evident from Book VI, Chapter 11. God should not have need, should He, of so many thousands of spirits to make so many thousands of earth bodies revolve about this single sphere of ours in twenty-four hours (which does not represent so much as a tiny mote of dust in comparison with such a multitude of light-giving and opaque spheres)? Rather, could not this sphere of ours complete this motion in twenty-four hours through its own unique and individual soul or intelligence and in the same way could not the other stars be capable of carrying out their own individual motions? We leave this to the judgment of others.

The paradox involving the very rapid turning of the earth bodies.

Number XII, page 356. Theodidactus: "I understand and find everything in conformity with the light of reason, but I do not comprehend the rate of speed of the stars, especially those which are separated at so great a distance from the stars of the heavens of the first magnitude." Cosmiel: "It is as easy for God to direct the orbital course of these spheres in twenty-four hours through the help of His administering angels as it was for me, in so brief a space of time, to conduct you from the Dog Star to Polaris. For if a large stone were lifted up into the firmament and were then released towards the earth, according to Scheiner and Mersenne it would cover this distance in only six hours. Why, then, are you in doubt about the speed with which the stars are moved?"

Our Comment

Just as Theodidactus does not comprehend the furious rate of speed of the stars, so also have I certain reservations in this regard. Since the distance of the stars is so great that "no matter how many undreds of times higher a person might go in the heavens, never vould there be an end of new stars which succeed one another," as Kircher himself declares on page 412. Who can be persuaded to elieve in the existence of such an irrational and feverish motion in ll the stars unless, by chance, some special reason, which he did ot wish to divulge, had led him to assume this was the case. A erson should ponder whether or not it would be contrary to reason nd sound sense that so many stars, which are beyond number and vhose distances have no definable bounds and whose vastness is ndescribable, should all revolve in such a way in the space of twenty-our hours about our earthly sphere (which can easily accomplish all his by itself through its own rotation), without interruption and ndeed, so rapidly that even the lower or nearer stars to us must omplete, in the wink of an eye, 152 radii of the earth which is the quivalent of 139,720 German miles (according to Riccioli in his *Almagestum*, page 49). What shall we say about the increasingly furious ate of speed of other stars, being hundreds and hundreds of times arther removed from us, which ought to circulate thousands and housands of times more rapidly for this reason? On what reasonable grounds can anyone understand why, among so great a number of arth systems, our sphere should hold the central position? For if ve were standing on any one of these stars (which was revolving in he same way as one of the planets), we should likewise seem to be n the center of the world and all the rest of the stars would appear o be revolving around us. Just as Theodidactus' journey from the Dog Star to Polaris was imaginary, so also does this rotation of the tars seem to be a product of the imagination. Finally, what we hould conclude about the huge rock (that has already been discussed n Book IV, Chapter 2), which supposedly covers the distance from he fixed stars to our earthly sphere in six hours?

An objection to the forward motion of the stars.

The fact that the earth is not in the center of the world.

Number XII again, page 357. Theodldactus: "But how is that articular motion of the firmament produced by which it progresses ontrary to the orderly movement of the zodiac from west to east?" Cosmiel: "This motion is produced through the delineation of a piral line in an east/west direction whereby it does not precisely neet with the meridian in the space of twenty-four hours, but still hrough a multiplication of many days and years, it gradually takes n a perceptible increase in size, as you already are aware.

The so-called motion of the Eighth Sphere from west to east.

Our Comment

ndeed, as for those revolutions of the stars to which the author, Kircher, alludes in Number V and elsewhere and declares are carried

The movement of each system is rational.

A motion (or rather, a forward motion) of all systems both in the 24 hours from rising to setting and setting to rising is contrary to all reason.

out in their own orbits about centers of their own systems, to see the necessity and rationale of these movements, one must consult Chapter 2 of the preceding Book. But to assign another motion contrary to this supposed daily motion or the First so-called Mobile, is not well advised, as we have discussed in detail in the last Chapter of the preceding Book. Furthermore, no reason, necessity, usefulness, etc. for these two motions can be indicated or understood, much less rendered true. You say: it is for the good of the earth. Gilbert replies *De Magnete*, Book VI, Chapter 4 "It is ridiculous for a philosopher to hold that all stars, both fixed and wandering, turn for no other purpose than for the convenience of the earth." Furthermore, Galileo in Dialogue I of *Systema Cosmicum*, page 39 states, "If the stars are so arranged to be the earth's servants, then all encompassing and most perfect nature has produced and governs all these splendid celestial bodies, immutable, immortal, and divine as they are, for no other purpose than for its servitude, which otherwise they call its filthy dregs or bilge. Furthermore, Father Anton de Rheita in Book IV of his *Oculus Enoch et Eliae* does not hold that the satellites of Saturn and Jupiter are in any way disturbed in their courses around Saturn and Jupiter because of our earth. For he says on page 182, "It is as far from the truth to believe that the rotational courses of these aforementioned satellites are carried out for the sake of our earth alone as to think that the monthly rotations of our moon about its center, the earth, take place for the sake of Jupiter or Saturn." This point of view René Descartes in Part 3 of his *Principia Philosophiae* extended to other lesser bodies than the stars, when he said, "It is far removed from the truth to suppose that all things are made as they are for our own sakes without their being of any other use. It would be clearly rediculous and inexact from the point of view of physics to make this supposition. Indeed, we have no doubt that there are many things which are in existence now or were in existence in the past and now are no longer, which never were seen by any man or intelligent being nor were created for any purpose for anyone." If one accepts this statement, how much less should one conclude the same to be the case with the countless, tremendously huge bodies of the stars! Then, if this purpose of serving the earth is removed, the daily motion of the stars about the earth is rendered completely useless, superfluous, and quite irrational, as well as the contrary motion which they call elsewhere the wavering motion ("trepidatio") or backward progression ("retrogradio"). Thus they neither have nor can have any mutual interaction between one another.

Number XII (again) page 362. Theodidactus: "I remember, when

I was on the Dog Star's lunar sphere, seeing a very bright star accompanied by a wondrous escort of nebulous stars. Even here, in this place, I look upon it." Cosmiel: "How far away do you think it is from you?" Theodidactus: "Not very far." Cosmiel: "Well, you should know that it is as far from you as you are distant from the earth in this place where you are now; but in order that you may look upon the incomprehensible works of God, I will take you there." Theodidactus" "So saying, he carried me off into the most remote and deep shadows of the world. In the Space I traversed, I discovered countless spheres that had never been seen by human eye, (one should undoubtedly understand these spheres to be planets, which, together with their satellites, belong to individual so-called systems, for the fixed stars of this place heretofore have been invisible to us from the earth) and finally he brought me to a new lunar sphere through the immense expanse of ethereal air we traversed, where I saw a new spectacle of stars before my eyes." Cosmiel: "These are the stars which you thought were nebulae from your previous position and which are as far distant from the Dog Star sun as the earth is from the canopy of the firmament. Here you no longer see your sun nor any of those spheres you have seen thus far. For all have disappeared into nothingness because of the incredible distance. Some of the spheres seen here are solar, others lunar, and some are stars which have never been viewed by mortal eyes. Some of these are as distant from others as this sphere on which you stand is distant from stars of the second magnitude, that is, these these bodies are three times farther removed from the earth than stars of the first magnitude in the firmament are; and if I should carry you a hundred times higher, there would be no end of new stars, one following upon another. From this you can easily perceive that the universe is not bounded by such restricted limits as certain men seem to believe."

Theodidactus is escorted to a star which he saw in the innermost recess of the heavens from the Dog Star's lunar sphere.

From the lunar or planetary sphere of this place he sees a new spectacle of stars from which our sun is no longer visible.

Some are suns, others are planets.

If one could be transported one hundred times higher, there still would be no end of stars.

Our Comment

Since, from this third station among the stars of Ethereal Space our sun can no longer be perceived, even though an abundance of other solar and lunar spheres are observable, (Kircher repeats this statement in Dialogue 2, page 412), "And even if a person should go up hundreds and hundreds of times times higher than this, he could never hope to find an end to the number of stars of this kind, nor could he compare or intellectually comprehend the diverse relationships of these earth systems, their disposition, order, motion, forces, etc." From this it seems to follow that the number of these systems

According to Kircher, the number of earth systems is infinite.

is infinite. To begin with, we shall assume that in the first station of the co-called firmament there are to be found at least ten fixed stars which are situated around this earth of ours. They are the reason it is regarded as the center of everything. In view of the extremely great distance of the second station, there most certainly will be 100, and in the third station, 1,000 (for according to Book I, Chapter 29, the number of stars that can be seen by the naked eye is calculated by astronomers to be 1,100. The rest, beyond the third station, cannot be seen by the eye alone), and in the fourth station, 10,000, the fifth, 100,000, and the sixth 1,000,000 etc. If this progression is repeated a hundred times and again another hundred, so great a number of world systems results as to extend to 200 ciphers, one after the other. Thus, in accordance with man's capacity for creative thinking, the number can properly be called infinite (as for the absurd assumption that this infinity, as it were, of world systems revolves each day about the sphere of our earth, this we pass over in silence). One can further pose the question as to whether an infinite group of world systems, in each of which (according to Number V) individual motions are carried out, can be regarded as a single world or rather as an infinite or countless number of worlds. When a solar body (according to Number VII), is accompanied by many satellites, however, and is created for the sake of lighting and warming them (so that they may not perish in eternal shadow and cold, according to Number III) and when no connection or interrelationship of all the systems can be seen other than in the so-called Ethereal Air (which, however, does not exist), it seems to follow that each solar body constitutes an individual world system or a single world, just as Aristarchus and others affirmed in the preceding Chapter.

The question is asked whether such an infinite number of world systems can be regarded as one world.

Number XII (again), page 363. Cosmiel: "Just as God Himself is great and does not know the bounds of His greatness, so He wished the world to be immense insofar as its natural condition permits in this present state of things; but only so great, to be sure, as it could and ought to be, not because He could not produce a greater world, but because the capacity of the created body, which is limited by itself and its own nature, could not sustain a greater extension. Similarly, many of your philosophers and theologians, when they ask whether God can perform the act of creation infinitely, rightly and with wisdom reply that He cannot, not because they think that the infinite power of the Creator has been exhausted, but rather because of what is encompassed in the nature of matter."

Kircher declared that God made the world as great as the capacity of created matter allowed.

Our Comment

The Reverend Father Kircher – above in Number IX – indeed does not believe in the existence of Empty Space, but perhaps thinks along the lines of certain theologians (Book I, Chapter 36) who hold that Space is created together with body and cannot exist except in a body of substance. Thus throughout this entire dialogue he calls that intermediate Space between the stars, Ethereal Air, and on page 436 states that no one can imagine that Space exists where there is nothing. Now, indeed, while he states above "that the matter of the world could not sustain greater extension by reason of its own nature," he also concedes that a person can imagine Space to exist where there is nothing.

While Father Kircher concedes the extension of matter, he also admits the existence of Space.

Now whatever is extended must necessarily be extended outwards (which in this place can be nothing other than surrounding Space or Vacuity. The latter was in existence, moreover, before matter itself was created and the extension of matter took place). According to Kircher, the matter or substance of the world, however, was extended, but only so far as its natural condition permitted. Therefore (since there was no other matter, but only nothingness) it follows that this extension was carried out into Empty Space. For if Space, and indeed Empty Space, had not been there, something else necessarily would have been there. But if there were something there, how could the matter of the world have been extended or received into something else? Necessarily, therefore, the so-called world substance was extended, not into something, but into surrounding Empty Space, as we explained in detail in Book II, Chapter 7.

From this, then,it follows that the opinion of René Descartes, when he writes about Empty Space beyond this world (however great its extent may have been described to be by Kircher), is incorrect for while he is denying its existence, at the same time he maintains that the world is infinite, as is evident in Book I Chapter 35. Indeed, God alone is infinite and no two things can be infinite at the same time, wherefore our author rightly replies here that God does not perform the act of creation infinitely, but that all things created are individual and exist in number and quantity, having no need of being held together, directed, supported, and preserved: and thus they are so distinct and separate throughout that surrounding, Infinite Space (ERRATUM, read "illud" for "aliud") that they do not hinder one another in their individual courses nor are they proportionate in any way to its size or extent since at any time they can be separately dispersed into very small pieces or reduced into nothingness.

God alone is infinite.

Space is the universal container of all things.

Furthermore, even though God does not perform the act of creation Infinitely, still, one must understand that insofar as the Immensity or Infinity of the Omnipotent Creator is concerned, there is limitlessness, but not insofar as the infinite number of things He has created. Indeed, we cannot deny that God either has created or can create an Infinity of creatures (so far as number is concerned), as well as an infinite Infinity of Infinites.

An infinity of creatures is to be understood as not being relative to Immensity but to Infinite Number.

Since, therefore, God creates no infinite Immensity, but all things in number or quantity, the world, then, just as our author Kircher describes it, is so vast that it is infinite according to man's intellectual reckoning, consisting of innumerable earth systems which, by reason of the very great distance between each of them, should be regarded as individual and separate, so that one is able to exist and to continue to exist without the other. One asks whether it is incompatible with Divine Omnipotence, Glory, and Majesty, to consider this world, as Kircher has done, to be the same as many others, so that, as in Dialogue 7 which has already been mentioned, the solar sphere which we call Aldebaran together with its satellites, the solar body of Arcturus, with its satellites as well as the spheres of Capella, Vega, and the stars of Ursa Major, etc., all solar bodies equally accompanied by a crowded array of spheres, single worlds in their own right, are regarded as separate and distinct world systems? Or would this contradict the Sacred Scriptures inasmuch as one reads in some places there that systems of this kind constitute one world? Indeed, should men be forbidden to have opinions like this? If such is the case, then we rightly should cease pursuing our argument and accept this infinite mass of innumerable systems as being one world, however vast it may seem and incompatible with reason. But if, in fact, this is not so, then nature, the mind, and reason seek further knowledge and wish to discover what is beyond. For instance, what if Cosmiel had carried Theodidactus hundreds and hundreds of times higher, and had finally been able to transport him up to a region where all solar and lunar spheres had vanished and no other torch in the heavens gave out light? The following Chapter is concerned with this question.

God does not create Immensity.

One asks whether there are many worlds. Or would this contradict the Sacred Scriptures?

CHAPTER 5

THE LIMIT OR OUTERMOST BOUNDARY OF THE STARS

Since almost all agree that the stars are finite in extent, one may ask what bounds them? Regarding this question and its answer, so many and such diverse statements are found that in no way can they, in fact, be reconciled. Some are of the opinion that they are limited by the outermost heaven, namely, the First (so-called Mobile); others by waters above the heavens; others by the Empyrean Heaven; others by Nothingness; and still others by Vacuity, etc., all of which we examined in Book I, Chapters 31, 32, 33, 34, and 35.

In Book II, however, we discussed rather thoroughly the fact that Being is immense, is necessarily everywhere, and exists in Infinity in such a way that it can and will always be able to exist. Thus, it must follow that that universal container of all things which we simply term Space, (the bounds of whose immensity cannot be given nor the limit beyond which it does not exist cannot be defined nor whose bodies cannot exist in greater number, although there are many, to be sure, of which we know nothing), always bounds the stars, wherever they may be.

Space also exists beyond the fixed stars.

Thus, however great Father Kircher may have described the world to be in the preceding Dialogue (provided that it is regarded as a unit which cannot embrace innumerable or countless bodies), still it cannot be compared with Immensity, because all that man designates by this term is no more than a point. But if, indeed, countless bodies should be present, they still would exist in number and quantity. Since, moreover, according to Book II, Chapters 9 and 10, all number is finite and according to Chapter 12 of the same Book, all quantity is finite, therefore, even though there were no end of worlds existing in Immense Space, all of these worlds would not create an Immensity for they would exist in number. Indeed, if, as was postulated in the aforementioned Chapter 10, any one of the worlds consisted of as many grains of sand (and even many tens of thousands more) as a number containing fifty-three ciphers in a series, nonetheless all would be contained in a number together with so many tens of thousand grains, nor would the number of worlds, increased by the number of sand granules, make an Immensity, but only a number increased by fifty-three ciphers placed in a series.

As for the question whether or not there is a boundary to stars or world systems, I must confess that I am hardly qualified to give an answer since arguments both pro and con can be set forth.

Is there a boundary to stars? Negative arguments.

Negative arguments are to be found in Book I, Chapters 34 and 35, where, for the sake of brevity, I refer my reader.

Plutarch subscribes to this point of view in Book I, Chapter 3 of his *De Placito Philosophorum* when he says, "Anaximander of Miletus said that Infinity is the foundation of everything, for from this all is created and into this all is ultimately returned; for this reason, an infinite number of worlds is created and is again returned into this from which they arise. He errs, however, according to Plutarch, because he does not say what Infinity is, whether it is Air, or Water, or Earth, or some other body. He is in error, therefore, in postulating the existence of matter, without describing its effective cause. For if Infinity is nothing other than matter, this in itself cannot participate in action unless an effective cause is present." So says Plutarch. Does he criticize Anaximander with justice or not? We leave this to others to decide.

Furthermore, it is the opinion of theologians that the world certainly is a single body since the Sacred Scriptures do not make mention of many worlds. If, moreover, the world is a single body, it necessarily must have an end (for the world is not regarded as infinite) and at the same time the fixed stars must be bounded by it because the Scriptures include these likewise in this world.

Affirmative arguments.

The opinion of Father de Rheita.

The Scriptures do not mention all things.

The affirmative arguments are (1) the opinion of Father Rheita cited above in Book VI, Chapter 15, "that there may be many things existing in nature (like the wandering stars or planets) of which no mention is made in the Scriptures." It does not follow from this that these bodies do not exist nor that beyond the Infinite tens of thousands of miles in the most remote recesses of Infinite Space, there is absolutely nothing. (2) The Peripatetics say that one sees the statement of Riccioli set forth in Book IX, section 4, Chapter 29, number 19, page 458: "If the Earth is moved according to Copernicus through its annual orbit, the enormous space between Saturn and the fixed stars would be empty of fixed stars and completely unoccupied. It is absurd, however, when a huge space like this is completely unoccupied." If this is the case, it would follow that Infinite Space beyond the so-called starry sphere is not unoccupied nor empty of stars. (3) If there is nothing else beyond the fixed stars in the Infinite and all these stars belong to this world as is commonly described, it would follow that the world is contained in this Infinite Space as an atom without quantity, without place either here or there, below, or above, and that all of Space, aside from this point, is infinitely empty of everything. They do not assume this to be so, however, but rather that many worlds of this kind exist because people commonly do not imagine Nothingness as being alone without an accompanying body. What is more, since it had seemed good to God that there be something here in the immensity of Space, accordingly it would seem evil that Nothing should be elsewhere. Thus if a person were to set an imaginary boundary or limit to these stars, in his theorizing he would progress, as it were, beyond the hand of God whose reach is never curtailed and would conclude that God is confined to this or that boundary beyond which He does not exercise His power. Rather

In Immense Space the world would not be regarded as an atom.

A single world is nowhere.

one should conclude that an infinite God governs infinitely; if, however, there is nothing beyond the stars, consequently there will be no ruling Power in that infinite abyss of Space. (4) The existence of a multitude of creatures does not detract from but rather makes divine Majesty and Immensity shine more brightly. For why should there be a single world of boundless height, depth, and breadth, which in no way can be compared with this Immensity? One tiny fish, indeed, can be set in comparison with the number of such creatures required to fill the vast sea and the total computed very easily; the number of worlds needed to fill Immensity, however, cannot be calculated. As God did not create the whole sea that one single fish might swim in it, nor all the air that one little bird or fly might fly around in it, so also does the Omnipotence of God seem to shine out more in a multitude of worlds endowed with so much variety that no single one can be compared with another nor a similar one found; rather all are different, one from the other, in many thousand ways. And as this Infinity of God is evident in all creation, it is apparent that this enormous multitude of stars, (whose number no mortal man can tell: whether it is bounded by ten times a thousand stars or rather one hundred times or one thousand times etc., this number), presents a very clear picture of the immense power of this great Architect."

An infinite God governs infinitely.

The Omnipotence, Glory, and Honor of the Creator is apparent from the infinite multitude of cratures.

Many such arguments and similar ones both pro and con can be set forth; judgment lies with him who has recourse to Infallible Truth, which I would willingly embrace. As for the statements in the preceding Chapter 3, however, these are not published as Infallible Truths nor are they mathematically proven and for this reason we do not wish to set them forth as sources of contention to be fought over. Rather as a cultivator of the mathematical sciences who is aware of certain facts beyond the province of popular astronomy, we would be introduced into the mysterious recesses of penetrating speculation.

The judgment of whoever is free is not to be disputed.

Meanwhile, no one will deny that in the Sacred Scriptures the stars are called the army of God and that the Immense Expanse of Heaven has been designed to accommodate the camp of the infinite celestial band. "He is the God of Zeboath, namely the leader of the heavenly armies by whose word were the heavens made and all its host." Psalms XXXIII, 6.

The stars are the army of God.

"I, even my hands, have stretched out the heavens, and all their host have I commanded." Isaiah XLV, 12.

God commands the host.

"All nations before Him are as nothing and they are counted to Him less than nothing, and vanity. To whom, then, will ye liken God? or what likeness will ye compare unto Him? (verses 17 and

All nations are as nothing.

God can be likened to no one. Lift up your eyes on high.

18). It is He that sitteth upon the circle of the earth, and the inhabitants thereof are as grasshoppers, (part of verse 22). Lift up your eyes on high and behold who hath created these things, that bringeth out their host by number: He calleth them all by name, by the greatness of His might, for that He is strong in power; not one faileth."

God addresses the celestial host by name and gives them names.

"He telleth the number of the stars; He calleth them all by their names." Psalm CXLVII. (ERRATUM read CXLVII for CXLVI.).

The heavens show the handiwork of God and the heavens declare his glory.

"The heavens declare the glory of God, and the firmament sheweth His handiwork." Psalm XIX, 1. "Thou, even Thou, art Lord alone; Thou hast made heaven the heaven of heavens, with all their host, the earth, and all things that are therein and Thou preservest them all; and the host of heaven worshippeth Thee." Nehemiah IX, 6.

The army of heaven adores God.

"Where wast thou when I laid the foundations of the earth? When the morning stars sang together, and all the sons of God shouted for joy?" Job XXXVIII, 7.

The stars praise God.

The ministers of God are spirits and flaming fires.

"Bless the Lord, O my soul O Lord my God Thou art clothed with honor and majesty; who coverest thyself with light as with a garment; who stretchest out the heavens like a curtain; who maketh His angels spirits; His ministers a flaming fire." Psalms CIV, 4.

When we consider, therefore, the infinite number of fixed stars and in addition the many thousands of their satellites, as Kircher holds, together with their companions scattered in the Immense Expanse of the heavens as in a field, we are compelled to admit that the camp of the heavenly host requires an immense Container or Space in which so great an army of spheres, being distributed appropriately, can be accommodated and so that its Leader be worthy of the great array which fittingly should not have a limit to its number.

A Leader of Immensity should have an infinite number of servants.

Rather, just as the Leader is immense, so also is the number of his flaming servants infinite, for it seems degrading that immense Majesty should be surrounded by a finite number of servants.

The flaming soldiery does not exist in imagination.

This heavenly host, therefore, does not exist in imagination as Intelligences, but rather as an array of soldiers arranged in battle formation before our eyes. It is meant to frighten evil and impious men who engage in depravity with the terror of lightning, just as it openly reveals to the good and god-fearing the unseen god, Zeboath,

Omnipotence is reflected in the bodies that have been created.

sanctified and compassionate, so that the ineffable Omnipotent Majesty, directly perceived through those bodies which have been created, may shine clearly in our eyes and encourage us to correct our evil acts.

If, when the night is calm and particularly when a breeze is blowing, we look up at the sky after a rain (for at that time we see many stars and perceive their twinkling forms more clearly and in greater number because through the motion of the air caused by the wind, the stars' rays fall upon that diaphanous flowing medium, namely the air, and in different ways are broken up in this, causing their twinkling to spread out and appear more extensive than usual to the observer), and consider that Immense Expanse filled with countless banners and standards of the heavenly host, and if we examine them together with their satellites with the eye of the mind as well as the body (ERRATUM, dele ";"), we see invisible Zeboath, clothed in light, as by a garment sprinkled over with the brightest of diamonds.

God can be seen with the eye of the mind as well as the body.

All other knowledge is reserved for our blessed and eternal life after Resurrection. "For in this mortal life", as the apostle says, "we have only partial knowledge. After perfect understanding comes, however, then that which has been incomplete will be completed. For now we see as through a darkening mirror and an allegory, but then we will observe everything face to face." Meanwhile may Power, Honor, and Glory be to God, the Father, Son, and the Holy Ghost, the Trinity, the Creator and Preserver of all things, for Eternity.

That which we know is incomplete but will be complete after Resurrection.

The End

Finished at Magdeburg
the 14th of March, 1663.

INDEX OF SUBJECTS

ARCHIVES INTERNATIONALES D'HISTOIRE DES IDÉES

*

INTERNATIONAL ARCHIVES OF THE HISTORY OF IDEAS

---

22. W.N. Hargreaves-Mawdsley: *The English Della Cruscans and Their Time, 1783-1828.* 1967 ISBN 90-247-0198-8
23. C.B. Schmitt: *Gianfrancesco Pico della Mirandola (1469-1533) and his Critique of Aristotle.* 1967 ISBN 90-247-0199-6
24. H.B. White: *Peace among the Willows.* The Political Philosophy of Francis Bacon. 1968 ISBN 90-247-0200-3
25. L. Apt: *Louis-Philippe de Ségur.* An Intellectual in a Revolutionary Age. 1969 ISBN 90-247-0201-1
26. E.H. Kadler: *Literary Figures in French Drama (1784- 1834).* 1969 ISBN 90-247-0202-X
27. G. Postel: *Le Thrésor des prophéties de l'univers.* Manuscrit publié avec une introduction et des notes par F. Secret. 1969 ISBN 90-247-0203-8
28. E.G. Boscherini: *Lexicon Spinozanum.* 2 vols., 1970 Set ISBN 90-247-0205-4
29. C.A. Bolton: *Church Reform in 18th-Century Italy.* The Synod of Pistoia (1786). 1969 ISBN 90-247-0208-9
30. D. Janicaud: *Une généalogie du spiritualisme français.* Aux sources du bergsonisme: [Félix] Ravaisson [1813-1900] et la métaphysique. 1969 ISBN 90-247-0209-7
31. J.-E. d'Angers: *L'Humanisme chrétien au 17e siècle.* St. François de Sales et Yves de Paris. 1970 ISBN 90-247-0210-0
32. H.B. White: *Copp'd Hills towards Heaven.* Shakespeare and the Classical Polity. 1970 ISBN 90-247-0250-X
33. P.J. Olscamp: *The Moral Philosophy of George Berkeley.* 1970 ISBN 90-247-0303-4
34. C.G. Noreña: *Juan Luis Vives (1492-1540).* 1970 ISBN 90-247-5008-3
35. J. O'Higgens: *Anthony Collins (1676-1729), the Man and His World.* 1970 ISBN 90-247-5007-5
36. F.T. Brechka: *Gerard van Swieten and His World (1700- 1772).* 1970 ISBN 90-247-5009-1
37. M.H. Waddicor: *Montesquieu and the Pilosophy of Natural Law.* 1970 ISBN 90-247-5039-3
38. O.R. Bloch: *La Philosophie de Gassendi (1592-1655).* Nominalisme, matérialisme et métaphysique. 1971 ISBN 90-247-5035-0
39. J. Hoyles: *The Waning of the Renaissance (1640-1740).* Studies in the Thought and Poetry of Henry More, John Norris and Isaac Watts. 1971 ISBN 90-247-5077-6
    For Henry More, *see also below under Volume 122 and 127.*
40. H. Bots: *Correspondance de Jacques Dupuy et de Nicolas Heinsius (1646-1656).* 1971 ISBN 90-247-5092-X
41. W.C. Lehmann: *Henry Home, Lord Kames, and the Scottish Enlightenment.* A Study in National Character and in the History of Ideas. 1971 ISBN 90-247-5018-0
42. C. Kramer: *Emmery de Lyere et Marnix de Sainte Aldegonde.* Un admirateur de Sébastien Franck et de Montaigne aux prises avec le champion des calvinistes néerlandais.[Avec le texte d'Emmery de Lyere:] *Antidote ou contrepoison contre les conseils sanguinaires et envinemez de Philippe de Marnix Sr. de Ste. Aldegonde.* 1971 ISBN 90-247-5136-5

ARCHIVES INTERNATIONALES D'HISTOIRE DES IDÉES
*
INTERNATIONAL ARCHIVES OF THE HISTORY OF IDEAS

---

43. P. Dibon: *Inventaire de la correspondance (1595-1650) d'André Rivet (1572-1651).* 1971 ISBN 90-247-5112-8
44. K.A. Kottman: *Law and Apocalypse.* The Moral Thought of Luis de Leon (1527?-1591). 1972 ISBN 90-247-1183-5
45. F.G. Nauen: *Revolution, Idealism and Human Freedom.* Schelling, Hölderlin and Hegel, and the Crisis of Early German Idealism. 1971 ISBN 90-247-5117-9
46. H. Jensen: *Motivation and the Moral Sense in Francis Hutcheson's* [1694-1746] *Ethical Theory.* 1971 ISBN 90-247-1187-8
47. A. Rosenberg: *[Simon] Tyssot de Patot and His Work (1655–1738).* 1972 ISBN 90-247-1199-1
48. C. Walton: *De la recherche du bien.* A study of [Nicolas de] Malebranche's [1638-1715] Science of Ethics. 1972 ISBN 90-247-1205-X
49. P.J.S. Whitmore (ed.): *A 17th-Century Exposure of Superstition.* Select Text of Claude Pithoys (1587-1676). 1972 ISBN 90-247-1298-X
50. A. Sauvy: *Livres saisis à Paris entre 1678 et 1701.* D'après une étude préliminaire de Motoko Ninomiya. 1972 ISBN 90-247-1347-1
51. W.R. Redmond: *Bibliography of the Philosophy in the Iberian Colonies of America.* 1972 ISBN 90-247-1190-8
52. C.B. Schmitt: *Cicero Scepticus.* A Study of the Influence of the *Academica* in the Renaissance. 1972 ISBN 90-247-1299-8
53. J. Hoyles: *The Edges of Augustanism.* The Aesthetics of Spirituality in Thomas Ken, John Byrom and William Law. 1972 ISBN 90-247-1317-X
54. J. Bruggeman and A.J. van de Ven (éds.): *Inventaire* des pièces d'Archives françaises se rapportant à l'Abbaye de Port-Royal des Champs et son cercle et à la Résistance contre la Bulle *Unigenitus* et à l'Appel. 1972 ISBN 90-247-5122-5
55. J.W. Montgomery: *Cross and Crucible.* Johann Valentin Andreae (1586–1654), Phoenix of the Theologians. Volume I: Andreae's Life, World-View, and Relations with Rosicrucianism and Alchemy; Volume II: The *Chymische Hochzeit* with Notes and Commentary. 1973 Set ISBN 90-247-5054-7
56. O. Lutaud: *Des révolutions d'Angleterre à la Révolution française.* Le tyrannicide & *Killing No Murder* (Cromwell, *Athalie*, Bonaparte). 1973 ISBN 90-247-1509-1
57. F. Duchesneau: *L'Empirisme de Locke.* 1973 ISBN 90-247-1349-8
58. R. Simon (éd.): *Henry de Boulainviller* - Œuvres Philosophiques, Tome I. 1973 ISBN 90-247-1332-3

    For Œvres Philosophiques, Tome II *see below under Volume 70.*
59. E.E. Harris: *Salvation from Despair.* A Reappraisal of Spinoza's Philosophy. 1973 ISBN 90-247-5158-6
60. J.-F. Battail: *L'Avocat philosophe Géraud de Cordemoy (1626-1684).* 1973 ISBN 90-247-1542-3
61. T. Liu: *Discord in Zion.* The Puritan Divines and the Puritan Revolution (1640-1660). 1973 ISBN 90-247-5156-X
62. A. Strugnell: *Diderot's Politics.* A Study of the Evolution of Diderot's Political Thought after the *Encyclopédie.* 1973 ISBN 90-247-1540-7

ARCHIVES INTERNATIONALES D'HISTOIRE DES IDÉES
*
INTERNATIONAL ARCHIVES OF THE HISTORY OF IDEAS

63. G. Defaux: *Pantagruel et les Sophistes.* Contribution à l'histoire de l'humanisme chrétien au 16ᵉ siècle. 1973 ISBN 90-247-1566-0
64. G. Planty-Bonjour: *Hegel et la pensée philosophique en Russie (1830-1917).* 1974 ISBN 90-247-1576-8
65. R.J. Brook: *[George] Berkeley's Philosophy of Science.* 1973 ISBN 90-247-1555-5
66. T.E. Jessop: *A Bibliography of George Berkeley.* With: *Inventory of Berkeley's Manuscript Remains* by A.A. Luce. 2nd revised and enlarged ed. 1973 ISBN 90-247-1577-6
67. E.I. Perry: *From Theology to History.* French Religious Controversy and the Revocation of the Edict of Nantes. 1973 ISBN 90-247-1578-4
68. P. Dibbon, H. Bots et E. Bots-Estourgie: *Inventaire de la correspondance (1631–1671) de Johannes Fredericus Gronovius* [1611–1671]. 1974 ISBN 90-247-1600-4
69. A.B. Collins: *The Secular is Sacred.* Platonism and Thomism in Marsilio Ficino's *Platonic Theology.* 1974 ISBN 90-247-1588-1
70. R. Simon (éd.): *Henry de Boulainviller.* Œuvres Philosophiques, Tome II. 1975 ISBN 90-247-1633-0
    For Œvres Philosophiques, Tome I *see under Volume 58.*
71. J.A.G. Tans et H. Schmitz du Moulin: *Pasquier Quesnel devant la Congrégation de l'Index.* Correspondance avec Francesco Barberini et mémoires sur la mise à l'Index de son édition des Œuvres de Saint Léon, publiés avec introduction et annotations. 1974 ISBN 90-247-1661-6
72. J.W. Carven: *Napoleon and the Lazarists (1804–1809).* 1974 ISBN 90-247-1667-5
73. G. Symcox: *The Crisis of French Sea Power (1688–1697).* From the *Guerre d'Escadre* to the *Guerre de Course.* 1974 ISBN 90-247-1645-4
74. R. MacGillivray: *Restoration Historians and the English Civil War.* 1974 ISBN 90-247-1678-0
75. A. Soman (ed.): *The Massacre of St. Bartholomew.* Reappraisals and Documents. 1974 ISBN 90-247-1652-7
76. R.E. Wanner: *Claude Fleury (1640-1723) as an Educational Historiographer and Thinker.* With an Introduction by W.W. Brickman. 1975 ISBN 90-247-1684-5
77. R.T. Carroll: *The Common-Sense Philosophy of Religion of Bishop Edward Stillingfleet (1635-1699).* 1975 ISBN 90-247-1647-0
78. J. Macary: *Masque et lumières au 18ᵉ [siècle].* André-François Deslandes, Citoyen et philosophe (1689-1757). 1975 ISBN 90-247-1698-5
79. S.M. Mason: *Montesquieu's Idea of Justice.* 1975 ISBN 90-247-1670-5
80. D.J.H. van Elden: *Esprits fins et esprits géométriques dans les portraits de Saint-Simon.* Contributions à l'étude du vocabulaire et du style. 1975 ISBN 90-247-1726-4
81. I. Primer (ed.): *Mandeville Studies.* New Explorations in the Art and Thought of Dr Bernard Mandeville (1670-1733). 1975 ISBN 90-247-1686-1
82. C.G. Noreña: *Studies in Spanish Renaissance Thought.* 1975 ISBN 90-247-1727-2
83. G. Wilson: *A Medievalist in the 18th Century.* Le Grand d'Aussy and the Fabliaux ou Contes. 1975 ISBN 90-247-1782-5
84. J.-R. Armogathe: *Theologia Cartesiana.* L'explication physique de l'Eucharistie chez Descartes et Dom Robert Desgabets. 1977 ISBN 90-247-1869-4

ARCHIVES INTERNATIONALES D'HISTOIRE DES IDÉES
*
INTERNATIONAL ARCHIVES OF THE HISTORY OF IDEAS

85. Bérault Stuart, Seigneur d'Aubigny: *Traité sur l'art de la guerre.* Introduction et édition par Élie de Comminges. 1976 ISBN 90-247-1871-6
86. S.L. Kaplan: *Bread, Politics and Political Economy in the Reign of Louis XV.* 2 vols., 1976 Set ISBN 90-247-1873-2
87. M. Lienhard (ed.): *The Origins and Characteristics of Anabaptism / Les débuts et les caractéristiques de l'Anabaptisme.* With an Extensive Bibliography / Avec une bibliographie détaillée. 1977 ISBN 90-247-1896-1
88. R. Descartes: *Règles utiles et claires pour la direction de l'esprit en la recherche de la vérité.* Traduction selon le lexique cartésien, et annotation conceptuelle par J.-L. Marion. Avec des notes mathématiques de P. Costabel. 1977 ISBN 90-247-1907-0
89. K. Hardesty: *The 'Supplément' to the 'Encyclopédie'.* [Diderot et d'Alembert]. 1977 ISBN 90-247-1965-8
90. H.B. White: *Antiquity Forgot.* Essays on Shakespeare, [Francis] Bacon, and Rembrandt. 1978 ISBN 90-247-1971-2
91. P.B.M. Blaas: *Continuity and Anachronism.* Parliamentary and Constitutional Development in Whig Historiography and in the Anti-Whig Reaction between 1890 and 1930. 1978 ISBN 90-247-2063-X
92. S.L. Kaplan (ed.): *La Bagarre.* Ferdinando Galiani's (1728-1787) 'Lost' Parody. With an Introduction by the Editor. 1979 ISBN 90-247-2125-3
93. E. McNiven Hine: *A Critical Study of [Étienne Bonnot de] Condillac's* [1714-1780] *'Traité des Systèmes'.* 1979 ISBN 90-247-2120-2
94. M.R.G. Spiller: *Concerning Natural Experimental Philosphy.* Meric Casaubon [1599-1671] and the Royal Society. 1980 ISBN 90-247-2414-7
95. F. Duchesneau: *La physiologie des Lumières.* Empirisme, modèles et théories. 1982 ISBN 90-247-2500-3
96. M. Heyd: *Between Orthodoxy and the Enlightenment.* Jean-Robert Chouet [1642-1731] and the Introduction of Cartesian Science in the Academy of Geneva. 1982 ISBN 90-247-2508-9
97. James O'Higgins: *Yves de Vallone* [1666/7-1705]: *The Making of an Esprit Fort.* 1982 ISBN 90-247-2520-8
98. M.L. Kuntz: *Guillaume Postel* [1510-1581]. Prophet of the Restitution of All Things. His Life and Thought. 1981 ISBN 90-247-2523-2
99. A. Rosenberg: *Nicolas Gueudeville and His Work (1652-172?).* 1982 ISBN 90-247-2533-X
100. S.L. Jaki: *Uneasy Genius: The Life and Work of Pierre Duhem* [1861-1916]. 1984 ISBN 90-247-2897-5; Pb (1987) 90-247-3532-7
101. Anne Conway [1631-1679]: *The Principles of the Most Ancient Modern Philosophy.* Edited and with an Introduction by P. Loptson. 1982 ISBN 90-247-2671-9
102. E.C. Patterson: *[Mrs.] Mary [Fairfax Greig] Sommerville* [1780-1872] *and the Cultivation of Science (1815-1840).* 1983 ISBN 90-247-2823-1
103. C.J. Berry: *Hume, Hegel and Human Nature.* 1982 ISBN 90-247-2682-4
104. C.J. Betts: *Early Deism in France.* From the so-called 'déistes' of Lyon (1564) to Voltaire's 'Lettres philosophiques' (1734). 1984 ISBN 90-247-2923-8

ARCHIVES INTERNATIONALES D'HISTOIRE DES IDÉES
*
INTERNATIONAL ARCHIVES OF THE HISTORY OF IDEAS

105. R. Gascoigne: *Religion, Rationality and Community.* Sacred and Secular in the Thought of Hegel and His Critics. 1985 ISBN 90-247-2992-0
106. S. Tweyman: *Scepticism and Belief in Hume's 'Dialogues Concerning Natural Religion'.* 1986 ISBN 90-247-3090-2
107. G. Cerny: *Theology, Politics and Letters at the Crossroads of European Civilization.* Jacques Basnage [1653-1723] and the Baylean Huguenot Refugees in the Dutch Republic. 1987 ISBN 90-247-3150-X
108. Spinoza's *Algebraic Calculation of the Rainbow & Calculation of Changes.* Edited and Translated from Dutch, with an Introduction, Explanatory Notes and an Appendix by M.J. Petry. 1985 ISBN 90-247-3149-6
109. R.G. McRae: *Philosophy and the Absolute.* The Modes of Hegel's Speculation. 1985 ISBN 90-247-3151-8
110. J.D. North and J.J. Roche (eds.): *The Light of Nature.* Essays in the History and Philosophy of Science presented to A.C. Crombie. 1985 ISBN 90-247-3165-8
111. C. Walton and P.J. Johnson (eds.): *[Thomas] Hobbes's 'Science of Natural Justice'.* 1987 ISBN 90-247-3226-3
112. B.W. Head: *Ideology and Social Science.* Destutt de Tracy and French Liberalism. 1985 ISBN 90-247-3228-X
113. A.Th. Peperzak: *Philosophy and Politics.* A Commentary on the Preface to Hegel's *Philosophy of Right.* 1987 ISBN Hb 90-247-3337-5; Pb ISBN 90-247-3338-3
114. S. Pines and Y. Yovel (eds.): *Maimonides* [1135-1204] *and Philosophy.* Papers Presented at the 6th Jerusalem Philosophical Encounter (May 1985). 1986 ISBN 90-247-3439-8
115. T.J. Saxby: *The Quest for the New Jerusalem, Jean de Labadie* [1610-1674] *and the Labadists (1610-1744).* 1987 ISBN 90-247-3485-1
116. C.E. Harline: *Pamphlets, Printing, and Political Culture in the Early Dutch Republic.* 1987 ISBN 90-247-3511-4
117. R.A. Watson and J.E. Force (eds.): *The Sceptical Mode in Modern Philosophy.* Essays in Honor of Richard H. Popkin. 1988 ISBN 90-247-3584-X
118. R.T. Bienvenu and M. Feingold (eds.): *In the Presence of the Past.* Essays in Honor of Frank Manuel. 1991 ISBN 0-7923-1008-X
119. J. van den Berg and E.G.E. van der Wall (eds.): *Jewish-Christian Relations in the 17th Century.* Studies and Documents. 1988 ISBN 90-247-3617-X
120. N. Waszek: *The Scottish Enlightenment and Hegel's Account of 'Civil Society'.* 1988 ISBN 90-247-3596-3
121. J. Walker (ed.): *Thought and Faith in the Philosophy of Hegel.* 1991 ISBN 0-7923-1234-1
122. Henry More [1614-1687]: *The Immortality of the Soul.* Edited with Introduction and Notes by A. Jacob. 1987 ISBN 90-247-3512-2
123. P.B. Scheurer and G. Debrock (eds.): *Newton's Scientific and Philosophical Legacy.* 1988 ISBN 90-247-3723-0
124. D.R. Kelley and R.H. Popkin (eds.): *The Shapes of Knowledge from the Renaissance to the Enlightenment.* 1991 ISBN 0-7923-1259-7

ARCHIVES INTERNATIONALES D'HISTOIRE DES IDÉES
*
INTERNATIONAL ARCHIVES OF THE HISTORY OF IDEAS

125. R.M. Golden (ed.): *The Huguenot Connection.* The Edict of Nantes, Its Revocation, and Early French Migration to South Carolina. 1988 ISBN 90-247-3645-5
126. S. Lindroth: *Les chemins du savoir en Suède.* De la fondation de l'Université d'Upsal à Jacob Berzelius. Études et Portraits. Traduit du suédois, présenté et annoté par J.-F. Battail. Avec une introduction sur Sten Lindroth par G. Eriksson. 1988 ISBN 90-247-3579-3
127. S. Hutton (ed.): *Henry More (1614-1687). Tercentenary Studies.* With a Biography and Bibliography by R. Crocker. 1989 ISBN 0-7923-0095-5
128. Y. Yovel (ed.): *Kant's Practical Philosophy Reconsidered.* Papers Presented at the 7th Jerusalem Philosophical Encounter (December 1986). 1989 ISBN 0-7923-0405-5
129. J.E. Force and R.H. Popkin: *Essays on the Context, Nature, and Influence of Isaac Newton's Theology.* 1990 ISBN 0-7923-0583-3
130. N. Capaldi and D.W. Livingston (eds.): *Liberty in Hume's 'History of England'.* 1990 ISBN 0-7923-0650-3
131. W. Brand: *Hume's Theory of Moral Judgment.* A Study in the Unity of *A Treatise of Human Nature.* 1992 ISBN 0-7923-1415-8
132. C.E. Harline (ed.): *The Rhyme and Reason of Politics in Early Modern Europe.* Collected Essays of Herbert H. Rowen. 1992 ISBN 0-7923-1527-8
133. N. Malebranche: *Treatise on Ethics* (1684). Translated and edited by C. Walton. 1993 ISBN 0-7923-1763-7
134. B.C. Southgate: *'Covetous of Truth'.* The Life and Work of Thomas White (1593–1676). 1993 ISBN 0-7923-1926-5
135. G. Santinello, C.W.T. Blackwell and Ph. Weller (eds.): *Models of the History of Philosophy.* Vol. 1: From its Origins in the Renaissance to the 'Historia Philosphica'. 1993 ISBN 0-7923-2200-2
136. M.J. Petry (ed.): *Hegel and Newtonianism.* 1993 ISBN 0-7923-2202-9
137. Otto von Guericke: *The New (so-called Magdeburg) Experiments* [Experimenta Nova, Amsterdam 1672]. Translated and edited by M.G.Foley Ames. 1994 ISBN 0-7923-2399-8
138. R.H. Popkin and G.M. Weiner (eds.): *Jewish Christians and Cristian Jews.* From the Renaissance to the Enlightenment. 1994 ISBN 0-7923-2452-8
139. J.E. Force and R.H. Popkin (eds.): *The Books of Nature and Scripture.* Recent Essays on Natural Philosophy, Theology, and Biblical Criticism in the Netherlands of Spinoza's Time and the British Isles of Newton's Time. 1994 ISBN 0-7923-2467-6

KLUWER ACADEMIC PUBLISHERS – DORDRECHT / BOSTON / LONDON

GPSR Compliance

The European Union's (EU) General Product Safety Regulation (GPSR) is a set of rules that requires consumer products to be safe and our obligations to ensure this.

If you have any concerns about our products, you can contact us on ProductSafety@springernature.com

In case Publisher is established outside the EU, the EU authorized representative is:

Springer Nature Customer Service Center GmbH
Europaplatz 3
69115 Heidelberg, Germany

Batch number: 10755284

Printed by Printforce, the Netherlands